The Scientific Legacy of Beppo Occhialini

Leonardo Gariboldi · Massimo Gervasi ·
Giorgio Sironi · Aldo Treves · Pasquale Tucci
Editors

The Scientific Legacy of Beppo Occhialini

Formative Years and the Return to Italy in 1950

Second Edition

Springer

Editors
Leonardo Gariboldi
Department of Physics
University of Milan
Milan, Italy

Massimo Gervasi
Department of Physics
University of Milano-Bicocca
Milan, Italy

Giorgio Sironi
University of Milano-Bicocca
Milan, Italy

Aldo Treves
Department of Physics and Mathematics
University of Insubria
Como, Italy

Pasquale Tucci
Milan, Italy

ISBN 978-3-031-37033-5 ISBN 978-3-031-37034-2 (eBook)
https://doi.org/10.1007/978-3-031-37034-2

This Springer imprint is published by the registered company Springer Nature Switzerland AG
The registered company address is: Gewerbestrasse 11, 6330 Cham, Switzerland

Paper in this product is recyclable.

Foreword

Giuseppe Occhialini, friendly named as Beppo, was an outstanding physicist, a key figure in the last century for the development of nuclear and particle physics. It is important to remember him for his innovative research and discoveries but also as a man.

This is why in 2006, in occasion of the centennial of his birth, SIF collaborated with Springer to publish the book *The Scientific Legacy of Beppo Occhialini*, edited by P. Redondi, G. Sironi, P. Tucci and G. Vegni. Moreover, to honor the memory of Giuseppe Occhialini, the Italian Physical Society (SIF), together with the Institute of Physics (IOP) in 2007 have decided to assign every year the Occhialini prize to a physicist operating mainly in Italy or in the United Kingdom or Ireland in recognition of relevant results in physics in the last 10 years.

In 2022, on the occasion of the 108th SIF Congress, taking place in Milano, SIF published and gave as a gift to the Congress participants, a special supplement of Giornale di Fisica, *Occhialini's Memoirs*, edited by L. Gariboldi, M. Gervasi, G. Sironi, P. Tucci. This supplement contains notes and documents of Beppo Occhialini found in Milan. Indeed, at the end of the 1980s, after the retirement of Beppo Occhialini, Giorgio Sironi moved from the first floor of the Astrophysics building in the backyard of the Milano Physics Department, in via Celoria, to the ground floor, in Beppo's former room. Giorgio Sironi found the room full of documents left by Beppo.

Something similar occurred with Guido Vegni, for a long time assistant professor of Beppo.

Because of this new available material, it has been decided to produce now this revised and extended version of the book published in 2006.

I heartedly thank Leonardo Gariboldi e Pasquale Tucci who undertook successfully the work of preparing this new book. I recommend reading this book because it is really inspiring for its scientific and historical content.

Bologna, Italy

Angela Bracco
Italian Physical Society, President

Società Italiana
di Fisica

Preface

A Brief History of the Dawn of the Occhialini-Dilworth Archives

At the end of the 80's, after the retirement of Beppo Occhialini, Giorgio Sironi moved from the first floor of the Astrophysics building in the backyard of the Milano Physics Department, in via Celoria, to the ground floor, in the former Beppo's room. Giorgio Sironi found the room full of documents left by Beppo. As the Department had no other place to store them, they were put in safe file cabinets in the rear part of the room, while Giorgio Sironi placed his desk at the room front. All the documents have been stored in this room up to December 1993 when Beppo passed away.

Something similar went on with Guido Vegni, for a long time assistant professor of Beppo. He had his room at the so-called "Capannino of Particle Physics" where Beppo and collaborators studied nuclear emulsions and bubble chamber pictures, early systems for digitizing and analyzing pictures and, together with Emanuele Quercigh, methods to produce intense magnetic field pulses.

When Beppo passed away Guido Vegni and Giorgio Sironi gathered all the documents left by Beppo. Once again the Physics Department said that there was no room nor special interest to preserve Beppo's documents. So Giorgio and Guido decided to transfer them to the Archives of the Physics History, that Pasquale Tucci was creating in the University rooms near, but separated from, the Astronomical Observatory in Brera building. These documents formed the first nucleus of what will become the Occhialini-Dilworth Archives.

New documents arrived in 2005 from the country house in Marcialla (Garfagnana, Tuscany) where Connie Dilworth moved after her retirement from the University at the end of the 90's. Before leaving Milano she collected in fact all the documents which remained in the flat where she and Beppo used to live, in Viale Argonne, and in her room at the theoretical astrophysics group in via Aselli. These documents were brought in Marcialla and there remained. When Connie passed away, in 2004, Etra Occhialini, daughter of Beppo and Connie, invited Tucci and Sironi to Marcialla and in 2005 handed them a large collection of documents. In a couple of days they were brought to Milano and gradually added to the already existing Beppo documents forming the Occhialini-Dilworth Archives (see: Occhialini E., Tucci P., "The Occhialini-Dilworth Archives", in *The Scientific Legacy of Beppo Occhialini*, edited by Redondi P., Sironi G., Tucci P. and Vegni G. (SIF/Springer-Verlag, Bologna/Berlino) 2006, pp. XXXIX–XLI).

At that time, after the acquisition of the new documents, the Archives were supposed to be completed, but it was not so. Few months later, as matter of fact, Etra called Sironi saying that she had still a folder with documents which had not been handed over yet. They were drafts of memoirs of Beppo's scientific life that he was preparing. They included outspoken judgements on various colleagues, which may have appeared hurting for somebody. Being doubtful on what to do, Etra handed over the folder to Sironi asking him to take a decision on it.

After reading the new documents, Sironi said they were extremely important and had to be preserved, but with the attention to treat them as confidential documents. Etra accepted the hint and left the folder to Sironi. Sironi made a copy for himself and handed over the original documents to Tucci to be preserved. By common decision, the folder left by Etra was sealed inside an envelope marked "riservato" ("confidential"). Waiting for future decisions, the envelope was not added to the Occhialini-Dilworth Archives.

In 2013 when Tucci retired and left the Brera rooms, the full archives were transferred to the Library of the Physics Department in Via Celoria, but the envelope with the Occhialini memoirs remained in the hands of Tucci. The death of Vegni in 2016 and of Etra Occhialini in 2019 made any decision even more difficult.

At the beginning of 2022 Sironi became convinced that the time to publish Beppo's memoirs has arrived. He spoke about this opportunity with some colleagues, who encouraged its publication. Afterward, Tucci, in February 2022, sent the original documents to the BICF Library (Biology, Informatics, Chemistry, Physics Library) of the University of Milan, where the Occhialini-Dilworth Archives are kept.

A Brief History About the Origin of the Occhialini "Memoirs"

To understand the composition and structure of these "Memoirs" we have to recall how they were prepared. Toward the end of his life, at the end of the 80's Beppo retired in Paris, Boulevard Raspail, guest of Marianne Labeyrie, sister of Jacques Labeyrie, an old friend of Beppo. Here Sironi had many opportunities of visiting Beppo, having talks with him and accompanying him for short walks along the boulevard. He was still a bright person, eager for the last news in physics and astrophysics, ready to joke on his health conditions. Once Beppo told Giorgio "Look at those nannies pushing prams with babies, in the past I used to rush to overcome and look at them, now I have to let them overcome me".

In Paris Beppo began to dictate in French his memoirs to Marianne. Every time a well organized set of Beppo talks was ready Marianne sent it to Italy to Connie who, also via back and forth exchanges with Marianne, translated, rearranged and typed it in English.

These exchanges between Marianne and Connie involved, among others, Giorgio Sironi and Dario Maccagni, as postmen. Unfortunately this process was halted by the death of Beppo in December 1993. So the memoirs are unfinished but the material we have covers over 30 years of Beppo scientific life. They are contained in the "riservato" folder Etra Occhialini gave Giorgio Sironi.

The documents in that folder can be divided into three parts.

- GO1, containing the final pages rearranged and typed by Connie Dilworth, in English;
- GO2, containing Connie's notes and drafts, in English;
- GO3, containing the original sheets written by Marianne, in French.

It is obvious that there are similarities among the information provided by GO1, GO2 and GO3, being GO2 the elaboration and translation in English of GO3 and of further exchanges between Marianne and Connie, and GO1 the typing and further elaboration made by Connie. All the three documents are essential to get an accurate reconstruction of Beppo's point of view. This is the main reason for taking the opportunity to present here the complete text faithful to the original one contained in the folder "riservato". The text here published has been prepared by Pasquale Tucci.

With the present publication the Italian Physical Society, and the Physics Department of Milano-Bicocca, the Occhialini Department, intend to offer the interested reader the possibility of studying and analyzing the original documents of Beppo Occhialini memoirs, compare with other documents of the Occhialini-Dilworth Archives and carry on specific studies on Beppo life and thinking. A next opportunity for further studies could be the remembrance of the thirtieth anniversary of Beppo's death, in 2023.

Massimo Gervasi
Università degli Studi di
Milano-Biccocca
Milan, Italy

Giorgio Sironi
Università degli Studi di
Milano-Biccocca (retired)
Milan, Italy

Acknowledgments

The originals of the transcribed texts are in the Occhialini-Dilworth Archives kept at the BICF (Library of Biology, Computer Science, Chemistry, Physics) of the University of Milan. The Archives material was donated by Etra Occhialini, the only daughter of Dilworth and Occhialini and was rearranged and inventoried by Agnese Mandrino, archivist of the Brera Astronomical Observatory.

Permission for consultation was granted to the authors of the transcription by Dr. Martella, head of BICF.

The authors are available to whichever heir they were unable to find for any clarification regarding the transcription of Archives papers.

Occhialini's Memoirs, a cura di L. Gariboldi, M. Gervasi, G. Sironi, P. Tucci, Supplemento Speciale del Giornale di Fisica, Vol. 63 (2022). Reproduced/reprinted with permission from Società Italiana di Fisica, Bologna, Copyright © 2022.

The Scientific Legacy of Beppo Occhialini, edited by P. Redondi, G. Sironi, P. Tucci and G. Vegni (SIF, Bologna; Springer, Berlin, Heidelberg, New York) 2006. Reproduced/reprinted with permission also from Società Italiana di Fisica, Bologna, Copyright © 2006.

Introduction

It's been over 30 years since Occhialini is no longer among us and almost 20 for Dilworth. The more one studies and analyzes their published works and their manuscripts kept in the Occhialini-Dilworth Archives at the University of Milan, the more evident become some little-known features of their personality.

Occhialini's contributions to the study of cosmic rays are well known: from 1931 to 1934 he worked in Cambridge (UK), at the Cavendish Laboratory directed by E. Rutherford. There, together with P. Blackett, he developed the controlled cloud chamber, a technique that Occhialini had learned from B. Rossi. From 1945 to 1948 Occhialini worked at the Wills laboratories in Bristol, directed by the arteriosclerotic (so named by Occhialini in his private correspondence with Dilworth) A.M. Tyndall,[1] with initial funding obtained from the Royal Society through P. Blackett.[2] There he worked in a group led by C.F. Powell.

In Bristol, Occhialini proposed to llford to produce a new type of photographic plate with a high concentration of silver bromide.

We know from historians' and from his colleagues' several papers and reminiscences that relationships, both human and scientific, between Occhialini and Blackett were marked by a great mutual friendship and esteem. Blackett was gentle and full of life and humor behind his shyness and reserve.[3]

On the other hand we know that there was no feeling between Occhialini and Powell. In a private sheet Occhialini claims that he had signed a "mercenary contract" with Bristol.[4]

---

[1] Archivio Occhialini-Dilworth, Serie 7M File 101.

[2] "Blackett m'a envoyé à Bristol et m'a remis de l'argent de la Royal Society pour cette mission aussi je devais rester accrocher jusqu'à la fin." Archivio Occhialini-Dilworth, Serie 7M File 101. Gariboldi, Leonardo; Tucci. Pasquale (2022). "Occhialini's Memoirs" Giornale di Fisica 2022 63: 1–75, pp. 34–35.

[3] Hodgkin. A. et al. (1975). "Memorial Meeting for Lord Blackett, O.M., C.H., F.R.S. at the Royal Society on 31 October 1974" *Notes and Records of the Royal Society of London* 1975 *29 (2)*: 135–162, pp. 145–146.

[4] Archivio Occhialini-Dilworth, Serie 7M File 101.

In Bristol Occhialini met Connie Dilworth: their human and scientific partnership allowed the couple to introduce remarkable innovations first in Brussels and then in Italy, initially in Genoa and afterward in Milan.[5]

In the mid-50s, Occhialini and Dilworth, as several other "cosmiciens", were faced with a crossroads. The new accelerating machines allowed the production of elementary particles in a predictable and more controlled manner than those detectable in cosmic rays. As Leprince-Ringuet summed it up in his closing speech at the Congress on cosmic rays in 1953: should we continue to worry about any new results from cosmic rays or was it better to turn to machines?[6] The technique of nuclear emulsions could be used for detection of particles produced by new machines and many cosmic ray scholars converted to them. But there was another possibility for the "cosmiciens": the study of the cosmological origin of cosmic rays.[7] In 1960 Occhialini and Dilworth went to the MIT as visiting Professors; there they interacted with B. Rossi. Once back to Italy they gave rise to the birth of cosmic and space physics, a research sector that saw them engaged until their retirement.

In each of the phases of his scientific activity Occhialini, and later together with Dilworth, left deep traces and got relevant scientific results.

But they did not exhaust their desire to investigate and understand the world and the people around us even when they retired. A novelty, in fact, is the recent acquisition by the Occhialini-Dilworth Archives of a substantial series of manuscripts and typescripts, dated 1992/1993, donated by their only daughter—Etra (Connie Pooh for the parents' friends)—who recently passed away. The events of the acquisition by the Archives are documented.[8] Here it is worth pointing out how what we have called "Occhialini's Memoirs" reveal a trait of the personality of the two that the previous works had only partially highlighted. From 1992, and until a few months before his death, Occhialini decided, together with Dilworth, to entrust the memory of the events of his life to written texts. The demise of Occhialini interrupted the work. But even in their incompleteness, the memoirs reveal unknown personality traits.

---

[5] Gariboldi, Leonardo; Tucci, Pasquale (2006). "Giuseppe Paolo Stanislao Occhialini (1907–1993). "A short biography" in Redondi, Pietro; Sironi, Giorgio; Tucci, Pasquale; Vegni, Guido (editors) (2006) The Scientific Legacy of Beppo Occhialini (Bologna: Springer-Verlag Berlin Heidelberg, 2006), pp. XI–XXXVII.

Gariboldi, Leonardo (2022c). "The Institute of Physics in the Post-War Period. Part 2: Some Highlights on Research in the Post-war Period" in Gariboldi, Leonardo; Bonolis, Luisa; Testa, Antonella (2022). The Milan Institute of Physics. A Research Institute from Fascism to the Reconstruction (Cham: Springer International Publishing, 2022), pp. 221–261, pp. 233–242.

[6] Leprince-Ringuet, Louis (1953). "Discours de clôture" in Proceedings of the *International Congress on Cosmic Radiation organized by the University of Toulouse, Bagnères de Bigorre, 6–11 July 1953.*

[7] Connie Dilworth, Note in Archivio Occhialini-Dilworth, Serie 7M File 101.

[8] See Occhialini, Etra; Tucci, Pasquale (2006). "The Occhialini-Dilworth Archive" in Redondi, Pietro; Sironi, Giorgio; Tucci, Pasquale; Vegni, Guido (editors) (2006). The Scientific Legacy of Beppo Occhialini (Bologna: Springer-Verlag Berlin Heidelberg, 2006), pp. XXXIX–XLI; and Gervasi, Massimo; Sironi, Giorgio (2022). "A brief history of the dawn of the Dilworth-Occhialini Archives" In "Occhialini's Memoirs" Giornale di Fisica (Supplemento) 2022 63: 1–73, pp. 1–3.

The complete transcription of the Memoirs, in their original language—French and English—is included in the present text.

There are several reasons to print a new edition of this book. The University of Milan and its Department of Physics, now named after Aldo Pontremoli, the founder of the first Physics Institute in Milan in 1924, celebrate their first hundredth anniversary in 2023–24. Such an anniversary is not only a stimulus to reflect on the future of current research and on the interactions of the university with all members and institutions of the society at large. It is also an occasion to consider a historical path and the main actors and events, which played a fundamental role in this century.

Among the main physicists working and teaching at the University of Milan, Occhialini can be considered the most known, if not the most relevant at all. From the early 50s, when he joined the University of Milan and the Milan division of the newly established National Institute of Nuclear Physics (INFN), Occhialini was the leader of a research group which stimulated international collaborations with other European universities and led to the development of cosmic ray physics on balloon and to the birth and development of space physics until his retirement. In Milan he was soon joined by Dilworth, with whom he formed a very active husband-and-wife team whose scientific and private lives almost coincided. As recalled above, the scientific and private papers left by Occhialini and Dilworth both at university and at home were donated by their daughter. The Occhialini-Dilworth Archive is hosted by the BICF (Biology-Informatics-Chemistry-Physics) Library of the University of Milan.

In 1998 the University of Milan "Bicocca" was established. It was born as a branch of the University of Milan. The Department of Physics was one of the founding departments of the university. At that epoch, almost the full astrophysics group decided to move to the new university. Among its members were present also some of the former students of Occhialini who continued its activity research and were still active at the University of Milan. Considering that many of the scientific research lines, still ongoing, had been started by Occhialini, upon its foundation, the members chose to name the Physics Department of the University of Milan "Bicocca" after him.

Still today many of the research activities of the Physics Department—from the study of Cosmic Rays and astro-particles to the space physics investigation, from the high energy astrophysics to the microwaves cosmology—derive from the proposals made for the first time in Milan by Occhialini.

The birth of the University of Milan Bicocca took place only 5 years after Occhialini's death, when his memory was still alive. In fact, Giuseppe Occhialini passed away in Paris on December 30, 1993. Therefore, the year 2023 is the 30th anniversary of the death of Giuseppe Occhialini. Various events, including exhibitions, editorial and multimedia releases, commemorate Occhialini on the occasion of this anniversary. The release of the present volume is part of this series of events. The contributions from the first edition that focus on Occhialini's activity before the 1950s are republished. To them we have added Occhialini's Memoirs, Occhialini's autobiography, and Dilworth's biography.

This new edition benefitted from the collaboration of two of Occhialini's associates, Giorgio Sironi and Aldo Treves. The last words of this introduction are left to them.

The first time I met Beppo was in November 1960 at his lectures of Fisica Superiore (Advanced Physics) at the Physics Institute of the University of Milan. In the following months my acquaintance with Beppo gradually improved and grew in the following years until 1993 when he passed away in Paris, just a few days I met him for the last time in Boulevard Raspail.

When talking with Beppo I was always fascinated by his capability of talking about, and finding deep links among apparently distant fields: physics, technical problems, literature, philosophy and everyday life habits.

This opened my mind and for this gift I received from him I can only say "Thanks Beppo ….."

Giorgio Sironi, Milano March 2nd, 2023.

In the occasion of the publication of a new edition of *The Scientific Legacy of Beppo Occhialini* with the addition of his memoirs, it was natural to recollect my interaction with Beppo, which started in 1966, when I asked for thesis in his group, and ended with his death. I was a student of his course of Advanced Physics in which I learned more than in all others of the Physics curriculum of the University of Milan. The course was obviously planned by him. He never lectured, that was the job of his assistants (I became one of them in a couple of years). At the end of the course he examined personally each student, and the exam itself was a supreme lesson of Physics. When my thesis was finished, together with my colleagues in the same situation, and I recall in particular G. Bignami and D. Maccagni, we were invited in his apartment to discuss our thesis presentations, and that was the occasion for Beppo to convince us of our ignorance, and to learn a lot, not only on Physics, but on the many subjects on which he was an expert. The contribution of C. Dilworth to lead us in the direction that Beppo had in mind, at times was essential. One of the reasons of Beppo's fascination was that he listened attentively, and one felt respected, even if the talk was between as student and a famous professor.

I had just entered in his group when the terrible flood of Florence occurred. Beppo considered the event a personal calamity. Everybody was impressed, and there was a rush to go to Florence and help. I was assigned to the Gabinetto Vieusseux. When I referred to Beppo I was fulgurated by his detailed familiarity of the place, of which I knew practically nothing. Long discussion regarded progress in Astronomy, in particular quasars and pulsars, on which Beppo was curious and informed, even in some way detached, implying that this was not his business. Other arguments could be art, cinema, or crime stories. Among the members of these long discussions I can mention in particular two young (at the time) astrophysicists of the group L. Maraschi and E. G. Tanzi.

Often the conversations regarded the Physics in his years in Cambridge. This is amply described in his memoirs. Let me add an element which I remember vividly. Beppo was describing the reaction of the Italian Physics community to his discovery with Blackett of the positron, and he referred of a telegram he had received just after the publication, complaining to have lost the original. The text was "All'esploratore buca Dirac: Alala'. I fisici romani" (To the explorer of Dirac hole. The Roman physicists: Alala' was a fascist form of salutation).

Beppo discussed often of the political opinions of the Italian physicists in the thirties, a theme that is present in the memoirs. He said the large majority were fascist, very few were not fascist, and two were antifascist: Bruno Pontecorvo, and Sergio Debenedetti.

I remember Beppo for his capacity as a scientist, as a teacher, often using non conventional ways, but especially for his humanity. Thinking of him, it comes to my mind a motto of Italian Humanists: *Nihil humani mihi puto alienum.*

Aldo Treves, University of Insubria, Como.

Leonardo Gariboldi
Massimo Gervasi
Giorgio Sironi
Aldo Treves
Pasquale Tucci

Signed photograph of the 1933 Solvay conference, given to Occhialini. (Copyright: University of Milano-Bicocca, Department of Physics "Giuseppe Occhialini")
Seated (left to right): Erwin Schrödinger, Irène Joliot, Niels Henrik David Bohr, Abram Ioffe, Marie Curie, Paul Langevin, Owen Willans Richardson, Lord Ernest Rutherford, Théophile de Donder, Maurice de Broglie, Louis de Broglie, Lise Meitner, James Chadwick.
Standing (left to right): Émile Henriot, Francis Perrin, Frédéric Joliot, Werner Heisenberg, Hendrik Anthony Kramers, E. Stahel, Enrico Fermi, Ernest Thomas Sinton Walton, Paul Dirac, Peter Joseph William Debye, Nevill Francis Mott, Blas Cabrera, George Gamow, Walther Bothe, Patrick Blackett, M.S. Rosenblum, Jacques Errera, Ed. Bauer, Wolfgang Pauli, Jules-Émile Verschaffelt, M. Cosyns, E. Herzen, John Douglas Cockcroft, Charles Drummond Ellis, Rudolf Peierls, Auguste Piccard, Ernest O. Lawrence, Léon Rosenfeld.
Absents: Albert Einstein and Charles Eugène Guye

Historiographical Notes to Occhialini's Memoirs

The folder holding the sheets here transcribed is kept in the Occhialini-Dilworth Archives at the BICF Library (Biology, Informatics, Chemistry, Physics Library) of the University of Milan. The folder was found by Etra Occhialini (1951–2019), the only child of Dilworth and Occhialini, when, after Dilworth's death, she emptied the house in Marcialla, in the province of Florence, where the mother lived after her retirement from the University of Milan.

The folder consists of a hundred sheets: some are type-scripted; others are handwritten by Constance Charlotte Dilworth (1924–2004), —Connie for her friends—; others are handwritten by Occhialini's friend Marianne Labeyrie. As a whole they constitute what we have called "Occhialini's Memoirs". They cover the two English periods of Giuseppe Paolo Stanislao Occhialini (1907–1993) —Beppo for his friends—in Cambridge between 1931 and 1934 and in Bristol between 1945 and 1948. Some portions of the Memoirs include the Brazilian period in between the English periods, and essentially deal with Occhialini's delicate position when Brazil joined the Allied Coalition against the Axis and he officially became an enemy alien.

The Memoirs were written between an unspecified day in 1992 and April 16, 1993. They do not have the organic form of a historical reconstruction but are Beppo's memories and clarifications of various episodes of his scientific life unconnected to each other. Connie was the real driving force behind the Memoirs. Beppo, who at that time lived in Paris, dictated to Marianne Labeyrie his considerations on various topics, without a specific order or scheme. They were sent to Connie who, starting from Beppo's recalls, prepared a handwritten draft, divided into topics, where her considerations were added to Beppo's ones. Afterward they were sent to Beppo who dictated to Marianne Labeyrie his comments which came back to Connie so that she could prepare an orderly and coherent manuscript, that presumably Connie herself typed.

All of Beppo's considerations are off the cuff, without reference to specific dates or documents. Nor was Connie interested in defining what Beppo had left vague. From the historiographical point of view it is easy to see how some of Beppo's statements, based only on his memory, are at odds with documents. As highlighted by many historians of science, a historical reconstruction based only on memories tends to favor a linear and rational path toward what the scientist thinks are his/her best "their" results. Beppo's aim was not to offer a historical reconstruction of the events that concerned him. Among other things, we do not even know if these Memoirs were intended for printing. It was instead a way of reaffirming the importance of some of his achievements that had been overlooked by both colleagues and historians.

In the Memoirs, the admiration, the esteem, almost the veneration that Beppo had for Blackett are evident. While the description of Rutherford highlights how the great scientist enjoyed making ironic and scathing judgments about those who proposed theories and experiments that he was not able to understand.

Occhialini's judgement on Heisenberg, Rosbaud, Houtermans is affected by the climate after World War II when admiration for the scientists was mixed with the suspicion that some of them could have collaborated with the Nazis.

Powell, on the other hand, was little esteemed, also in his entourage, because he was a communist and accused of not having participated in the war effort. The only one who defended him was his old professor Tyndall, director of the department. But Occhialini's account reveals above all Powell's lack of interest in the technological innovations—for example, the concentrated emulsions that Occhialini had suggested to Ilford—and introduced into the study of cosmic rays. When Occhialini decided to go to the University of Brussels, Blackett tried to keep him in England, in Manchester, but by now, the relationships with England had broken down in Bristol, where Occhialini had the feeling of being a mercenary who had sold himself for a few pounds a month.

The sheets contained in the folder, especially the manuscript ones, were rather messy. We have tried to maintain the order in which they came to us. But for some groups of sheets we preferred to give them a logical and chronological arrangement in order to make their reading easier.

The text of both the manuscripts and the typescripts have been reported in full and, when readable, we reported the corrections that had been made on them. They are indicated in our transcription with a horizontal bar on the single or group of words.

To our knowledge Connie's final draft of the manuscripts has been lost, except for some parts which are in the Occhialini-Dilworth Archives. Sometimes the difference between Connie's draft and the typescript is interesting. It is the reason why we have inserted both the manuscript and the typescript of the same subject.

Small changes have been made to punctuation, to make the reading of the text smoother.

The surnames, often distorted, have been corrected and, when possible, we give the year of birth and death in footnotes.

Few footnotes clarify some passages that are difficult to understand.

Biographical Notes on Dilworth and Occhialini

Giuseppe "Beppo" Occhialini and Constance "Connie" Dilworth were a husband-and-wife team of physicists who played a prominent role in the history of cosmic ray physics. We shall offer here a sketch of their biographies and refer the readers to the works in the bibliography for more details.

Constance Charlotte Dilworth was born in Streatham Manor, Leigham Avenue, London county, on February 5, 1924, the daughter of George Darwell Dilworth and Charlotte Cant Price. She studied at King's College, University of London, from 1941. She got her B.Sc. in 1944 and her M.Sc. in 1945.

During World War II, Dilworth worked at the Admiralty Research Laboratory. After her master degree, she worked as researcher in Cecil Frank Powell's team at H.H. Wills Laboratory in Bristol. After a first study on the effects of electric current on the surface films of semiconductors, she mostly worked on nuclear emulsions, a research tool studied by Powell since the late 1930s. Powell's team was then joined by Occhialini, whose collaboration with Cecil Waller of Ilford Ltd. favored the production of a new kind of emulsions, which allowed the discovery of the π meson. From now on, Dilworth's scientific career was strictly connected to Occhialini's one. In Bristol, Dilworth contributed in particular to the invention of a method of uniform development of the emulsions.

Dilworth followed Occhialini to Brussels when he left Bristol in 1948. With Max Cosyns, they established a new research team working on nuclear emulsions by Ilford and Kodak.

Dilworth's collaborations with Italian physicists became more intense when Occhialini became a professor at Genoa and Milan universities. Several young Italian physicists worked with her in Bruxelles and then in Milan on the development of the nuclear emulsions technique. In Bruxelles, Dilworth was particularly engaged in the study of the Auger effect in the μ-meson capture by heavy nuclei, the study of the magnetic deflection of long tracks in electron-sensitive emulsions, and the study of cosmic ray showers.

When Giovanni Polvani succeeded in establishing a local division of the newborn National Institute of Nuclear Physics, Dilworth could join Occhialini's team in Milan as an INFN researcher. She collaborated to the organization of flights of emulsion stacks on balloons at high altitude, which culminated with the G-Stack Collaboration in 1954. Later stacks were instead exposed to artificially produced particle beams at the Bevatron and at CERN. Dilworth collaborated to the K^--Collaboration (the study of K-mesons), the studies on the τ-θ puzzle (one of the first examples of parity non-conservation) and on the K_μ mesons, the hyperfragments, the decay schemes of the Σ-hyperon.

In 1957–59 Dilworth was entrusted with a teaching position at the Specialization School in Atomic and Molecular Physics at Milan University. She was a professor of Radioactivity from 1960 (full professor from 1967) and then of Modern Physics ("Fisica Superiore" in the Italian university system).

While many collaborators continued to work on artificially produced particles, in particular at CERN, Occhialini and Dilworth turned their attention to space physics. In 1959–60 they both went to the MIT to work with Bruno Rossi. Dilworth was a staff member in the Laboratory for Nuclear Science and studied meteorological and geophysical researches on cosmic rays, in particular the devices to study interplanetary plasma.

Back to Milan, they established a cosmic physics team, which started an international collaboration with Jacques Labeyrie of the Centre d'Études Atomiques in Saclay and then also with Munich (the Mi-Mo-Sa collaboration). The Milan-Saclay collaboration launched devices on balloons and satellites to study the different components of primary cosmic radiation. From 1964, Dilworth was engaged in the European Space Research Organization; she acted as chairperson of the Space Committees. She collaborated to the launches of the first-generation satellites: the HEOS-A (1968) to study interplanetary magnetic fields and solar wind, and the TD-1 (1972) to study UV, X, γ, heavy nuclei, and solar radiation. Then she collaborated on the COS-B project, the satellite carrying the Gamma Ray Telescope (1975), which obtained the first complete γ-map of the Galaxy, and on the EXOSAT satellite (1983) to study several astrophysics topics. Dilworth also coordinated the establishment of the Gornergrat national observatory for infrared astronomy.

In 1985 Dilworth retired from scientific life and went to live in their country house in Marcialla di Certaldo, in Tuscany. She died on May 17, 2004.

Dilworth was honored with a degree by Bern University in 1954 and University of London in 1981. She was a member of the Italian Physical Society, the Royal Astronomical Society, the American Geophysical Union, and the Lombard Institute of Sciences and Humanities.

Giuseppe Paolo Stanislao Occhialini was born in Fossombrone, Pesaro province, on December 5 1907, the son of Raffaele Augusto Occhialini and Etra Grossi. Occhialini's father was a physics professor at Sassari (1921–24), Siena (1924–28) and Genoa (1929–51) universities.

Occhialini attended High School ("Liceo Scientifico") in Florence; he then studied physics at Florence University where he enrolled in 1927. In 1930 he graduated and became assistant at the Institute of Physics directed by Antonio Garbasso.

The scientific interests of the Institute of Physics shifted to nuclear physics and cosmic rays after the arrival of Bruno Rossi in 1927 and Gilberto Bernardini in 1928. In this way Occhialini became an expert of Rossi's innovation: the coincidence circuit. Rossi planned with Patrick Blackett the admission at the Cavendish Laboratory in Cambridge of one of his collaborators to learn the cloud chamber technique. In the summer 1931, Occhialini joined the Cavendish team, then directed by Ernest Rutherford.

The Cavendish Laboratory continued the tradition of research made by producing visible phenomena with scintillators, photographic plates, and cloud chambers; the Geiger-Müller counters were instead not in common use.

Blackett and Occhialini developed the controlled cloud chamber, i.e., a cloud chamber whose expansion was triggered by a coincidence circuit of counters placed above and under the chamber. The integration of the two techniques allowed a sudden and great increase in the number of useful images with particle tracks. In 1933, after the announcement of the discovery of the positron by Carl Anderson (1932), they published their discovery of the electron-positron couple production and the confirmation of Dirac's challenged theory of particle-antiparticle creation and annihilation.

Back to Italy, Occhialini was not able to continue a scientific activity of such a high level. The National Council of Researches did not offer him the support to build a cloud chamber. In 1937 he was invited by Gleb Wataghin to join him in organizing a school of physics at São Paulo University, in Brazil. Occhialini raised a group of young Brazilian physicists—Marcelo Damy De Sousa Santos, Mario Schönberg, Ugo Camerini, and César Lattes among the others—in cosmic rays research. From Brazil, Occhialini was active in organizing an international action to set Houtermans free from the imprisonment in the Soviet Union.

In March 1942 Brazil joined the allied nations fighting against Italy. Occhialini was called back to Italy but the English government refused to allow him the passage through the Atlantic. In June 1942 he was removed from São Paulo University. He worked as an alpine guide on the mountains between São Paulo and Rio de Janeiro. After Italy signed the armistice in September 1943, he worked in the Biophysics Laboratory, directed by Carlos Chagas. Blackett tried to persuade the Foreign Office into believing Occhialini's anti-Fascism and to employ him in the war efforts. At the same time, Occhialini refused an invitation from Ohio University and a Rockefeller fellowship. Eventually Occhialini landed unconditionally in Cardiff on January 23rd, 1945. He got a short temporary job at the Research Laboratories of the General Electric Company in Wembley because of the refusal of his employment from the Aliens War Service "Aliens War Service" (in blu): cosa si intende? Department. Occhialini accepted the invitation from the director of the H.H. Wills Laboratory and arrived in Bristol in September 1945 to work with Cecil Frank Powell's team on nuclear emulsions.

The collaboration of the Bristol team with the Ilford Ltd led to the production of boron-loaded plates, which were an impressive step forward in the nuclear emulsions technique. The exposition of Ilford C2 plates on the Pic-du-Midi allowed to identify the tracks of a meson decaying into another meson (the π-μ decay). The new π meson was identified as the nuclear meson predicted by Yukawa's theory. The discovery was announced in 1947. Occhialini, Dilworth and Ron Payne also studied the optimal way to develop the thick emulsions and invented the Temperature Development processing method. Blackett's and Powell's researches carried out with Occhialini were awarded with the Nobel Prize for Physics, respectively, in 1948 and 1950.

In June 1948, Occhialini decided to leave England (he refused Blackett's invitation to join him in Manchester); he went with Dilworth to work in Brussels at the Centre of Nuclear Physics of the Free University of Brussels where he stayed until the end of the 1950s. Dilworth and Occhialini got married in Brussels on October 4, 1950. The next year, their daughter Etra Mary Giovanna Occhialini was born -Connie Pooh for her parents' friends. While he continued to work in Brussels, he won the public exam for the chair of Modern Physics for Cagliari University in 1950. However, he was called by Genoa University to the same chair, which had been held by his father.

After his father's death, he left Genoa to Milan University in 1952. In Milan he could profit from Giovanni Polvani's scientific policy, that succeeded in establishing a local division of the National Institute of Nuclear Physics. At the same time, the Brussels laboratory hosted many young Italian physicists, in particular from Genoa and Milan. They continued to make advances in the nuclear emulsion technique.

In the 1950s, Occhialini was one of the most relevant organizers of international collaborations to fly emulsion stacks on balloons at high altitude, such as the G-Stack collaboration, or to expose emulsion stacks to beams of artificially produced particles, such as the K^--collaboration.

In 1959–60, Occhialini spent a sabbatical year with Dilworth at the MIT in Bruno Rossi's laboratory. Once back in Milan, he established a new research team that in 1968 was established as the Laboratory of Cosmic Physics and Related Technologies. The Milan team was engaged in international collaborations, at first in particular with the French group based in Saclay and led by Labeyrie. He was a member of the Council and of the Technical and Scientific Committee of the European Space Research Organization and can be considered one of the founding fathers of the European space physics. His most relevant contribution was the organization of the launch of COS-B satellite to study γ-astrophysics. The European research programs of launching of instruments on balloon or satellite saw Occhialini's constant contribution until his retirement in 1985.

In the 1980s Occhialini was also engaged in dating techniques with the analysis of fission tracks in rocks.

After his retirement, Occhialini spent most of his time in Marcialla di Certaldo, a hamlet in Tuscany, and in Paris. He died in Paris on December 30, 1993.

Leonardo Gariboldi
Università degli Studi di Milano
Milan, Italy
Pasquale Tucci
Università degli Studi di Milano—Retired
Milan, Italy

References

Belloni L., Dilworth C. The Restructuring of Physical Sciences in Europe and the United States 1945–1960 (World Scientific, Singapore) 1989, pp. 732–744.

Bignami F. *Nature*, **367** (1994) 515.

Bignami F. *Biograph. Mem. Fellows R. Soc. London*, 48 (2002) 331.

Fiorini E. *Il Nuovo Saggiatore* 1, N. 1 (1985) 18.

Gariboldi L. *Boletim da Sociedade Brasileira da Física*, **4** (2004).

Gariboldi L. *Il Nuovo Saggiatore*, **20**, N. 3–4 (2004) 16.

Gariboldi L. *History of Physics and Astronomy in Italy in the 19th and 20th Centuries. Sources, Themes and International Context (La Serigrafica, Milan)* 2003, pp. 137–150.

Gariboldi L. *Quaderni di Storia della Fisica*, **14** (2006) 95.

Gariboldi L. *The Global and the Local: The History of Science and the Cultural Integration of Europe* (Polish Academy of Arts and Sciences, Cracow) 2007, pp. 480–486.

Gariboldi L. *L'eredità di Fermi, Majorana e altri temi* (Bibliopolis, Naples) 2007, pp. 207–210.

Gariboldi L. *Scientific research in World War II: What scientists did in the war* (Routledge, New York) 2008, pp. 31–43.

Gariboldi L. *Relatività, quanti, caos e altre rivoluzioni della fisica* (Guaraldi, Rimini) 2010, pp. 329–336.

Gariboldi L. *Dizionario biografico degli italiani*, **79** (2013) 79.

Gariboldi L. *Giornale di Fisica*, **56** (2015) 251.

Gariboldi L., Tucci P. *Il Nuovo Saggiatore*, **23**, N. 3–4 (2007) 23.

Gariboldi L., Tucci P. *Il cammino dell'uomo alla scoperta dell'universo: Beppo Occhialini, scienziato europeo* (Printing, Bologna) 2007, pp. 69–81.

Gariboldi L., Verzeroli M. *Atti del XL Convegno annuale SISFA* (Pisa University Press, Pisa) 2021, pp. 133–139.

Occhialini G. *Scienziati e Tecnologi Contemporanei*, Vol. **2** (Mondadori, Milan) 1974, pp. 322–324.

Polvani G. *Rendiconti del Seminario Matematico e Fisico di Milano*, **39** (1959) 11.

Redondi P., Sironi G., Tucci P., Vegni G. (Editors), *The Scientific Legacy of Beppo Occhialini* (SIF/Springer-Verlag, Bologna/Berlino) 2006.

Russo A. *Sapere*, **4** (1996) 62.

Scarsi L. *Nuovo Cimento C*, **20** (1997) 613.

Tagliaferri G. *Rendiconti dell'Istituto Lombardo Accademia di Scienze e Lettere*, **128** (1997) 231.

Telegdi V. *Proc. Am. Philos. Soc.*, **146** (2002) 218.

Tucci P. *Il contributo italiano alla storia del pensiero—Scienze* (Istituto della Enciclopedia Italiana, Roma) 2013, pp. 721–725.

Tucci P., Gariboldi L. *The Global and the Local: The History of Science and the Cultural Integration of Europe* (Polish Academy of Arts and Sciences, Cracow) 2007, pp. 507–514.

Verzeroli M. *Nuovo Cimento C*, **44** (2021) 1.

Contents

Occhialini's Memoirs

L. Gariboldi, M. Gervasi, G. Sironi, and P. Tucci

Rutherford

Rutherford [Typescript]

Rutherford

In the two years I spent in Cambridge, most of the time at the Cavendish, I had two, maybe three, personal encounters with Rutherford.[1] My first encounter was the day after my arrival, on the stairs of the Cavendish. Blackett[2] introduced me. I filled one of my rare letters to my father with a description of the event, and of the sense of awe that I felt. I spoke of him as a 'big carnivore'.

Then I met him at the seminars and at the Royal Society meetings to which he almost always went. The public at the Cavendish was that of the 1932 group photo in which I hid myself at the top right, a shy habit I always had. I remember the public divided into three groups in their response to Rutherford's comments.When Ruther-

[1]Ernest Rutherford (1871–1937).
[2]Patrick Blackett (1897–1974).

L. Gariboldi (✉)
Dipartimento di Fisica "Aldo Pontremoli", Università degli Studi di Milano, Milan, Italy
e-mail: leonardo.gariboldi@unimi.it

M. Gervasi
Dipartimento di Fisica "Giuseppe Occhialini", Università degli Studi di Milano-Bicocca, Milan, Italy

G. Sironi · P. Tucci
Università degli Studi di Milano-Bicocca (retired), Milan, Italy

L. Gariboldi et al. (eds.), *The Scientific Legacy of Beppo Occhialini*,
https://doi.org/10.1007/978-3-031-37034-2_1

ford cracked a joke, one group laughed excessively, another with more composure, the third never laughed. Chadwick[3] and Dee[4] belonged to the third group.

In his comments there was an undercurrent of malice. He never hesitated to express an opinion, even if it was, to say the least, not diplomatic. When Millikan[5] came to the Cavendish and showed Anderson's[6] first photos, Rutherford gave me the impression of a Master of the Hunt as he called all the young Wilson Chamber specialists one by one by name to ask questions. Poor Millikan found himself attacked by all the young hounds. Rutherford did not ask Dee or Blackett to comment. The scene was distressing. Millikan had talked 'muy pomposo' but at the end he was destroyed. It seemed to me contrary to the laws of hospitality.

I regret that I did not write down all his comments. I remember after a talk[7] of Ellis[8] in which he contradicted Lise Meitner,[9] the catty phrase "Even a lady can drop a brick." The phrase about the theoretician, Eddington:[10] "We might hope the guesses will prove inspired" and about Dirac:[11] "When Dirac, who has already greatly improved, will know some physics…".

When Bainbridge[12] told us of the mass measurements with his spectrograph, Rutherford expressed his appreciation, but also his certitude that the precise measurements would be given by the analysis of nuclear reactions. Bainbridge had been in Cambridge for a month and I had got to know him well. I asked him what he thought of Rutherford's comment. He replied: "If we had been in the States I would have said—Sez you!"

The Cavendish was full of Rutherford anecdotes. One of them illustrated well his barely concealed despise for technique. It goes like this, (maybe the dates are wrong): Professor Rutherford coming to Cambridge meets C.T.R. Wilson[13] busy shaping a Cloud chamber piston: Sir Ernest coming back from Canada finds Wilson busy shaping a cloud chamber piston: Lord Nelson comes to the Cavendish and meets again Wilson, busy shaping a cloud chamber piston. Wilson's comment on this mean story was a meek 'Rutherford was right, but you see, it was not the same piston.'

There was something of the cruelty of a pagan god in Rutherford, there was gaiety in his massacres.

---

[5] Robert Millikan (1868–1953).

[6] Philip Anderson (1923–2020). Dilworth-Occhialini refer to photos taken by Anderson with a Wilson cloud chamber. In those photos the track of a new particle, the positron, was visible, whose mass was equal to the electron one but its charge was the opposite.

[7] See Jensen C., Aaserud F., Kragh H., Rüdinger E., Stuewer R. H., *Controversy and Consensus: Nuclear Beta Decay 1911–1934* (Birkhäuser, Basel) 2000, pp. 55–93. https://doi.org/10.1007/978-3-0348-8444-0_3.

[8] Charles Ellis (1895–1980).

[9] Lise Meitner (1878–1968).

[10] Arthur Eddington (1882–1944).

[11] Paul Dirac (1902–1984).

[12] Kenneth Bainbridge (1904–1996).

[13] Charles Wilson (1869–1959).

Blackett's first experience in the Cavendish reflects this despise for technique. Rutherford had been struck by Shimizu's[14] simple cloud chamber with a continuously moving piston (like a sewing machine needle). He asked Blackett to construct for him such a simple chamber, capable of showing α-particles and protons. It was Napoleonic strategy, for him tactics did not exist. Blackett started to build a very complicated automatic apparatus. Rutherford went around the Cavendish exclaiming: "If Blackett goes on delaying I'll have the carpenter make a wooden chamber and it will work." I forget who told me this, but Blackett explained to me: "Rutherford was very impatient, he had not understood the problem so there was a period of coolness between us while not a single photo of a track came out of the chamber, until one Monday morning I closed the 9 switches of the apparatus. After that, in 40 days I took more photos than had been taken till then anywhere in the world."

Geiger[15] came to the Cavendish to talk about his work with proportional counters. Afterwards I found Rutherford alone and approached him to ask him if the separation of the α-particle 'kicks' could be due to the intermediate charge of the α-particle at the end of its range. He said no, but that he had thought it might be that at first sight, and concluded "your brain works the right way". It was a back-handed compliment, but it showed that he had no cruelty towards a young man and for that I had endless gratitude.

On the other hand, the story of the Wilson chamber in wood shows his limitation in a changing world. He accepted proportional counters and electronics without enthusiasm, and his young men were crushed by his personality.

At a dinner, he defined the annular magnet for the fine structure of α-particles as a pretty Yo-Yo. He seemed to have an almost religious despise for people like Geiger, Wilson, Blackett maybe Cockcroft[16] and Walton[17] who were the Héraults.[18]

I never understood why Geiger counters were not adopted by the Cavendish. When I came there one man, H. Webster,[19] was working with Geiger counters with Chadwick. When at the end of his stay he left to go to South America, he was close to the discovery of the neutron. After that, the only Geiger counters made in the Cavendish were very small ones with walls of aluminium foil that I made for Ellis and another I made for Rutherford and Bowden.[20]

Rutherford's scorn for technique seems to have spread through the world. After the war, at the Rochester conference, a rapporteur showed only the curves he had obtained, saying that the experimental apparatus was 'not interesting'. I have always thought that a bigoted attitude like that of the Victorians who celebrated the birth of children while refusing to mention how they are conceived.

---

[14] Takeo Shimizu (1898–1976).

[15] Hans Geiger (1882–1945).

[16] John Cockcroft (1897–1967).

[17] Ernest Walton (1903–1995).

[18] Connie Dilworth didn't know who the Héraults were, as we can see on p. 72.

[19] Hugh Colins Webster (1905–1979).

[20] Bertram Vivian Bowden (1910–1989).

When Cockcroft and Walton presented their work to the Royal Society, Rutherford gave a long commentary. It was very moving, he described how hard work was in his youth, the lack of voltmeters to measure the voltage of the accumulators, the counting of scintillations, the coincidence system for determining the efficiency of the method and, in a manner which made me think of the Merry Wives of Windsor, he finished with this phrase: "And now with this apparatus Cockcroft and Walton go on disintegrating elements all their life without a thought." The text is Rutherford's, the underlining is mine, but it was there in the emphasis with which he pronounced it.

I wondered if he had some resentment, if he thought he should have signed the paper. Cockcroft had no inhibitions with me. The design of the machine and the use of low voltage had come out of a personal contact of Cockcroft with Gamow,[21] before Walton, who unconsciously was in the way of inventing the betatron, had joined forces with Cockcroft. Without a fault.

Some years later Lawrence[22] received the Nobel prize and Cockcroft and Walton were ignored by the Nobel committee. One version recounts that Cockcroft when questioned said the Nobel prize should be shared with Walton and this had paralysed the committee.

So much for Rutherford's presences at the Royal Society, but sometimes absence is more important than presence. On the 17th February 1933 when Blackett gave the evidence for the positive electron to the Royal Society, neither Rutherford nor Chadwick was present. That seemed to prove they did not believe in it.

Rutherford did not believe in all the results of his lab. The analysis of Gray[23] and Tarrant[24] had proved that γ-rays from Thorium after scattering had energy in the range between 500.000 and 1.000.000 eV. The Cavendish, following Rutherford's flag tended to the conclusion that the scattered radiation of Gray and Tarrant was due to multiple scattering over-corrected in the diffusion chamber they used. (I remember pleading with Gray and Tarrant to use Geiger counters in coincidence as that brilliant Chinese physicist Tchao[25] had done in Pasadena.)

Nobody around Rutherford believed in Dirac's theory. As I learnt afterwards from Chadwick, he thought that on the basis of the mass difference there could be place for a charge difference associated with the neutron.

Chadwick, I got to know better later and grew to like him. I never got close to Rutherford, and I certainly was irritated by his absence from that Royal Society meeting. He had been one of the idols of my youth. My father had brought me up in a religion in which Rutherford was a god who was always right.

[21] George Gamow (1904–1968).

[22] Ernest Orlando Lawrence (1901–1958).

[23] Louis H. Gray (1905–1965).

[24] Gerald T.P. Tarrant (1906–1965).

[25] Chung-Yao Chao (1902–1998).

In the last days before the Royal Society presentation, the paper took its final form with occasional discussions with Dirac. Blackett showed surprise at Dirac's common sense and understanding of experimental phenomena. It is certain that Dirac was more respected in Rome at the laboratory of Fermi[26] and Rasetti.[27]

At this time I was learning German using the Bible as text. My first composition was the Genesis in German, substituting Rutherford for "God".[28]

It was a great success with Mrs. Blackett and drew a sad smile from Blackett. In this Genesis I described the spirit of Rutherford which moved over the waters: Chadwick discovers that the radiation Bothe,[29] Becker[30] and Joliot[31] was composed of neutrons "und Rutherford sagt dass es nicht gut war". Then Cockcroft and Walton propose to disintegrate light elements with protons of low energy "und Rutherford sagt dass es gut war". Now Blackett proposed that the Dirac particle exists and that its existence explains the effect Tchao-Gray-Tarrant, and Joliot's tracks moving towards the source "und Rutherford sagt dass es nicht gut war".

I sent this composition to my father. I remember he did not appreciate it.

Blackett defined the situation in a lapidary manner: "The work at the Cavendish was first class, but the pigeon hole of fundamental ideas belonged to the old ones—to Rutherford and Chadwick."

Rutherford was not the only one who did not believe in the positive electron. Blackett showed me a postcard, saying that if I collected autographs I might like to keep it. The postcard said "Congratulations on your work, but even if the positive electron exists, I don't believe it, and I do not want to believe in the Dirac theory"—signed Niels Bohr.[32] I kept that postcard for many years, but I left it in Brazil with friends who lost it. The text I have given is very close to the original. The emphasis on the fact that Bohr did not believe in it and that he was sure that in the future he would be proved right, is not my invention.

That Bohr had objections does not seem to have influenced the Nobel committee who seven months later awarded Dirac the Nobel prize. Still, Rutherford and Chadwick were in good company.

[26] Enrico Fermi (1901–1954).

[27] Franco Rasetti (1901–2001).

[28] In a file named "The Book", published in the present publication on p. 69, Dilworth and Occhialini spoke about the role of God and his relationship with the Archangels. The description of Rutherford as a God is linked to their considerations about the Book of Genesis.

[29] Walther Bothe (1891–1957).

[30] Herbert Becker (1899–1980).

[31] Frédéric Joliot (1900–1958).

[32] Niels Bohr (1885–1962).

[Cambridge]

*Before and After the 17***th** *February 1933 [Typescript]* [33]

Before and after the 17th February 1933

On the 22nd December 1932, I heard from Bernardini[34] that I had lost my job as assistant at the university because I had not been there present on the 15th October (beginning of the academic year) as the law required. As usual I did not know of this law, and having asked for leave to finish my research in Cambridge and, through my professor Garbasso[35] for an extension of my CNR fellowship, I thought I had no reason to worry.

After all these years I can think that the fascist State was right, but then I felt the situation was tragic. The only place where one could do research was the university, but things had changed since I was first appointed. Now to be appointed of the university staff one had to belong to the fascist party. I was convinced that my being sacked was a blackmail by the Secretary of the University to force me to take the party card. My relations with my professor had cooled (I did not know that he was ill) and I was persecuted by letters from my friends urging me to leave England and to try to make peace with the university authorities and with the fascist party. Under these conditions I could only work like a madman in the hope that my luck would turn.

I was completely broke. I owed my landlady two months of board and heating. I had given up my bedroom to my Canadian friend Ouellet[36] and slept on a couch in the common sitting room. (I had tried sleeping in the garden shed but a bad cold brought me back into the house.)

My shoes were worn out, and also the only pair of good trousers I used to go to the Royal Society. I bought a second-hand pair from my French friend Rosier. I went to the lab. in tennis shoes and white trousers, carrying a tennis racket on the handlebar of the bicycle (to complete the uniform). The bicycle was borrowed from the bicycle stand of the Biology lab. There were a lot of bicycles in that stand, some of them dusty, as if their owners were out of Cambridge. I marked them with a piece of wire and checked they were always in the same place. One of those I would use to go to the Cavendish.

Sometime in January 1933, (I do not remember the exact date), after three days in which the chamber was full of mist and we had not been able to get tracks, I went to the lab. on a Saturday morning to see if the situation had improved. I intended just to look at a few expansions with the counter control. I do not know why, I loaded

[33] The Typescript Has the Same Content as Dilworth's Manuscript Which Begins with "On the 22nd December 1932, ..." whose transcription is after this typescript.

[34] Gilberto Bernardini (1906–1995).

[35] Antonio Garbasso (1871–1933).

[36] Cyrias Ouellet (1906–1994).

the cameras and adjusted the lamps to avoid scattered light on the glass. I borrowed Karen's[37] Basque beret to complete the optical screen and set the magnetic field to the maximum of 3000 gauss with a delay of 2.5 min. I waited for the chamber to trigger, watching the thermometer (of motor-oil) which was rising dangerously. I had my hand on the switch of the magnet current when the chamber triggered. There was some mist but the heat of the event created a clear zone in which I thought I saw the event I had been waiting for.

It was 11 am when Karen and I arrived bicycling at Blackett's house. I told him there was something important on the two plates, that I had not had the courage to develop them. For a sort of superstition I refused to go with him to the lab. for the development. The photo of the shower is well known, the shadow of Karen's beret can be seen partially on the right. The extraordinary thing is that we told no one about it, we kept it to ourselves. I do not know when Rutherford saw the photo. When Blackett came home for lunch he asked me to describe what I had seen during the flash. Karen completed my sketch by telling him I had said there were five 'useful' tracks.

I had broken the rules by going to the lab on a Saturday, alone with a woman. What was to have been a simple control of the operation of the chamber I had turned into an act of research. Nevertheless God forgave my sins and gave me the privilege of a first sighting. I resisted the desire to confirm that glimpse by developing the plates. Had I done so I would have faced Blackett with a fait accompli. After all his work it was right that he should have the responsibility of seeing the evidence of what I had guessed. There was a mixture of positive and negative particles and the positive ones did not look like protons.

It was an act of love to refuse to go back with him to the Cavendish so that he could meet the particle alone. Several, maybe ten times in my life, I did something like that, in my youth in caves, on mountains, before a nuclear emulsion. Blackett was the only one who remembered and understood the meaning of the gesture.

The day after there began the discussions with Dirac, and the presentation at the Royal Society was fixed for the 17th February. In the paper there was half a page concerning the effect found by Tchao and by Gray and Tarrant, and the 'backward moving' tracks of Joliot. I had asked for this reference in the paper but at one stage Blackett took it out. I was surprised when, in the final edition, Blackett proposed to put it in again. That made me understand that there was a critical attitude in the Cavendish towards the paper, and Blackett was replying to it with the affirmation that the Tchao-Gray-Tarrant effect was real. (In my autobiography in the STET[38] there is a Freudian slip when I said that in the paper there were suggestions difficult to accept. I should have said, difficult to accept in a Cavendish influenced by Rutherford and Chadwick.)

---

[37] Around those years two women were in the staff of the Cavendish lab: Mackenzie and Salaman. We have been unable to find who of the two was Karen.

[38] Occhialini G. P. S., *Scienziati e Tecnologi Contemporanei*, Vol. **2** (Mondadori, Milano) 1974, pp. 322–324.

Not long ago, Yang[39] wrote to me enclosing an article in which he complained that we had put Gray and Tarrant on the same level as Tchao. I think we were justified in doing so. Tchao came to England in 1932 on his way to Germany. He stopped in Cambridge to see Gray and Tarrant who treated him as a friend. They brought him to meet me and I received him like a brother. That is to say that among the experts in his field he had a triumphal welcome. He told me, with gentle regret, that his work had not been appreciated in Pasadena. We spent two days together talking about counters and anticoincidence. Many years later, not knowing if he were still living, I proposed him for the Wolf prize.

The Royal Society paper gave me a new identity. I thought it would be fun to have more than one initial, like Blackett. I started with two, G.P. but Blackett said that with the O for Occhialini that stood for General Post Office, so I added an S, the initial of my sporting pseudonym. My father was furious, he did not understand my sense of humour. I think now he was right. Then in 1952, when I was in my hometown, I asked to see the birth register, and there was written Giuseppe Paolo Stanislao![40]

So came the day of days, the 17th February. The presentation went very well, though the title of the paper hid some of the penetrating radiation in the photo. Neither Rutherford nor Chadwick was there. Like Bohr, they did not believe in the theory of Dirac.

The next morning at 8 o'clock my landlady brought my breakfast, with a newspaper, saying scornfully: "Two months you haven't paid your board and now you let your friends learn the good news from the newspapers". I was not happy about the tam-tam of the newspapers, I thought that without the press publicity Anderson might have delayed his paper. His paper was written the 26th February and published with the wonderful photo on the 1st March.

In Rome they believed, in spite of Bohr's opinion. They sent me a telegram based on my love of caving: "Congratulations for exploration of Dirac hole".

My financial situation improved. The CNR awarded me a second fellowship and the Illustrated London News[41] paid £ 100 for the beautiful photo of the shower. Blackett shared the money with me.

The secretary of the 'Fascio' of the Italian colony in London (he was called Carmagna) came to Cambridge to propose that I go to London to meet the comrades. Behind that was a hint that I would have been given the party card. Looking back now, I must admit I was tempted, but I did not go. I was too busy. Mrs Blackett came to the conclusion that I did not do it because I was afraid of her judgement. Maybe she was right.

---

[39] Chen-Ning Yang (1922-*alive*).

[40] In the manuscript "Notes", published in the present publication, Dilworth is doubtful about this reconstruction. These are her words: "Is it possible that your father added three names after the war? It took a long time to get your birth certificate for your wedding because, your father said, the archives of Fossombrone had been destroyed during the war." It should be noted that, as Occhialini said, his father was furious when he learned that Beppo had added S. to his name since, from the formal point of view, Occhialini Giuseppe Paolo and Occhialini Giuseppe Paolo Stanislao would be considered two different authors.

[41] *Illustrated London News*, 11 March 1933.

His Excellency Marconi,[42] President of the Italian CNR asked me for written report. Blackett was not in Cambridge and the request had been urgent, by telegram. I composed a very careful document which, to my memory, contained the essentials. I was troubled in sending it because I felt I had not the right to [do] so without the approval of Blackett. I was right to be worried because my letter, sent personally to Marconi to be included in the general report of the year, was published in full in the Ricerca Scientifica, organ of the CNR.[43] I read the article and showed it to Blackett. There was no problem. I had betrayed neither friends nor institutions.

The University of Florence had sacked me because they had not understood the importance of the work I was doing. I had not wanted to give them details because I felt it was necessary to be discreet about the work in the Cavendish in view of the international competition (Millikan, Anderson, Joliot, Bothe and Meitner). After the 18th February I started to write a letter to my professor, Garbasso, to explain the situation. Before I could finish it I heard from Bernadini that Garbasso was ill. I telephoned to ask what it was. It was leukemia. I had a sense of relief, the nature of the sickness was such as to eliminate any responsibility for my ungrateful silence. I sent a letter to Garbasso in which I explained everything. The letter arrived a few days before he died. He called the people of the lab to justify to them my attitude. So I was pardoned, but the scars of this wound remained for years.

Later I found among my father's correspondence, a letter from Garbasso which explained his attitude. He thought I was being exploited by Blackett and that idea was shared by the people of the lab; not by my father. It was not correct. Blackett, to whom I attributed greatness in my STET autobiography, was the reason for my firm allegiance to England in the following years. It has been said under his influence. Scientifically, that is so, but politically I always kept an affectionate independence, with a doubt that he was a communist. This doubt was eliminated in December 1945 when we discussed and took decisions together on the necessity of helping Powell and Bristol. Blackett was left-wing labour, very cultivated, and I was a barbarian without political culture, grown up in the atmosphere of the Florentine renaissance, fundamentally an 'anti-fascist' which at the time I realised with shame implied an absence of any ideological identity. Now I am proud to have kept my independence.

Editorial note: the following paragraph was marked by Dilworth who wrote CANCEL? *on its side.*

My association with Blackett and Powell produced in the fifties a situation in which I was refused a visa to go to a Rochester conference. When the American consul asked me if I was or had been a communist, I replied "Me no, what about you?" Pauli[44] told Ferretti that I was an anarchist, not a communist, and it was a pity, because had I been one, I would have created confusion and destruction in

[42] Guglielmo Marconi (1874–1937).

[43] Occhialini G., *La Ricerca Scientifica*, **1** (1933) 372.

[44] Wolfgang Pauli (1900–1958).

the Party. At the time of the flights of the G-Stack, Bonetti,[45] Connie[46] and I were refused permission to enter the military launching field of Novi Ligure. It would be interesting to know who was the informer. The American consul in Milan, when he met me at a ceremony, proffered his excuses, explaining that his attitude had been forced by an important file, and that I must have had powerful enemies.[47]

In March, Blackett was elected a Fellow of the Royal Society and received funds to build a new Wilson Chamber with an electromagnet which was available to him somewhere in England. This magnet was, if I remember correctly, capable of giving a permanent field of 12000 gauss, as compared with the 3000 gauss for a limited time in the Oerlikon solenoid of the old Wilson chamber.

Editorial note: the text in italics in the following paragraph was crossed out by Dilworth who wrote to P.10 *on its side.*

Blackett was very occupied by the design of this new chamber. In fact I may say that the last days in which we worked together ended on the 17th February 1933. *My memory of the details of the period that followed, from 1933 to 1937 is not so clear. I had been protected by the atmosphere of the Cavendish, starting work at 8 am and finishing at 6 pm, in a kind of religious cloister. I had not realised that on the* 18th *February, an important chapter of my life closed without a possible substitution.*[48]

I did not want to abandon the old Wilson Chamber. Curiously enough, by the way, it never had a name. I had christened all the counters. There was Daria, Lillibet and Marlène. One day Rutherford and to my terror asked why Blackett succeeded in removing the implication of polygamy by explaining that the counters were temperamental. (It must be said that the log book of experiments with the counters was defined by Blackett as practically obscene.)

Some of the photos in the old chamber contained heavy tracks which at the time seemed mysterious and Dirac suggested they might be monopoles. Blackett found the explanation later, in October '33 when I was in Italy after the death of my mother.[49] He attributed them to α-particles from contamination.

Since Blackett was busy about the new chamber I asked him if he could arrange with Chadwick to borrow a neutron source to verify Joliot's tracks. Blackett left me to organise the work. I thought Chadwick was simply lending us the Polonium-Beryllium source, but Chadwick, as I realised later, considered it a collaboration between him and me, I being the junior partner and Blackett to be excluded. (I did obtain triple signatures on the paper.)

The source was available only at night, being used during the [day] for the proportional amplifiers so, contrary to the rules of the Cavendish, I had the honour of

[45] Alberto Bonetti (1920-*alive*).

[46] Constance Charlotte Dilworth (1924–2004).

[47] In the manuscript "Notes", published in the present publication, Dilworth wrote: "The paragraph I mark "CANCEL?" seems to me to interrupt the flow. Also it does not have into account the Brussels association which undoubtedly contributed—essentially for Bonetti and me.".

[48] In the manuscript "Notes", published in the present publication, Dilworth wrote: "The remarks on the period 1933–37 I have crossed out here because I think it fits better at the end (P.10)."

[49] Occhialini's mother was Etra Grossi (1887–1933).

spending nights in the lab with Chadwick. It was an exciting, ambiguous relationship. His coldness and dry humour had always intimidated me, and it took me several nights to overcome this. I had thought he had come to accept the validity of the Dirac theory, but it was not so. There was no place in his vision of the world for the creation of electron-positron pairs by γ-rays.

Blackett had recommended that I do what Chadwick told me to, so when on the second night Chadwick insisted on putting a layer of paraffin (blackened to prevent scattering of light) in contact with the cylinder, I accepted though the effect I was looking for required a heavy element. So for three nights with the paraffin, we got no effect. The source was very weak so what I had expected to do in a few nights took much longer. We had a track every 15 expansions and could take only 50 expansions in a night. I had to take down the chamber when it got dirty several times. Still Chadwick did not believe in the Dirac theory.

Then I obtained a γ-ray source. Thorium C from the Royal Society. As Chadwick exclaimed 'they were coming from every direction'. Still, he was not entirely convinced. He remembered that a young man called Alexander[50] during his stay in the Cavendish had taken very good photos with a γ-ray source and had seen no positrons. I insisted on seeing Alexander's photos. There they were the positrons. Chadwick was very depressed. He said, "I should have discovered the damned particle three years ago."

In August I went to a Conference in Zurich, intending to go on from there to Italy. I received 50 Swiss francs from Scherrer[51] for my contribution to the conference and met several people, among them Rosbaud[52] and a young Indian, Bhabha,[53] who wanted to talk about showers. Also there was Racah[54] who explained to me the situation in Florence. The university had readmitted me on the staff under the condition that I take the card of the Fascist party before the 15th September 1933. I decided I could not accept this situation, but also I could not face arguments with my colleagues and above all with my father and mother, so I decided to return to England to finish the γ-ray work.

On the boat crossing the Channel I found Vicky Weisskopf[55] on his way to Cambridge and we spent the night together at the Holborn Hotel. We had to share a room since the hotel was full. I had been very impressed by Joliot's talk at Zurich on the positive electrons produced by the bombardment of aluminium by neutrons, so I asked Weisskopf to try to think of an explanation for this effect. I proposed a 'brain-storming' session in our hotel room, but Weisskopf wanted to sleep. He cursed my insomnia so heartily that for years after no one at a conference would share a

[50] In the manuscript "Notes", published in the present publication, Dilworth wrote: "The bit about Alexander's photos has been put together by me from some cryptic notes of yours and my memory of your telling the story.".

[51] Paul Scherrer (1890–1969).

[52] Paul Rosbaud (1896–1963).

[53] Homi Bhabha (1909–1966).

[54] Giulio Racah (1909–1965).

[55] Viktor Weisskopf (1908–2002).

room with me. I sometimes wondered if the malediction of Weisskopf had aroused suspicions that I had homosexual tendencies. I still believe that Weisskopf with his brilliant mind would have arrived at the interpretation of the production of positrons as due to induced radioactivity. As I told him once, I think he traded a Nobel prize for a night's sleep.

In September I was called to Genova, my mother was ill. She died before I could see her.

There I learnt that she had taken the initiative of registering me as a member of the Fascist party in my hometown, with the help of the local secretary of the party who was a childhood friend of mine. It was difficult to get back to England because, being in my 25th year I was liable for military service, but I succeeded.

When I got back, I found Blackett had since three months transferred to London to Birkbeck College. He had taken the Wilson chamber with him and proposed I help him get it working again. It was a difficult time. I went frequently to Cambridge for personal reasons not, as Blackett seemed to think, because I preferred working with Chadwick.

My memory of the details of the period that followed, from 1933 to 1937, is not so clear. I had been protected by the atmosphere of the Cavendish, starting work at 8 am and finishing at 6 pm in a kind of religious cloister. I had not realised that on the 18th February an important chapter of my life closed without possible substitution.

"On the 22nd December 1932 ..." [Dilworth's Manuscript][56]

On the 22nd December 1932, I heard from Bernardini that I had lost my job as assistant at the University because I had not been there present on the 15th October (beginning of the academic year) as the law required. A usual, I did not know of this law and having asked for leave my research in Cambridge and, through my professor Garbasso, for an extension of my CNR fellowship, I thought I had no reason to worry.

After all these years I can think that the Fascist state was right, but then I found the situation tragic. The only place where we could do research was the University, but things had changed since I was first appointed, now to be a member of University staff one had to belong to the Fascist party. I was convinced that my being sacked was a blackmail by the Secretary of the University to force me to take the Party card. My relations with my professor cooled (I did not know he was ill) and I was persecuted by letters from my friends urging me to leave England and to try to make peace with the University authorities and the fascist Party. Under these conditions I could only work like a madman in the hope that my luck would turn.

I was completely broke. I owed two months for boarding and heating to my landlady. I had given up my bedroom to my Canadian friend Ouellet and slept on a

[56] Transcription of Dilworth's manuscript which begins with "On the 22nd December 1932, ...". It is the manuscript version of "Before and after the 17th February 1933".

couch in the common setting room. (I had tried sleeping in the garden shed, but a bad cold brought me back into the house.)

My shoes were worn out, as also the only pair of good trousers I used to go to the Royal Society. I bought a second-hand pair from a French friend Rasier. I went to the lab. in tennis shoes and shirts, carrying a tennis racket on the handlebar of the bicycle (to complete the uniform). The bicycle was borrowed from the bicycle stand at the Biology lab. There were a lot of bicycles in that stand, some of them dusty, as if their owners were out of Cambridge. I marked them with a piece of wire and checked they were always in the same plane. One of these I would use to go to the Cavendish.

Sometime in January 1933, (I do not remember the exact date) after three days in which the chamber was full of mist and we had not been to get tracks, I went to the lab. on a Saturday morning to see if the situation had improved. I intended just to look at a few expansions with the counter control. I do not know why, I loaded the two cameras and adjusted the lamps to avoid diffusion on the glass; I borrowed Karen's Basque beret to complete the optical screen and set the magnetic field to a the maximum of 3000 gauss with a delay of 2 1/2 min. In the darkened room I waited for the chamber to trigger, watching the thermometer (of moler-oil) which was raising dangerously. I had my hand on the switch of magnet current when the chamber triggered. There was some mist but the heat of the event created a clear zone in which I thought I saw the event I had been working for.

It was 11 am when Karen and I arrived, bicycling, at the Blackett's house. I told him there was something important on two plates, that I had not had the courage to develop them. For a sort of superstition I refused to go with him to the lab for the development. The photo of the shower is well known, the shadow of Karen's beret can be seen partially on the right. The extraordinary thing is that we told no one about it, we kept it to ourselves. I do not know when Rutherford saw the photo. When Blackett came home for lunch he asked me to describe what I had seen during the flash. Karen completed my sketch by telling him I have said there were 5 "useful" tracks.

I had broken the rules by going to the lab on a Saturday, alone with a woman. What was to have been a simple control of the operation of the chamber I turned into an act of research. Nevertheless God forgave my sins and gave me the privilege of a first sighting. I resisted to the desire to confirm that glimpse by developing the plates. Had I done so I would have faced Blackett with a fait accompli.

After all his work it was right he should have the responsibility of seeing the evidence of what I had guessed. There was a mixture of positive and negative particles and the positive ones did not look like protons.

It was an act of love to refuse to go back with him to the Cavendish, so that he could meet the particle alone. Several, maybe ten, times in my life I did something like that, in my youth, in caves, in mountains, before a nuclear emulsion. Blackett was the only one who remembered and understood the meaning (sacralité) of the gesture.

The day after, began the discussions with Dirac, and the presentation at the Royal Society was fixed for the 17th February In the paper there was half a page concerning the effect found by Tchao and by Gray and Tarrant, and the "backward moving" tracks of Joliot. I had asked for this reference in the paper but at one stage Blackett eliminated it. I was surprised when in the final edition Blackett proposed to put it in again. That made me understand that there was a critical attitude in the Cavendish towards the paper and Blackett was replying to it with the affirmation that the Tchao—Gray & Tarrant effect was real.[57]

Not long ago Yang wrote to me, enclosing an article in which he complained that we had put Gray and Tarrant on the same level as Tchao. I think we were justified in doing so. Tchao came to England in 1932 on his way to Germany. He stopped in Cambridge to see Gray and Tarrant who treated him like a friend. They brought him to meet me and I received him like a brother. That is to say that among the experts in his field he had a triumphal welcome. He told me, with gentle regret, that his work had not been appreciated in Pasadena. We spent two days together talking about counters and anticoincidence.

Many years later, not knowing if he were still living I proposed him for the Wolf prize.

The Royal Society paper gave me a new identity. I thought it would be fun to have more than one initial like Blackett. I started with two, G. P. Occhialini, but Blackett said that stood for General Post Office so I added an S, the initial of my sporting pseudonym. My father was furious, he did not understand my sense of humour, I think now he was right. Then in 1952 when I was in my home town I asked to see the Births Register and there was written Giuseppe Paolo Stanislao. (Believe it or not, it is true!)

So came the day of days, the 17th February. The presentation went very well though the title of the paper hid some of the penetrating radiation in the photo. Neither Rutherford nor Chadwick was there. Like Bohr they did not believe in Dirac's the theory of Dirac.

The next morning at 8 o'clock my landlady brought my breakfast, with a newspaper, saying scornfully "Two months you haven't paid your board and now you let friends learn the good news from the newspapers!" I was not happy about the tam-tam of the newspapers, I thought that without the press publicity Anderson might have delayed his paper. His paper was written the 26th February and published with the wonderful photo in the 1st March.

In Rome they believed, in spite of Bohr's opinion. They sent me a telegram based on my love of caving "Congratulations for explanation of Dirac hole."

My financial situation improved. The CNR awarded me a second fellowship and the illustrated London News paid £100 for the beautiful photo of the shower. Blackett shared the money with me.

[57] Occhialini's Note reported by Dilworth: "In my autobiography in the STET there is a Freudian slip when I said that in the paper there were suggestions which were difficult to accept. I should have said, difficult to accept in a Cavendish influenced by Rutherford and Chadwick."

The secretary of the "fascio" of the Italian colony in London (he was called Camagna) came to Cambridge to propose that I go to London to meet the comrades. Behind that was the hint that I would have been given the Party card. Looking back now I must admit I was tempted, but I did not go. I was too busy. Mrs. Blackett came to conclusion that I did not do it because I was afraid of her opinion. Maybe she was right.

His Excellency Marconi, President of the Italian CNR asked me for a written report. Blackett was not in Cambridge; and the request had been urgent, by telegram. I composed a very careful document which, to my memory, contained the essentials: I was troubled in sending it because I felt I had not the right to do so without the approval of Blackett. I was right to be worried, because my letter, sent personally to Marconi to be included in the general report of the year was published in full in the Ricerca Scientifica, organ of the CNR. I read the article and showed it to Blackett, there was no problem, I had not betrayed neither friends nor institutions.

The University of Florence had sacked me because they had not understood the importance of the work I was doing. I had not wanted to give them details because I felt it was necessary to be discreet about the work in the Cavendish, in view of the international competition (Millikan, Anderson, Joliot, Bothe and Meitner). After the 18th February I started to write a letter to my Professor Garbasso to explain the situation. Before I could finish it, I heard from Bernardini that Garbasso was ill. I telephoned to ask what it was. It was leukemia. I had a sense of relief, the nature of the sickness was such as to eliminate any responsibility for my ungrateful silence. (These emotional impulses tend to influence the chronological precision of memory). I sent a letter to Garbasso in which I explained everything. The letter arrived a few days before he died. He called the people of the lab to justify to them my attitude. So I was pardoned, but the scars of this wound remained for years.

Later I found among my father's correspondence, a letter from Garbasso which explained his attitude. He thought I was being exploited by Blackett, and that idea was shared by the people of the lab.; not by my father. It was not correct. Blackett, to whom I attributed greatness in my STET autobiography, was the reason for my firm allegiance to England in the following years. It has been said that I was under his influence scientifically, that is so but politically I always kept an affectionate independence, with a doubt that maybe he was a communist. This doubt ceased to exist in December 1945 when we discussed, and took decision together on the necessity of helping Powell and Bristol. Blackett was left-wing Labour, very cultivated and I was a barbarian without political culture, grown up in the atmosphere of the Florentine renaissance, fundamentally an "anti-fascist" with which at the time I realised with shame implied an absence of any ideological identity. Now I am proud to have kept my independence.

[*Editorial Note: the text "Bruxelles?" was handwritten by Dilworth on the side of next paragraph.*]

[My association with Blackett & Powell produced in the fifties a situation in which I was refused the visa to go to a Rochester conference. When the American consul asked me if I was or had been a communist, I replied, "Me no, what about you?"

Pauli told Ferretti that I was not an anarchist, not a communist, and it was a pity that I was not in the Party because had I been I would have caused confusion and destruction.

At the time of the flight of the G-stack, Bonetti, Connie and I were refused permission to enter the military launching field at Novi Ligure. It would be interesting to know who informed the American consul in Milan, when he met me at a ceremony, proffered his excuses, explaining that his attitude had been forced by an important file, and that I must have had powerful enemies.]

In March, Blackett was elected a Fellow of the Royal Society and received funds to build a new Wilson Chamber with an electromagnet which was available to him somewhere in England. The magnet, if I remember correctly, was capable of giving a permanent field of 12000 gauss, to be compared with the 3000 gauss for a limited time in the Oerlikon solenoid of the old Wilson Chamber.

Blackett was very occupied by the design of this new chamber. In fact I may say that the last days in which we worked together ended on the 17th February 1933. My memory of the details of the period that followed, from 1933 to 1937 is not so clear. I had been protected by the atmosphere of the Cavendish, starting work at 8 am and finishing at 6 pm, in a kind of religious cloister. I had not realised that on the 18th Feb. an important chapter of my life closed without a possible substitution.[58]

I did not to abandon the old Wilson Chamber. Curiously enough, by the way, it never had a name. I had christened all the counters, there was Daria, Lillibet and Marlène. One day Rutherford saw the labels, and asked why. I was terrorised, but Blackett waived away the polygamic implication by explaining that the counters were temperamental. (It must be said that the log book of the experiments with the counters was defined by Blackett as practically obscene).

Some of the photos in the old chamber contained heavy tracks which at the time appeared mysterious, and Dirac suggested they might be monopoles. Blackett found the explanation later, in October '33 when I was in Italy after the death of my mother. He attributed them to α-particles from contamination.

Since Blackett was busy about the new chamber, I asked him if I could arrange with Chadwick to borrow a neutron source to verify Joliot's tracks. Blackett left me to organise the work. I thought Chadwick was simply lending us the Polonium-Beryllium source, but Chadwick, I realised only later, considered it a collaboration between him and me, I being the junior partner and Blackett to be excluded.

It was available only at night. The source was being used during the day for the proportional amplifiers. So contrary to the rules of the Cavendish, I had the honour of spending nights in the lab. with Chadwick. It was an exciting, ambiguous relationship.

His coolness and dry humour had always intimidated me and it took me several nights to overcome this. [I had thought Chadwick had come to accept the validity of the Dirac theory, but it was not so.] There was no place in his vision of the world for the creation of the electron-positron pairs by γ-rays. Blackett had recommended that I do what Chadwick told me to [do] so when on the second night Chadwick

[58] The paragraph form "My memory ..." to "...a possible substitution." was moved in this position by Connie Dilworth. See file "Notes".

insisted on putting a layer of paraffin (blackened to prevent scattering of light) in contact with the cylinder, I accepted, though the effect I was looking for required a heavy element. So for 3 nights with the paraffin we got no effect.

The source was very weak, so what I had expected to do in a few nights took much longer. We had a track every 15 expansions and could take at most 50 expansions per night. I had to take down the chamber, when it got dirty, several times. Still Chadwick did not believe in the Dirac theory.

Then I obtained a γ-ray thorium source from the Royal Society.

As Chadwick said: "They are coming from every direction." Still he was not convinced. He remembered that a young man called Alexander had taken very good photos with a γ-ray source and had seen no positrons. I insisted on seeing the photos. There they were. Chadwick was very depressed. He said "I should have discovered the damned particle three years ago."

In August I went to a conference in Zurich intending to go afterward to Italy. I received 50 Swiss Francs from Scherrer for my contribution to the conference, and meet there several people, among them Rosbaud, and a young Indian, Bhabha, who wanted to talk about showers. Also there was Racah who explained to me the situation in Florence. The University had re-admitted me on the staff under the condition that I take the card of the fascist party before the 15th September 1933. I decided I could accept this situation but not could face arguments with my colleagues, and above all with my father and mother so I decided to return to England to finish the γ-ray work.

On the boat crossing the Channel I found Vicky Weisskopf and we spent the night together at the Holborn Hotel, which was full, so we had to share a room. I had been very impressed by Joliot's talk at Zurich on the positive electrons produced by the bombardment of aluminium by neutrons, so I asked Weisskopf to try to explain it. I proposed a brain-storming session in our hotel room, but Weisskopf wanted to sleep. He cursed me and my insomnia so heartily that for years after no one at a conference would share a room with me. I sometimes wondered if the Malediction of Weisskopf had aroused suspicion that I had homosexual tendencies.

I still believe that Weisskopf with his brilliant mind would have arrived at the interpretation of the production of positrons as due to induced radioactivity. As I told him once, he traded a Nobel prize for a night's sleep.

In September I was called to Genova, my mother was ill. She died (of peritonitis) before I could see her.

There I learnt that she had taken the initiative of registering me as a member of the fascist party in my hometown with the help of the secretary of the party who was a childhood friend of mine.

It was difficult to get back to England because, being in my 25th year, I was liable to military service, but I succeeded. When I got back I found Blackett had transferred to London to Birkbeck College. He had taken the Wilson Chamber with him and proposed I help him get it working again.

It was a difficult time. I went frequently to Cambridge for personal reasons, not, as Blackett seemed to think, because I preferred working with Chadwick.

Notes [Typescript][59]

Notes:

P.4: Paolo Stanislao. Is it possible that your father added these names after the war? It took a long time to get your birth certificate for our wedding because, your father said, the archives of Fossombrone had been destroyed during the war.

P.6–7 The paragraph I mark CANCEL? seems to me interrupt the flow. Also it does not have into account the Brussels association which undoubtedly contributed—essentially for Bonetti and me.

P.7 The remarks on the period 1933–37 I have crossed out here because I think it fits better at the end (P.10).

P.9 The bit about Alexander's photos has been put together by me from some cryptic notes of yours and my memory of your telling the story.

Brazil—London—Bristol

Brazil—London—Bristol. a Brief Account [Typescript]

"Brazil—London—Bristol"
A brief account"

I found it difficult to concentrate on research in Brazil, the surroundings were strange, the Italian community mostly fascist and the news from the world outside most depressing. By 1940 I started to get my bearings and was working on a constant volume, automatic Wilson chamber, very cheap, the first of what was to be an array of 6 Wilson chambers on Itatiaia to investigate extensive showers. Pearl Harbour stopped that.

After the attack on Pearl Harbour, Brazil broke off diplomatic relations with the Axis Powers. The Italian government recalled its nationals of which I was one. I could not refuse for fear of putting my father, who was in Italy, in danger. I was preparing to embark when I was told that the British had refused the "navicert" for me, so I stayed in Brazil.

It was long after that I discovered the origin of this intervention of the British authorities in my case. It had been through the initiative of my friend Damy de Souza Santos[60] who spoke with his cousin Aranha, the Minister of Foreign Affairs.

While awaiting embarkment in Rio I had meet Germaine Krull[61] who, herself preparing to leave as volunteer with the Free French, changed my attitude to my own

[59] The Page Numbers Indicated in This Manuscript "Notes" Refer to the Dilworth's Manuscript Which Begins with "On the 22nd December 1932, …".

[60] Marcelo Damy de Sousa Santos (1914–2009).

[61] Germaine Krull (1897–1985).

problem. She convinced me that I could play a part in the war against fascism if I returned to Europe.

I returned to S. Paolo, to the University, but the action of the German submarines forced Brazil into the war. The University of S. Paolo gave me the option either to sign a declaration against the Axis or to leave the University. I refused to sign the declaration, my official reason being that, although everyone knew I was antifascist, I had never made a public declaration and to do it now would be an act of opportunism. My private reason was that I was afraid of some act of vengeance against my father. (He later told me I had been right, he would have been in danger.)

Hoping in the day that Italy would get out of the war, I went to live very economically in the mountains of Itatiaia where I had dreamt of installing six Wilson chambers. The day I heard that Italy was no longer an enemy of the Allies, I went down to Rio to keep the promise I had made to myself and to Germaine.

I contacted the British Embassy and, through my friend and protector Carlos Chagas, the Brazilian Ministry for Foreign Affairs, with a request to join the expeditionary corps. The Brazilians said no. Chagas encouraged me to stay in Rio to await the reply of the British Embassy. It arrived, positive, around March 1943.[62] I learnt later that I was to have been sent to Canada.

The invasion of Normandy changed many things, including the date of my departure. That became possible only in December. In the meantime the university of S. Paolo offered to take me back. I had the impression they had guessed the state of my finances and wanted to help me.

Before taking the boat I had the Wilson chamber in working order and left it in the hands of of two young men, Lattes[63] and Camerini.[64] I paid my debts and embarked with £5 in my pockets.

The arrival in London was the start of a long odyssey. The attitude of the British authorities had changed owing to recriminations from the Americans who suspected infiltration of Soviet spies in war research (the case of Fuchs).[65] After several attempts I was found a place in the General Electric Co. but this assignment was annulled after 5 d by MI5. The day after Easter, the local employment agency ordered told me to go to a restaurant in Soho where I had been found a job as a dishwasher.

Blackett, in whose house I had been living and who had shared my sorrows (he was one of the eight scientists who had guaranteed for me) decided I should try for a university job with no connection with war work.

With most of the scientists engaged in some kind of war-work, the Universities were kept running on a skeleton staff just sufficient to teach the students, who would then be available for the war effort.

[62] 1944. We assume it was after Italy signed the armistice, on September 3, 1943.

[63] César Lattes (1924–2005).

[64] Ugo Camerini (1925–2014).

[65] Klaus Fuchs (1911–1988).

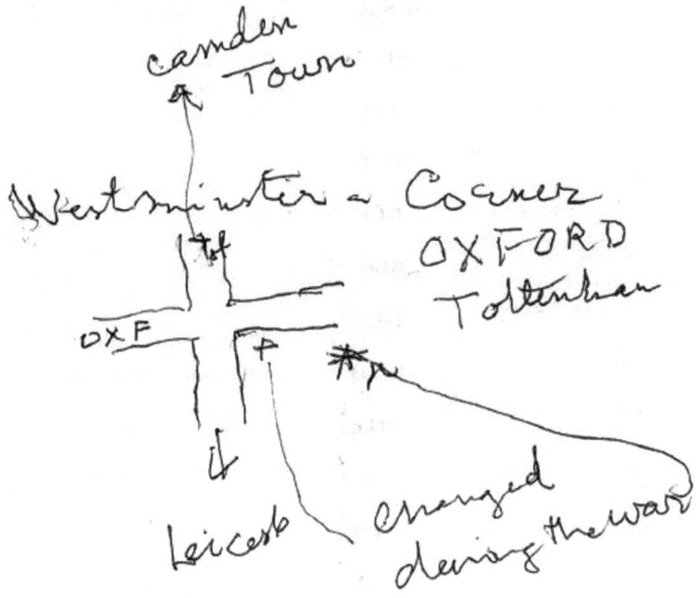

Fig. 1 Map drawn by Occhialini. Courtesy of the BICF library of the University of Milan

Blackett told me that in these conditions he felt that to send me to his University, Manchester, would not be correct. He would have been profiting by his knowledge of my situation to help his people. He suggested the Cavendish, but for some reason Bragg[66] was not interested. Then he thought of Bristol and suggested I go there to see what could be done in cosmic rays with the nuclear emulsion technique.

Since my arrival in London, Prof. Appleton[67] had arranged for me a grant as research fellow of DSIR, covered my expenses by taking me on as a research worker for DSIR. This was sufficient while I was a guest in Blackett's house. To cover expenses in Bristol, Blackett took me to a bank in Tottenham Court Road [*Editorial note: Map drawn by Occhialini in Fig.* 1.] and drew £50 from an account which he swore to me was not his personal account but that of a grant of the Royal Society. I felt I had an official mission.

In Bristol I found Powell[68] with a group of plates exposed to the Liverpool cyclotron on which he was studying 'butterflies' (angular distributions). I explained to him my cosmic ray mission. Another visitor at the time was Rosenblum,[69]

[66] Lawrence Bragg (1890–1971).

[67] Victor Appleton (1892–1965). From 1938 to 1948 he was Director of the Department of Scientific and Industrial Research (DSIR).

[68] Cecil Frank Powell (1903–1969).

[69] Salomon Rosenblum (1896–1959). Rapkine was the director of the Bureau Scientifique de la France Combattante.

member of the Rapkine Mission who wanted to use the Ilford plates for his research at Bellevue.[70]

Brazil—London—Bristol [Dilworth's Manuscript]

"Brazil—London–Bristol"

I found it difficult to concentrate on research in Brazil, the surroundings were strange, the Italian community mostly fascist and the news from the world outside most depressing. By 1940 I started to get my bearings and was working on a constant volume, automatic Wilson chamber, very economical, the first of what was to be an array of 6 Wilson chambers on Itatiaia to investigate big showers. Pearl Harbour stopped that.

After the attack on Pearl Harbour, Brazil broke off diplomatic relations with the Axis Powers. The Italian government recalled its citizens of which I was one. I could not refuse for fear of putting my father, who was in Italy, in danger. I was preparing to embark when I was told that the British had refused the "navicert" for me, so I stayed in Brazil.

It was long after that I discovered the origin of this intervention of the British authorities in my case. It had been through the initiative of my friend Damy de Souza Santos who spoke with his cousin Aranha, the Minister of Foreign Affairs.

While awaiting embarkment in Rio I had meet Germaine Krull who, herself preparing to leave as volunteer with the Free French, changed my attitude to my own problem. She convinced me that I could play a part in the war against fascism if I returned to Europe.

Returned to S. Paolo, to the University, but the action of the German submarines forced Brazil into the war. The University of S. Paolo gave me the option either to sign a declaration against the Axis or to leave the University. I refused to sign the declaration, my official reason being that, although everyone knew I was antifascist, I had never made a public declaration and to do it now would be an act of opportunism. My private reason was that I was afraid of some act of vengeance against my father. (He later told me I had been right, he would have been in danger.)

Hoping in the day that Italy would get out of the war, I went to live very economically in the mountains of Itatiaia where I had dreamt of installing six Wilson chambers. The day I heard that Italy was no longer an enemy of the Allies, I went down to Rio to keep the promise I had made to myself and to Germaine.

I contacted the British Embassy and, through my friend and protector Carlos Chagas, the Brazilian Ministry for Foreign Affairs, with a request to join the expedi-

---

[70] In the verso of the last sheet of the manuscript is inserted a draft of a letter, in Italian, for a payment of 300.000 Italian lira: "In response to his note of 17/03/93, and following today's telephone agreement with his colleague, I attach a check for 300,000 Italian lira. Hoping that one time you will not leave me unaware debtor for so many months." Although "debitore" in Italian is masculine the draft was written by Connie Dilworth.

tionary corps. The Brazilians said no. Chagas encouraged me to stay in Rio to await the reply of the British Embassy. It arrived, positive, around March 1943.[71] I learnt later that I was to have been sent to Canada.

The invasion of Normandy changed many things, including the date of my departure. That became possible only in December. In the meantime the University of S. Paolo offered to take me back. I had the impression they had guessed the state of my finances and wanted to help me.

Before taking the boat I had the Wilson chamber in working order and left it in the hands of of two young men, Lattes and Camerini. I paid my debts and embarked with £5 in my pockets.

The arrival in London was the start of a long odyssey. The attitude of the British authorities had changed owing to recriminations from the Americans who suspected infiltration of Soviet spies in war research (the case of Fuchs). After several attempts I was found a place in the General Electric Co. but this assignment was annulled after 5 d by MI5. The day after Easter, the local employment agency told me to go to a restaurant in Soho where I had been found a job as a dishwasher.

Blackett, in whose house I had been living and who had shared my sorrows (he was one of the eight scientists who had guaranteed for me) decided I should try for a University job with no connection with war work.

With most of the physicists engaged in some kind of war-work, the Universities were kept running on a skeleton staff just sufficient to teach the students, who would then be available for the war effort.

Blackett told me that in these conditions he felt that to send me to his University, Manchester, would not be correct. He would have been profiting by his knowledge of my situation to help his people. He suggested the Cavendish, but for some reason Bragg was not interested. Then he thought of Bristol and suggested I go there to see what could be done in cosmic rays with the nuclear emulsion technique.

Since my arrival in London, Prof. Appleton, Director of DSIR had covered my expenses by taking me on as a research worker for DSIR. This was sufficient while I was a guest in Blackett's house. To cover expenses in Bristol, Blackett took me to a bank in Tottenham Court Road and drew £50 from an account which he swore to me was not his personal account but that of a grant of the Royal Society. I felt I had an official mission.

In Bristol I found Powell with a group of plates exposed to the Liverpool cyclotron on which he was studying 'butterflies' (angular distributions). I explained to him my cosmic ray mission. Another visitor at the time was Rosenblum, member of the Rapkine Mission who wanted to use the Ilford plates for his research at Bellevue.

[71] 1944. We assume it was after Italy signed the armistice, on September 3, 1943.

Résumé Autobiographique. Brésil—Londres—Bristol [Occhialini's Comments (Labeyrie's Manuscript)]

Résumé autobiographique

Brésil—Londres—Bristol

Alas poor Yorick.[72]

Je reconnais que mon activité scientifique au Brésil a été négative. L'ambiance était difficile, les Italiens étaient fascistes, j'avais les larmes faciles pour tout ce qui passait dans le monde. En 1940 je commençais à m'acclimater, j'avais une chambre de Wilson automatique, un modèle merveilleux de chambre de Wilson en volume constant, très bon marché pour investiguer à Itatiaia le grandes showers (6 chambres étaient en projet).

Pearl Harbour a interrompu ça. Le Brésil a rompu ses relations avec l'Axe, je devais ainsi m'embarquer pour l'Italie pour demande du gouvernement italien prisonnier de la peur pour le destin de mon père en Italie. J'étais prêt à partir quand les Anglais ont nié le navicart. Je suis resté au Brésil. La raison de cette action anglaise pas sollicité par moi a été due à l'action de mon ami Damy de Souza Santos qui est intervenu avec son cousin Aranha, ministre des affaires étrangères. Pendant l'attente à Rio la rencontre avec Germaine Krull a changé mon attitude envers mon problème. Elle partait volontaire et m'a convaincu que j'aurais pu jouer un rôle dans la guerre contre le fascisme si je rentrais en Europe. Rentré à San Paolo à l'Université, l'action de sous-marins de l'Axe fasciste le Brésil a entré en guerre. L'Université de San Paolo m'a donné l'option entre abandonner ou rester dans l'Université en signant une déclaration de solidarité contre l'Axe. J'ai dû refuser comme d'ordinaire pour une raison publique. Tout le monde savait que j'étais antifasciste mais je ne m'étais pas déclaré et sa aurait été un geste d'opportunisme; la raison secrète était que j'avais peur de la vengeance sur mon père (j'avais raison).

Dans l'espoir que l'Italie sortie de la guerre je suis allé vivre très bon marché sur le haut d'Itatiaia ou j'avais rêvé d'installer la station de 6 chambres de Wilson. D'où pour la sortie de la guerre de l'Italie je suis descendu direction Rio pour maintenir ma promesse que j'avais faite à moi-même et à Germaine. J'ai contacté l'Ambassade Anglaise et à travers mon ami et protecteur Carlos Chagas le Ministère des extérieurs Brésilien pour faire partie de corps expéditionnaire; la réponse a été négative. Chagas m'a encouragé à rester à Rio en attendant la réponse vers Mars 1943. J'ai su après que ma destination ultime était Canada.

L'invasion a paralysé la date de mon départ qui devenait possible seulement en Décembre; avant cela l'Université de Sao Paolo avait offert de me reprendre. J'ai l'impression qu'ils devinaient mes conditions financières et qu'ils voulaient m'aider. A mon départ j'ai consigné à 2 personnes jeunes Lattes et Camerini la chambre de

---

[72] Written by Dilworth. The sentence comes from Shakespeare, Hamlet, Act 5 Scene 1, and refers to Hamlet who learns from the gravedigger that the skull belonged to Yorick.

Wilson automatique prête à fonctionner. J'ai payé mes dettes et je me suis embarqué avec £5 dans mes poches.

À l'arrivé à Londres a commencé une longue odyssée. La politique anglaise avait changé à cause des récriminations américaines pour le suspect que des espions s'étaient infiltrés (Fuchs). Après une série d'essais, une incursion à la General Elect. Cie annulé après 5 jours par MI5. Le jour après Pâques je recevais l'ordre d'aller laver la vaisselle dans Soho. Blackett, dans la maison duquel j'étais hospité et qui était l'un des 8 savants qui avait garanti pour moi et qui avait souffert ma passion a décidé que je devais essayer l'Université sans travail de guerre. Manchester était hors de question, mon arrivée là aurait été abusive. Toutes les Universités étaient vides. La Cavendish était possible et Bristol aussi.

Je connaissais le travail de Bristol et son excellence. Blackett m'a conduit à une banque à Tottenham Road, il a retiré £50 de frais de mission sur un compte qu'il m'à déclaré sur sa parole d'honneur qui n'était pas personnel mais qu'il était un "grant" de la Royal Society mis à sa disposition. Ce signifiait que j'allais à Bristol pour investiguer quel était la vraie possibilité des plaques nucléaires. À Bristol j'ai trouvé dans le laboratoire Powell avec un groupe de plaques exposées au cyclotron de Liverpool où ils étudiaient les Butterflies. (J'ai oublié de dire que de mon arrivée Professor Appleton directeur du DISR m'avait assuré comme chercheur et qu'ainsi les £50 ou £100 de Blackett était un additif à mes gages).

J'ai expliqué à Powell ma mission en rayons cosmiques. Ma accommodation le premiers mois était très primitive. Il y avait dans le laboratoire un membre de la mission Rapkine, Rosenblum. Rosenblum voulait appliquer les plaques Ilford a ses études à Bellevue.

Heisenberg

Heisenberg [Typescript]

Heisenberg

Hard things were being said about Heisenberg's[73] role in the Nazi war-effort, so when on the occasion of his first academic visit to England after the war, Mott[74] invited Powell and me to a party that evening to meet him, it is not surprising that our attitude was negative. Powell said "I have been in Poland, I have seen Cestokova, I cannot come". Mott then asked me and I too refused, but Mott insisted, "You have not seen the gas chambers, what is your reason? I replied that if I came, I would have to say things to Heisenberg that would put a strain on Mott's rules of hospitality. Mott said then that I should come and could say whatever I wished.

---

[73] Werner Heisenberg (1901–1976).

[74] Nevil Mott (1905–1996).

I felt trapped. I had not prepared a lecture to give to Heisenberg and I had put myself in a difficult situation. Then I remembered a conversation I had had with Rosbaud (The Griffon) who had given me a list of names which I memorised.

I came to Mott's house (across the courtyard in front of the Royal Fort) at about 9 o'clock that evening. Heisenberg shook hands so warmly that I started to have doubts about my decision to confront him. Then he reminded me of our last meeting in Manchester in March 1938 and the political discussion we had in Blackett's house *and his rapture about Ludlow which we had visited with Blackett and Mrs Blackett. I was there on holidays from Brazil.*[75]

In fact, Blackett had decided to have Heisenberg stay with him that time so he could report back to his German colleagues that the English physicists were leading their usual life. (He did not of course leave around his drawings of a *the precision* bomb-sight for *bombing airplanes*).[76]

At that time, Austria had been invaded and Heisenberg was worried about Schrödinger.[77] He asked if he could get information anonymously through the Italians. I telephoned Bernardini in Rome. He told me the situation was under control, Fermi had learnt from von Weizsäcker[78] at the Vatican Embassy that Schrödinger was safe and well.

This memory convinced me that Heisenberg could not have been a highly placed Nazi dignitary. He would not have needed to get the information through Rome had he been an important Nazi.

Nonetheless, while standing with him near Mott's fireplace, and noticing that the rest of the company was gradually retreating from around us, I decided to put my prepared question. "Professor Heisenberg" I said "since you are on good terms with Ronald Fraser[79] (the high commissioner) as proved by your presence here, why did you not delay your visit and *take the initiative to send*[80] send first some of the known anti-nazis such as —" and I recited the list of names of those whom Rosbaud had told me were the German physicists who could come to England with honour.

At this moment the charm was broken by Mott who joined us to say "I don't know what Occhialini has been telling you, what he says is on his own responsibility, you must not think everyone here agrees with him."

That was that. I felt that I had made a powerful enemy and that someone in Bristol would never have forgiven me.

Around 10.30 I was called to the lab. and I took my leave of Mott. Then the miracle happened, Heisenberg asked to come with me. There Connie made the ritual coffee and Heisenberg and I were left discreetly alone in an empty alcove while Heisenberg explained to me his position.

---

[75] The words in italics were added by Occhialini. These handwritten words are some of the few that the scientist added in the documents we are editing in the present publication.

[76] The words in italics were added by Occhialini.

[77] Erwin Schrödinger (1887–1961).

[78] Carl Friedrich Freiherr von Weizsäcker (1912–2007).

[79] Ronald Fraser (1888–1974).

[80] The words in italics were added by Occhialini.

In Manchester he had given me the impression of being a good German but politically naive. That opinion was confirmed. I perceived the emotion with which he declared that they did not know that Hitler and his crowd were assassins. I did not have the courage to say to him that, even if they had not been assassins, Nazism would still have been bad.

Then he asked about the work in the lab. I told him of the difficulty one group was having in applying the theory of scattering to measurement. Heisenberg mentioned the work of a German physicist called Molière,[81] and promised to send copies of Molière's papers when he got back to Germany.

Some days later, Molière's papers arrived at the Royal Fort and I passed them to Goldschmidt.[82] The measurement by scattering of the mass of the π and μ mesons proceeded.

[Concentrated Emulsions]

Concentrated Emulsions [Typescript]

Concentrated Emulsions

When I arrived in Bristol the lab. consisted of one large room with 3 microscopes plus a Y4 monocular and a small studio on the 4th floor. On the 3rd floor there was a dark room. The staff consisted of Heitler[83] (brother of the theoretician) and Guggenheimer,[84] and one microscopist, Mrs Andrews.[85]

Powell had demonstrated that the precision in measurement of length and angle of tracks in the emulsion was comparable to that in the Wilson chamber. In my opinion, the important step he had made in the technique was to place emphasis on the quality of the microscope optics (use of immersion objectives).

I was given a group of plates exposed to cosmic at the Jungfraujoch by Powell, Heitler and Fertel[86] and I started to study them. At the same time Powell suggested that I get the feel of the technique by measuring the distribution in length of the long tracks from Thorium stars. I did it with such zeal and application, changing frequently eyepiece and scale, that my distributions turned out to be narrower than Powell's own. In fact he asked my permission to use them instead of his in the paper then being prepared for publication in the Review of Scientific Instruments.

---

[81] Gert Molière (1909–1964).

[82] In 1947, two people named Goldschmidt were at the Bristol laboratory: Victor Moritz Goldschmidt (1888–1947) and Yves Goldschmidt-Clermont (1922–1988).

[83] Walter Heinrich Heitler (1904–1981). His brother was Hans Heitler.

[84] Kurt Martin Guggenheimer (1902–1975).

[85] M.L. Andrews (…-…).

[86] Geoffrey E.F. Fertel (1913–1949).

There was no acknowledgement, but a few days later the Director of the Laboratory (Prof. Tyndall)[87] called me to his study to offer me a job at £27 a month. Ecstasy!

I went ahead with the scanning of the plates exposed to cosmic rays but the results were disappointing.

One day, I don't remember the date, but maybe in June, a representative of Ilford, Chilton,[88] came to discuss improvement of the plates, and Rosenblum and I were invited to the discussion. Chilton remembered me, we had met at the Cavendish when he was a student there. That helped dispel my habitual shyness and I was able to explain to him that, in view of Blackett's mission to me for cosmic rays, the low density of grains on tracks of fast particles made them too difficult to follow, and I put the question as to the possibility of increasing the quantity of silver, on the basis of the experience of increasing pressure in a Wilson chamber. Rosenblum was silent. Powell said, with a bit of dry humour, that that would mean re-making the range-energy relation.

After that there was Hiroshima.

Shortly afterwards, Lettice Ramsey[89] sent me a book on Scientific Photography by Lawrence, an old Cambridge friend of mine. In it I found the temperature coefficient for the development process and the recipe for the Amidol developer. I immediately tried the effect of temperature, packing ice into condoms so as not to dilute the developer.

I was the only one who used the 3rd floor dark room. I often washed my underclothes there, and sometimes my sheets, before leaving the lab at my usual time, about 1 am. I had left the condoms full of water, but next morning they had disappeared. The lab. staff looked awry at me. "We thought someone had been rash" said Powell. Fröhlich[90] consoled me. I wanted to go and see Tyndall, but Powell said he had put things right.

That showed me for the first time how difficult it is to be accepted by the English as one of them.

The situation, as I see it now, was that the right wing of the lab. considered me an ally of communists like Powell or Blackett. The left considered me a trotskyist or reactionary. For the rest, I was an Italian fascist turncoat.

But the developed plates were splendid.

About this time, maybe in September, I opened a drawer in the microscope room and found a postal package marked 'Ilford'. I asked Powell, discretely, what it was. He had been too busy to open it. In it we found the first concentrated emulsions with a letter asking us to try them. There was no motive given (such as 'following your request') but a note that a similar package had been sent to the Cavendish. The letter was dated Monday and it was now Thursday. Powell had been busy with his lectures, with writing a paper, and with Marxist dialectics. He apologised saying there was no

---

[87] Arthur Mannering Tyndall (1881–1961).

[88] Leonard Vincent Chilton (1905–1997).

[89] Lettice Ramsey (1898–1985).

[90] Herbert Frölich (1905–1991).

hurry, and that after all, Ilford had "delivered the goods". Evidently he had forgotten my explicit request to Chilton.

It was 6 pm I took the plates and put them to soak in a Thorium solution, closed them in a box and left them in the dark room with a label "Do Not Touch". The following noon I developed the plates; they were dry by 4 pm and under the microscope: they were a revelation. Powell was flabbergasted. I told him we should notify Ilford, we had delayed four days, the Cavendish would be first. On my insistence, having telephoned Ilford but found it closed, we sent a telegram that same evening. It was Friday evening. Ilford was closed on Saturdays.

Powell received a letter from, I think, Harrison in which he said "Compliments, you beat the Cavendish by 24 h."

It was the beginning.

"When I Arrived in Bristol…" [Dilworth's Manuscript][91]

When I arrived in Bristol the lab. consisted of one room plus a small studio on the 4th floor and had 3 microscopes plus a Y4 monocular. On the 3rd floor was a dark room. The staff consisted of Heitler (brother of the theoretician) and Guggenheimer, and one microscopist, Mrs Andrews.

Powell had shown the precision in measurement of length and angle of tracks in the emulsion was comparable to that in the Wilson chamber. In my opinion, the important step he had made in the techniques was his emphasis on the quality of the microscope optics (use of immersion objectives).

I was given a group of plates exposed to cosmic rays by Powell, Heitler and Fertel at the Jungfraujoch and I started to study them. At the same time Powell suggested that I get the feel of the technique by measuring the distribution in length of the long tracks from Thorium stars. I did it with such zeal and application, changing frequently eyepiece and scale, that my distributions turned out to be more narrow than Powell's own. In fact he asked my permission to use my distribution instead of his in the paper then being prepared for publication in the Review of Scientific Instruments. There was no acknowledgement, but a few days later the Director of the Laboratory (Prof. Tyndall) called me to his study to offer me a position in the laboratory at £27 a month. Ecstasy!

I went ahead with the scanning of the two plates exposed to cosmic rays but the results were disappointing.

One day, I don't remember the date, but maybe in June, a representative of Ilford, Chilton, came to discuss improvement of the plates, and Rosenblum and I were invited to the discussion. Chilton remembered me, we had met at the Cavendish when he was a student there. That helped dispel my habitual shyness and I was able to explain to him that, in view of Blackett's mission to me for cosmic rays, the low

[91] Transcription of Dilworth's Manuscript Which Begins with "When I Arrived in Bristol…". Dilworth Titled the Correspondent Typescript "Concentrated Emulsions".

density of grains on tracks of fast particles made them too difficult to follow, and I put the question as to the possibility of increasing the quantity of silver, on the basis of the experience of increasing pressure in a Wilson chamber. Rosenblum was silent. Powell said, with a bit of dry humour, that that would mean re-making the range-energy relation.

After that there was Hiroshima.

Shortly afterwards, Lettice Ramsey sent me a book on Scientific Photography by Lawrence, an old Cambridge friend of mine. In it I found the temperature coefficient of the process of development and the recipe for the Amidol developer. I immediately tried the effect of temperature, packing ice into condoms so as not to dilute the developer.

I was the only one who used the 3rd floor dark room. I often washed my underclothes there, and sometimes my sheets, before leaving the lab at my usual time, about 1 am. I had left the condoms full of water, but next morning they had disappeared. The lab staff looked awry at me. "We thought someone had been rash" said Powell. Fröhlich consoled me. I wanted to go and see Tyndall, but Powell said he had put things right.

That showed me for the first time how difficult it is to be accepted by the English as one of them.

The situation, as I see it now, was that the right wing of the lab considered me an ally of communists like Powell or Blackett. The left considered me a trotskyist or reactionary. For the rest, I was an Italian fascist turncoat.

But the developed plates were splendid.

About this time, maybe in September, I opened a drawer in the microscope room and found a postal package marked 'Ilford'. I asked Powell, discretely, what it was. He had been too busy to open it. In it we found the first concentrated emulsions with a letter asking us to try them. There was no motive given (such as 'following your request') but a note that a similar package had been sent to the Cavendish. The letter was dated Monday and it was now Thursday. Powell had been busy with his lectures, with writing a paper, and with Marxist dialectics. He apologised saying there was no hurry, and that after all, Ilford had "delivered the goods". Evidently, he had forgotten my explicit request to Chilton.

It was 6 pm. I took the plates and put them to soak in a Thorium solution, closed them in a box and left them in the dark room with a label "Do Not Touch". The following noon I developed the plates; they were dry by 4 pm and under the microscope: they were a revelation. Powell was flabbergasted. I told him we should notify Ilford, we had delayed 4 days, the Cavendish would be first. On my insistence, having telephoned Ilford but found it closed, we sent a telegram that same evening, a Friday evening. Ilford was closed on Saturdays.

Powell received a letter from, I think, Harrison in which he said "Compliments, you beat the Cavendish by 24 h."

It was the beginning.

"Bristol..." [Occhialini's Comments (Labeyrie's Manuscript)]

Bristol[92]

Emulsions Concentrées[93]

Noël 1992

Dans mon opinion le step important de Powell dans la technique avait été l'emphase sur la nécessité d'une optique supérieure (immersion objectif) et compétition avec la chambre de Wilson dans la mesure des angles et des longueurs.

J'ai reçu des plaques exposées aux Rayons cosmique par Powell, Heitler, Ferstel et j'ai commencé à étudier full time ainsi que le check up sur la distribution à longueur des traces longues de Thorium (Revue scientifique discovery). Comme je dis avant cette distribution étais la preuve de la puissance de la mesure du range dans l'émulsion. Mes mesures portaient à une exactitude encore plus grande avec un mean spread supérieur. [*Editorial note: Occhialini's graph in Fig.* 2.]

Ces résultats sur une longueur de 62 microns avaient été obtenus en changeant continuellement d'oculaire et d'échelle avant le scepticisme général. Je me suis offert de faire un check indépendant sur une quantité inconnue. J'ai reçu 2 plaques qui portaient sur une longueur de 700 micron et dans lequel il y avait à mesurer 3 longueurs. Quand la chose a été finie il a été prouvé que mes mesures étaient plus exactes que celle qui étaient en train d'être publiées. Powell a proposé d'insérer anonymement mes distributions en les substituants aux mesures déjà prêtes pour publication.

J'étais d'accord. Aucun acknowledgement m'a été offert mais quelques jours après le Directeur de Laboratoire me convoquait pour m'offrir une [position] dans le laboratoire à £27 par mois. Extase. L'examen des 2 plaques des rayons cosmiques continuait avec des résultats décevants.

En—(date inconnue), peut-être an Juin, est arrivé un représentant d'Ilford pour parler d'amélioration des plaques, j'étais admis à la discussion avec Rosenblum.

Le représentant d'Ilford Chilton m'avait rencontré, étudiant au Cavendish et se souvenait de moi. Ça m'a tiré beaucoup de timidité et j'ai pu expliquer sur la base de ma mission pour compte de Blackett (plaques rayons cosmiques) que la densité des grains sur les traces des particules de haute vitesse rendait difficile de les suivre, et j'ai posé le problème d'augmenter la quantité d'argent sur la base de l'expériences en Chambre de Wilson de l'augmentation de pression. Rosenblum était silencieux. Powell a dit avec humour que sa aurait signifié refaire reconstruire la range energy relation. Après ça il y a eu Hiroshima.

[92] Written by Dilworth.

[93] Added by Dilworth.

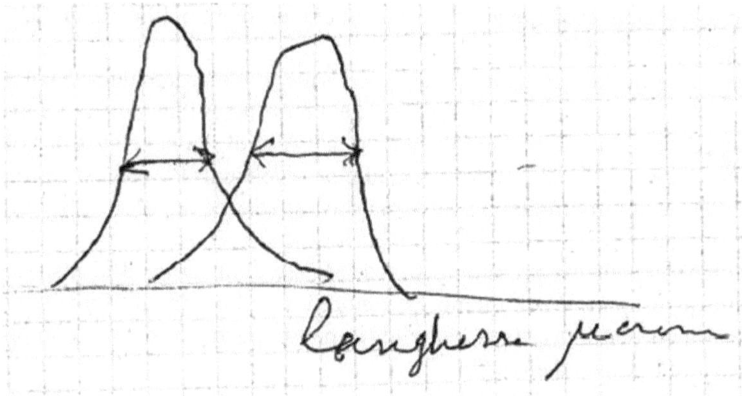

Fig. 2 The graphs and the legend on the abscissa were sketched and written by Occhialini. Courtesy of the BICF library of the University of Milan

J'ai reçu de Lattice Ramsey en cadeau un livre de Lawrence mon vieil ami à Cambridge, sur lequel j'ai trouvé les 2 indications sur le développement en température et l'emploi de l'amidol.

A ce moment j'ai fait le premier essai avec des préservatifs remplis de glace pour ne pas diluer le développeur. Le système a été foudroyant. J'étais le seul à employer la chambre noire au 3ème étage. Je faisais mon lavage de mes sous-vêtements et aussi des draps dans la chambre noire quand j'ai abandonné le labo comme c'était habitude à 1h la nuit, j'ai laissé les préservatifs plein d'eau, le matin suivant in n'étaient plus là. Le staff du Labo me regardait avec reproche. We thought someone had been rash, said Powell.[94]

Frölich m'a consolé, je voulais visiter Tyndall mais Powell m'a dit qu'il l'avait fait lui-même en ma défense. Ça m'a montré pour la 1ère fois qu'il est difficile d'être accepté par les Anglais comme un des leurs. La situation comme je le vois maintenant est que la droite dans le Labo me considérait comme un allié d'un communiste comme Powell ou comme Blackett. La gauche me considérait comme un trotskiste ou un réactionnaire. Pour d'autres j'étais un fasciste italien qui avait changé de wagon.

** les plaques étaient splendides

À ce moment peut-être à septembre j'ai ouvert le tiroir dans la salle des microscopes et il y avait un colis postal marqué Ilford. J'ai demandé discrètement à Powell, il avait été trop occupé pour l'ouvrir. On l'a ouvert. C'était les premières émulsions concentrées avec prière de les essayer. Il n'y avait aucune indication dans la lettre sur la raison de l'envoi mais qu'un paquet analogue avait été envoyé au Cavendish Laboratory. La date de la lettre était un lundi et on était jeudi à ce moment. Powell occupé avec ses courses, avec la rédaction des travaux en cours et aussi avec la dialectique marxiste, s'est excusé disant qu'il n'y avait pas d'urgence et qu'après

[94] This last sentence was written by Occhialini.

tout Ilford avait "delivered the goods". Il avait évidemment oublié ma demande, ma requête explicite à Chilton.

Il était 6h du soir j'ai obtenu les plaques je les aient immergées dans la solution de Thorium. Je les aient mises à sécher, je les aient enfermées dans une boîte avec une affiche dans la chambre noire "ne toucher à rien". À midi j'ai développé les plaques, elles étaient sèches à 4h. Au microscope elle se révélaient. Powell est resté sidéré. Nous pouvons le communiquer à Ilford, nous avons retardé de 4 Jours, le Cavendish va arriver premier.

On a téléphoné sur ma pression à Ilford. Mais c'était fermé. J'ai insisté sur un télégramme qui est parti le soir même. Le vendredi soir. Le samedi Ilford était fermé.

Powell a reçu une lettre dans laquelle, je pense Harrison, disait "compliments vous avez battu le Cavendish de 24h."

Ça c'est le commencement.

J'ai proposé que le travail de ces émulsions devait être publié, alors est arrivé une défense de publier parce que la chose était d'intérêt guerrier. Au Canada sont passé au travers aucune industrie. Un physicien Canadien français Demers[95] avait fait des émulsions concentrées merveilleuses et pour le "Secret Act" nous ne pouvions pas publier.

J'étais en désespoir, aussi pour le fait que il y avait un suspect dans l'air qu'il y avait été un——. Powell avait un ami communiste au Canada et c'était possible que cet ami lui avait communiqué le travail de Demers. D'un autre côté pour ce qui me regarde je pouvais testifier que Powell n'avait jamais été intéressé pour cette émulsion concentrée; son travail n'avait pas besoin. Ces émulsions étaient nécessaires pour une personne qui travaillait en rayons cosmiques. Mais signifiait que Powell n'était pas responsable.

À Bristol il n'y avait pas de possibilité de prendre des micro-photos. Crée dans la tradition de Cavendish de Fowler, Blackett, Wilson, le document photographique était important.

J'ai ainsi téléphoné à Lattice Ramsey à Cambridge pour savoir si elle pouvait trouver une caméra pour micro-photographie. Le matin après elle communiquait qu'elle en avait acheté une pour un Leitz attachement pour £32. C'était plus qu'un mois de mes gages. Elle demandait que j'aille la prendre. Je suis parti le vendredi soir pour Cambridge, j'ai pris la caméra attachement, je suis rentré le samedi à Bristol, je suis allé au Pub Albion et j'ai fait une petite enquête pour savoir si quelqu'un pouvait m'aider un Leica et plus important encore avec un peu de films.

J'ai fait amitié avec un type extraordinaire appelé Busby qui s'est mis à ma disposition. Il habitait près du Royal Fort. Nous sommes rentrés avec sa Leica. Nous avons photographié une série d'étoiles de Thorium et un proton (proton important).

[95] Pierre Demers (1914–2017).

à 2h.30 on avait fini avec les photos, Busby remportait sa précieuse Leica avec lui et je commençais à produire une série des cartes postales (c'était dimanche nuit) que Powell a trouvé le matin après sur son bureau.

1er janvier 1993

Appendice aux émulsions concentrées

Bloc

J'ai toujours trouvé l'obscurité et le bruit de l'eau constructive et j'ai toujours pensé que le gros patron qui s'était occupé d'émulsions nucléaires et s'était mis trop dans les mains des techniciens pendant qu'on attendait la permission de publier ou non le travail, on continuait à y ajouter des pièces au milieu d'une polémique qui regardai la question de sécurité et desquelles je me tenais de côté. Je réalisais que ma présence comme citoyen de l'Axe pouvait rendre la position de Bristol et éventuellement de Blackett, Cockcroft et Appleton délicate.

Pour la question qui regarde la mémoire une forte visualisation est nécessaire. C'est-à-dire pour que le souvenir reste et soit valable il doit y avoir une connexion dans la mémoire entre les faits de la conversation et la géométrie dans laquelle elle se produit.

Il y avait dans la chambre noire un placard bas très profond, en l'absence d'une table à moi je voulais l'employer pour tenir mes frusques pour lesquelles il n'y avait pas de place dans la maison de Freddy.

J'ai divisé avec 2 autres ma chambre pour dormir (on dormait 4 dans la même chambre). Ainsi j'ai demandé la permission de vider le placard. Je l'ai obtenu et j'ai commencé à range dans le couloir la quantité incroyable de choses qui étaient entassé de dedans à plusieurs niveaux. Dans l'angle gauche au fond j'ai trouvé un paquet postal envoyé par Ilford Ltd très vieux. Je ne sais pas si c'était 1941 ou 42. Je l'ai ouvert, j'ai trouvé dedans une boite de plaques d'Ilford Ltd avec une lettre de Monsieur——adressé à Powell dans la quel il communiquais que le plaques contenaient une émulsion enrichie de Bromure d'Argent qui contenait le double de la quantité normale. J'ai développé une des plaques divisées en 4 pour découvrir qu'elles étaient toutes complètement noires. Ce noir n'étant dû à des positives d'Argent.

À ce moment j'avais appris par le Photographic α Book la possibilité d'une émigration *froide oleochimique*[96] qui couvrait la surface et qui pouvait être éliminée avec un frottage délicat avec du méthylated spirit.

Cette information quelques mois après devait être capitale pour la découverte du PyMu quand j'ai récupéré de la poubelle ou elles avaient été jetées les plaques très alcalines chargées du Pic du Midi.

---

96 The words in italics were written by Occhialini.

J'écris ça à ce point parce que j'ai peur dans cette description de ne pas arriver en temps à la décrire au moment opportun qui regard les Rayons cosmiques.

Je n'ai jamais rien caché à Powell et le matin après je lui ai porté la boite avec la lettre et les plaques noires développées. Ça a porté à des conséquences publiques, une addition dans le brouillon du papier à Monsieur Block qui était probablement la personne qui avait envoyé la lettre d'Ilford (late Dr Block). Sur un livre grand marmorisé, le matin après, Powell me montrait sur une page qui avait été toujours blanche, un méssage avec la date du meeting avec Chilton qui contenait a peu près la phrase "suggested to Ilford the use of concentrated emulsion".

Comme beaucoup de choses qui se sont passées à Bristol, c'était ambigu. Il ne disait pas "I suggested", il ne disait pas "Beppo and I suggested". Ça reste un des points obscurs de cette question. J'avais parlé à Cecil de la nécessité d'augmenter la pression du gaz c'est-à-dire en termes d'émulsion d'Argent, et n'avait jamais dit que cette tentative avait été faite.

Est-ce qu'il avait oublié cela? Est-ce que dans l'attitude et le jugement négatif envers Ilford en n'ayant ouvert le paquet que j'avais découvert il l'avait ignoré complètement? Ou il avait tout simplement oublié.

J'ai une propension pour la 2eme explication le phénomène: de ne pas ouvrir les paquets que je pensais, envoyés par Chilton aurait arrêté l'investigation si Mr Block avait envoyé le paquet seulement à une personne. Il semble presque que quelqu'un chez Ilford avait prévu la possibilité d'un manque d'intérêts du côté Bristol quand il avait envoyé les plaques non seulement à Powell mais aussi à Lipsy et Gibson. Il y avait ainsi une double assurance.

Pour ce qui regarde l'épisode du "suggestion" (we suggested, je me souviens d'une petite conversation sceptique de mon côté la phrase "it is in the name of the firm partner") il y a encore un point, Connie m'a conté que Waller avait dit au congrès de Bristol que Chilton lui avait dit que Powell demandait des plaques concentrées.

Pour Waller mon nom ne signifiait rien et Powell était, en bon droit, la personne à Bristol avec laquelle Ilford négociait.

Je me propose d'investiguer cette situation quand je rencontrerai Powell au Purgatoire, mais j'ai peur que le pauvre Beppo en tâchant de redresser cette situation ajoute à sa réputation d'idiot du village cette du mythomane.

Un autre mystère pour les historiens est le problème Rotheblat qui déclare qu'il a contacté Ilford pour faire produire des plaques concentrées lui aussi. Ici le suspect que quelqu'un de l'énergie Atomique Anglaise ait cherché à l'insu du pauvre Pierre Demers de capitaliser sur son travail.

La conclusion que je dois tirer est ce que les personnes qui ont fait le travail n'ont pas été les spécialistes en plaques nucléaires mais 2 mercenaires Demers et Beppo, indépendamment l'un de l'autre.

Ce type de témoignage tends à défendre les intérêts personnels de Powell contre l'accusation d'être initié (cherche sur le vocabulaire français le signifié). C'est certain si Powell avait été informé par quelqu'un de ses amis du Canada de l'activité de Demers, son attitude avec moi aurait été différente. Il n'avait aucun intérêt en

Rayons Cosmiques. Sa contribution au travail est le dessin détaillé de la Chambre à scattering qui constituait son intérêt principal et son activité abandonnée dans la matinée du "marteau".

Résumé autobiographique
Brésil—Londres—Bristol

Comme je ne suis pas sûr de pouvoir avoir le temps de faire le récit, j'explique aussitôt que ça a été la matinée après le développement des plaques du Pic du Midi dans lesquelles j'avais trouvé aussitôt le 1er marteau et qui avait provoqué les hurlements historiques de Tyndall. "Powell you must publish it at once".

La langue anglaise est ambigüe dans ce même moment j'ai compris que l'idylle entre mois et Tyndall était finie et que le you signifiait tu et toi seul, en effet Powell très confus a excusé la manie d'un Professeur en âge avancé qui avait de remords envers son élève, assistant et qui le protégeait d'une manière qui étais embarrassante, ma réponse a été que j'étais d'accord et que moi aussi j?avais des liens de gratitude avec lui Powell, que j'appartenais à un peuple pour lequel l'amitié était un lien)

fin de la parenthèse[97]

parenthèse additionnelle aussi:

Il avait montré un gentil scepticisme à mes rêves de rendre l'émulsion un instrument de précision universelle. Il ne m'avait pas encouragé dans mes tentatives d'augmenter l'épaisseur au contraire il avait contribué à créer dans le staff une attitude négative en m'arrêtant dans mon action de défense contre la réputation d'être un "latin lover" qui remplissait la chambre noire de préservatives. S'il avait été au parfum il m'aurait encouragé en assumant une responsabilité conjointe dans la product promotion en se retranchant derrière mon innocence. Rien de cela. Il avait ses racines dans le travail réellement très beau des "Butterflies" pour lequel je ne sais pas s'il a trouvé des explications théoriques. Il se sentait persécuté et était très pessimiste pour ce qui regardait l'attitude des Académicien ses collègues dans l'Université Anglaise. C'est à ce moment que Blackett est entré dans le play à travers un weekend qui s'est passé à Londres.

---

4 janvier 1993
après midi

Jean François est venu déjeuner

Aujourd'hui c'est le 4 janvier et pendant la nuit j'ai réalisé l'inutilité de ce que j'écrivais, il n'y a aucune preuve de ce que j'ai écrit aussi si quelqu'un lit mon histoire de Bristol, il pensera que je suis un mythomane.

---

97 The parenthèse opens at "Comme je ne suis pas sûr...".

Une réponse à tous les "pourquoi?" est difficile à donner à qui n'a pas vécu la situation. En ayant appliqué à Houtermans la qualification de "Dimitri", j'ai accusé ses critiques de ne pas reconnaitre les "Dimitri" (j'ai défendu Houtermans dans les années dures de l'après-guerre) qu'il adorait en lisant Dostoevsky, il les rencontrait dans la vie réelle et ne les reconnaissait pas. Ainsi je peux dire que je me reconnais dans l'Idiot et je me trouve compréhensible.

Lattes un soir d'ivresse m'a accusé en public d'être un lâche parce que j'avais démontré que j'étais capable de mourir en défendant des amis mais que j'étais incapable de défendre mes intérêts.[98]

Il est possible qu'il ait raison, mais combien sont-elles les personnes qui sont capables de réclamer l'argent qu'ils ont prêté à un ami?

Le lien avec Powell était créatif. J'ai rencontré un être effacé, prisonnier d'une idéologie marxiste, pas sûr de lui-même comme une personne qui sortait d'un hôpital, je l'ai vu se transformer sous mes yeux.

"Concentr. Emulsion" [Occhialini's Comments (Labeyrie's Manuscript)][99]

16 avril 1993

Concentr. emulsion[100]

Remarques page 1 période 2
Correct, le progrès était important si on compare certaine des résultats obtenus par Indian working with dry objectives.

para 3. Powell au commencement n'avait pas de confiance dans mes mesures faites sur des traces dont la longueur étaient connues. Il m'a donné 2 plaques inconnues qui contenaient le matériel pour 3 travaux pour la Royal Society. Ici l'engagement était plus grand parce que le champ des longueurs arrivait à 1,2 mm (Proton projected from neutron Boron). J'ai eu le choix.

Ces 2 courbes ont été substituées, toutes celle pour la Royal Society, by Heitner, Guggenheimer and Powell. The différence in quality on results were stiriking. My maximum étaient dans la position juste mais visiblement plus étroit que ceux du Bristol team.

page 2
Je ne sais pas si Hiroshima a été avant o après.

[98] Dilworth-Occhialini expressed the same concept in the file "Politics and Personalties".

[99] Transcription of Occhialini's comments to Dilworth's manuscript which begins with "When I arrived in Bristol…". The pages numbers refer to Dilworth's manuscript.

[100] The title was added by Dilworth.

Ce n'était pas seulement la basse densité mais la valeur très élevée du grain du fond toutes le plaques étaient développées avec un développeur au métal Hydroquinol très alcalin, qu'on employait dilué de 1 à 5 et la base de ce choix était qu'il s'appelait développeur pour Rayons X.

Je n'ai jamais su qui avait inventé la dénomination qui était un exemple d'intoxication intellectuelle.

Se me souviens bien l'une des difficultés pour le jugement sur les plaques de Rayons cosmiques était dû à une longue exposition, il y avait du feeding et c'était ainsi un étrange mélange dans lequel on pouvait comprendre pourquoi le seul résultat de cette exposition avait été un fragment lourd.

Page 3

La réaction du labo et aussi de Powell explique pourquoi je n'ai pas employé la méthode en ses conséquences logiques. J'explique plus avant, que les plaques de 200 microns ont été volées et qu'ainsi l'exposition au Pic du Midi a été faite avec des plaques non adaptées en partie acide et en partie alcaline.

Page 3. 4e para.

Les plaques développées étaient splendides Powell gasped.

L'arrivée des plaques de Ilford, l'arrivée des émulsions concentrées a interrompu la querelle.

Je n'avais pas oublié la visite d'Ilford-Chilton. J'ai vu arriver le paquet le matin. Le soir n'était plus là. 2 jours après j'ai ouvert un tiroir et découvert le paquet non ouvert.

J'ai suivi l'action qui me montrait la manque d'intérêt de Powell et son manque de confiance en Ilford de "never delivered the goods".

Situation. Le paquet était arrivé 2 jours avant au moins, c'était certainement jeudi ainsi la lettre ne pouvait pas être daté lundi.

4eme page O.K. only out the box.

Pour l'histoire Ilford pour moi était Chilton au Cavendish's day. Il était mon interlocuteur, son souvenir de moi m'avait ému, ainsi je l'ai considéré comme l'interlocuteur. Je ne savais pas de l'existence de Waller et ça explique pourquoi le papier porte son nom.

Le papier dont la conception était complètement ma décision, je voulais faire une lettre à Nature, Powell demandait "ainsi nous écrivons une lettre à Nature en disant que nous avons demandé à Ilford de produire des plaques concentrées."

À ce moment j'ai exhibé la distribution de longueur des longues traces du Thorium qui, comparée avec la distribution que j'avais déjà faites avec le half tone montrait l'avantage des nouvelles plaques.

Quelques jours après Ilford envoyait des nouvelle plaques qui mi remplissaient de désespoir. Il y avait une qualité inférieure dans les nouvelles plaques. Ainsi j'ai paniqué et j'ai demandé à Powell de téléphoner à Chilton ou à Harrison ou à qui que ce soit chez Ilford pour dire que s'ils avaient changé les procédures, les résultats étaient désastreux. J'avais peur que s'ils avaient changé leur routine ou que leur succès était dû à la chance et que maintenant le secret était perdu. Powell a téléphoné à Ilford, a reçu l'assurance qu'ils seraient venus à nous visiter et c'est ainsi que nous avons rencontré Waller.

The Publication

The Publication [Typescript]

The Publication

It seemed to me evident that with this leap in quality the method could be useful. I felt that my leaving Brazil had been justified, and that I could be pardoned for not having taken part in the war. The important thing was, to my mind, to publish immediately. I persuaded Powell, in spite of his wife's complaint that he was losing weight due to my frenzied activity. In fact, Powell had a strange lassitude. The years of obscurity seemed to have drained him of initiative.

We proposed that the paper be authored, apart from ourselves, by a member of the Cavendish Laboratory, Livesey,[101] and by Chilton of Ilford. At that time I did not know that the real inventor of the concentrated emulsions was Waller, Chilton being merely an Ilford representative. I had met Waller when he came to Bristol to see the situation after we had complained that the second batch of emulsions was not of the same high quality (the third and successive ones were consistently good). Waller was so handsome, like Mountbatten, neatly dressed and with well-kept hands, that I was sure he must be a high-ranking member of the Ilford directorate, not a laboratory type. Only several years later, when Ilford found itself in disadvantage with respect to Kodak over the electron sensitive emulsions, could I repay my debt by testing and reporting to Waller on what were to become the final victors, the G5.

The paper was written for the Review of Scientific Instruments.[102] I repeated the old α-particle range distribution to show graphically that the precision was improved with respect to the old plates. We filled the paper with technical details. Powell put in a drawing of his new scattering chamber.[103]

I had been brought up in the tradition of the Cavendish, of Wilson, Feather,[104] and Blackett, that it was important to give photographic evidence, but in Bristol there was no means for taking microphotos. I telephoned to my friend Lettice Ramsey in Cambridge to ask her if she could find a microscope attachment. The morning after

[101] Derek Leonard Livesey (1923–1992).

[102] The paper was published in the *Journal of Scientific Instruments*: Powell C. F., Occhialini G. P. S., Livesey D. L. and Chilton L.V., "A New Photographic Emulsion for the Detection of Fast Charged Particles", *J. Sci. Instrum.* **23** (1946) 102–106.

[103] A propos of measurement of shrinkage. [Handwritten footnote by Dilworth].

[104] Norman Feather (1904–1978).

she replied that she had bought a Leitz attachment for £32 and that I should go and get it. It was more than a month's wages for me.

I went to Cambridge on a Friday and returned with the attachment on Sunday evening to Bristol. There I went to my local pub, the Albion, and asked if anyone knew a photographer, who had a Leica and film. There was, he was Busby, a great chap, who offered to help me. He lived near the Royal Fort,[105] so he went home and got his Leica loaded with film, the developing box and photographic paper of postcard format.

We photographed a series of Thorium stars and a proton, very important. At 2 am Busby left the Royal Fort with his Leica and developing kit. I remained washing and drying the photos, which I left on Powell's desk so he and Rosenblum could see them in the morning. Tired out, I came back to the lab only at midday.

That group of postcards gave me the chance to have the darkroom put in order. A good enlarger was bought, and photographic paper. The necessary Leica was lent by Powell's father-in-law, a very good man, a Jew who listened patiently to my railing against the Germans.

The photographing exploit preceded my visit to Blackett, a visit which had been delayed by two weeks because of his engagements. I spent a weekend with him in his home in Maida Vale near Swiss Cottage. (Moving house was a family sport of the Blacketts, the latest one being always the best.) On the Sunday we walked across the parks, like in the old Cambridge days.

I showed Blackett the photos of the Thorium stars and, especially, of the protons. He warned me that there was a Canadian physicist, Demers, who during the war had made concentrated emulsions. His work had not been published as it was part of the secret war work. We talked about Powell. I said that I was under the impression that he was persecuted. Blackett asked me if he was a member of the Communist Party. I replied that 'to the best of my belief and knowledge' he was not, but that Blackett's question was important because it proved to me, what I had never been sure, that he, Blackett, was not a member. He told me Lord Portal[106] had the same doubt ("He called me a traitor"). Coming back to Powell, he explained that the reason Powell was not a member of the Royal Society was not bad will, but the fact that there were others who had interrupted their research to take part in the war effort and so had priority.

We then talked again about Ilford and I explained to him how the new emulsions had come about, and Powell's lack of enthusiasm, proven by the fact that I had to develop them myself. I had realized also that scientific interests came in second place to a firm like Ilford. To follow a suggestion from outside could be a financial risk. By the end of the discussion Blackett had promised to try to arrange for a committee of scientists empowered to distribute funds to encourage the production and improvement of the emulsions. "You will not be a member of the committee"

[105] I do not know why he used the Albion instead of the Robin Hood which was near the Royal Fort. [Handwritten footnote by Dilworth].

[106] Charles Portal (1893–1971).

said Blackett "but being in Bristol you will be able to let me know if someone tries to muscle in."

So the Photographic Panel was born in a park in London. (Rochester[107] knows the date, but I have never had the opportunity of correcting his mistaken idea that the concentrated emulsions were a product of the Panel. It was vice versa).

Blackett saw in these emulsions a means of reinvigorating research in Europe. That evening he opened a bottle of wine for a toast to the hope that Italy could have a part in the boom.

Some months later Powell was elected Fellow of the Royal Society.

Back in Bristol I told Powell of Blackett's reaction. Now Bristol should find the funds to carry on. I guaranteed Blackett's friendly backing but did not mention the Panel. I urged the necessity of getting more people to exploit this new technique by forming a 'Foreign Legion'. He accepted the idea since few of the young British physicists had yet been released from their war-time positions. So it came about that Lattes, whom I had left with Camerini in Brasil, was offered a fellowship.

While all this was going on, we received an injunction from the Atomic Commission to suspend publication in view of the fact that Demers's work in Canada was still covered by the Official Secrets Act. This started a controversy, and there was in the air a vague suggestion that the Bristol initiative could have been a copy of Demers. Alan Nunn May,[108] a communist, (one of the first to be tried and condemned as an 'atomic spy') was a colleague and friend of Powell and had been working in Canada. The presumed complicity between Powell and Nunn May could have been the basis of the interdiction of publication.

An incident which tended to defend Powell against this accusation was the discovery I made when clearing out a cupboard in the darkroom. In my lodgings I shared a room with three other men and I needed a place to keep some of my things so I had asked for permission to use that cupboard. I started clearing it out and stacking in the corridor an incredible number of objects. At the bottom, on the left, I found a package from Ilford Ltd, very old, 1941 or 1942. I opened it and found a box of plates with a letter from Dr Block to Powell, informing him that these plates contained twice the normal content of AgBr. I cut one of the plates into four parts and developed them, but found them all completely black. It was not due to a deposit of silver. (I had learnt from the book on Scientific Photography of the possibility of the migration of a oleochemical fog which covered the surface and could be removed by delicately rubbing with methylated spirit. That information was to be precious in saving the π-μ plates).

I never hid anything from Powell. The next morning I took him the box with the letter and the black developed plates. A reference to them and to "the late Dr. Block" was inserted into the paper then being written.

The next day Powell showed me, in the big book he used as a diary, on a page which had always been blank, a note, with the date of the meeting with Chilton,

[107] George Rochester (1908–2001).

[108] Alan Nunn May (1911–2003).

"suggested to Ilford to use concentrated emulsion." Ambiguous, not 'I' nor 'we', just 'suggested'.

Powell's habit of not opening the parcels from Ilford possibly explains why Chilton sent Waller's plates to the Cavendish also.

Another mystery remains: Rotblat's[109] claim that he too had asked Ilford to produce concentrated emulsions. Here one could suspect that in the British Atomic Energy had tried to exploit poor Demers's initiative.

In the end the injunction against the paper was lifted and it was sent for publication. (Feb. 1946)

When Lattes arrived on his fellowship he suggested that Camerini could come on his own expenses. It was the beginning of the 'Foreign Legion'. Goldschmidt-Clermont came later and Franzinetti.[110]

The Publication [Dilworth's Manuscript]

The Publication

It seemed to me evident that with this leap in quality the method could be used. I felt that my leaving Brazil had been justified, and that I could be pardoned for not having taken part in the war. The important thing was, to my mind, to publish immediately. I persuaded Powell, in spite of his wife's complaint that he was losing weight due to my frenesy. In fact, Powell had a strange lassitude. The years of obscurity seemed to have drained him of initiative.

We agreed that the paper should be authored, by ourselves, a member of the Cavendish Laboratory, Livesey, and by Chilton of Ilford. At that time I did not know that the real inventor of the concentrated emulsions was Waller, Chilton being merely an Ilford representative. I had met Waller, when he came to Bristol to see the situation after we had complained that the second batch of emulsions was not of the same high quality. (The third, and successive emulsions were consistently good). Waller was so handsome (he resembled Mountbatten) style, with well kept hands and neatly dressed, that I was sure he must be a high-ranking member of the Ilford directorate, not a laboratory type. Only several years later, when Ilford found itself in disadvantage with respect to Kodak for the electron-sensitive emulsions, could I repay my debt from Brussels testing and reporting on what were to become the final victors, the Ilford G5.

The content of the paper, originally, hopefully, intended for Nature, but which ended in Reviews of Scientific Instruments, included my repeat of the α-range distribution to show the increased precision, several technical points, and the design of a scattering chamber that Powell had requested of Payne.

---

[109] Joseph Rotblat (1908–2005).

[110] Carlo Franzinetti (1923–1980).

In addition, there were the photographs. It was one of the traditions of the Cavendish that there should be a photographic evidence. Powell had not followed this tradition, only one photo had appeared in "Endeavour?, so in the lab there were not the necessary accessories.

I phoned to my photographer friend Lettice Ramsay in Cambridge. She found a Leica microscope attachment for £35, more than a monthly salary for me and bought it. On a Friday evening I went to Cambridge to get it. Back in Bristol, Saturday evening, I asked in my pub, the Albion, if anyone knew of a photographer. There was a one, Busby. (I never knew why Busby used the Albion. For me, it was the nearest put to the room I shared with 3 other men, but Busby lived near the Royal Fort, so his local pub should have been the Robin Hood, which later become my pub). Busby was magnificent. He came with me to the Royal Fort with his Leica, full of film, his developer tank and printing paper, postcard format. He left at 2 am and next morning Powell found the post-cards on his desk.

I was ready then for the weekend with Blackett. It had been delayed for two weeks because he was busy. I stayed in his house in Swiss Cottage. One of the chief amusements of the Blackett family seemed to be that of changing house. We walked in the park, as we used to do in Cambridge days. I showed him my postcards and suggested that something might be done to encourage Ilford. He said a Government committee could be empowered to contribute funds for this purpose. Thus the Panel was born.

Blackett explained to me that, as a foreigner, I could not be a member of the Panel, but he counted on me in Bristol to let him know if some character tried to "muscle in".

On the same occasion, he warned me of the Demers problem. Demers was a Canadian physicist who, as part of his war-work, had found the way to produce, in artisan fashion, concentrated emulsions. His work was covered by the Official Secrets Act and therefore not published.

In fact, we in Bristol were notified that we could not publish.

It was a difficult situation, particularly so because there was a suspicion that Powell could have had information from a communists Bristol colleague and friend, Alan Nunn May, of Demers work.

That it was not so, I knew, not only from Powell's lack of interest in the parcel from Ilford, but also from the discovery, in clearing out a cupboard in the dark room, of a package of plates of '42, unopened, from Ilford, with 2x the concentration of AgBr. These plates I developed. Thy were black. Appealing to Lettice, she suggested a treatment with———which later was precious in saving the Boron loaded plates from the Pic du Midi which contained the first π-μ decay.

What interested me in Blackett's discussion of the Demers affair, was his question if Powell were a Party member. I replied honestly that I did not know, but for me, the question itself was important. It eliminated a doubt I had that Blackett himself could have been a card-carrying member.

Eventually the prohibition to publish was withdrawn; and so began the era to which Blackett toasted the final evening of that weekend, "To Italy and Europe, ruined by the war, who can with this technique in cosmic rays, rebuild scientific research."

I came back to Bristol to tell Powell of Blackett's total backing of our enterprise. Then I insisted on the necessity of increasing our "fire power". On one hand, to invest in technical gadgets, for instance a photographic enlarger. It was bought, and the Leica to put in it was loaned by Isabelle's (Powell's wife) father—whom I had known in pre-war year. On the other hand we needed men-power. The young people were mostly still engaged in their war-time activity. Few of them had obtained license after Hiroshima. So I suggested to Powell that my young friends from Brazil could be useful.

Lattes was invited with a fellowship. He suggested that Camerini, who had private means, could come too.

So the "Foreign Legion" was born. It was later increased by Franzinetti and Goldschmidt-Clermont. All Jews.[111]

Editorial note: Among Dilworth's handwritten sheets of "The Publication" there are some deleted notes. They indicate the topics to be addressed, a consideration about the Jews and a block diagram of the topics to be covered.

Editorial note: block diagram in Fig. 3.

"Plaques Concentrées" [Occhialini's Comments (Labeyrie's Manuscript)]

27 dec. 1992

Plaques concentrées[(a)][112]
Publication[(a)]
Blackett[(a)]
Panel[(a)]
Lattes, Camerini[(b)]
Concentr. Emulsion[(b)]
Publication[(b)]
Photos[(b)]
Demers[(b)]
Blackett—Panel[(b)]
Lattes[(b)]
Camerini[(b)]

[111] [Handwritten note by Dilworth].

[112] The lines with [(a)] were written by Labeyrie, the ones with [(b)] by Occhialini.

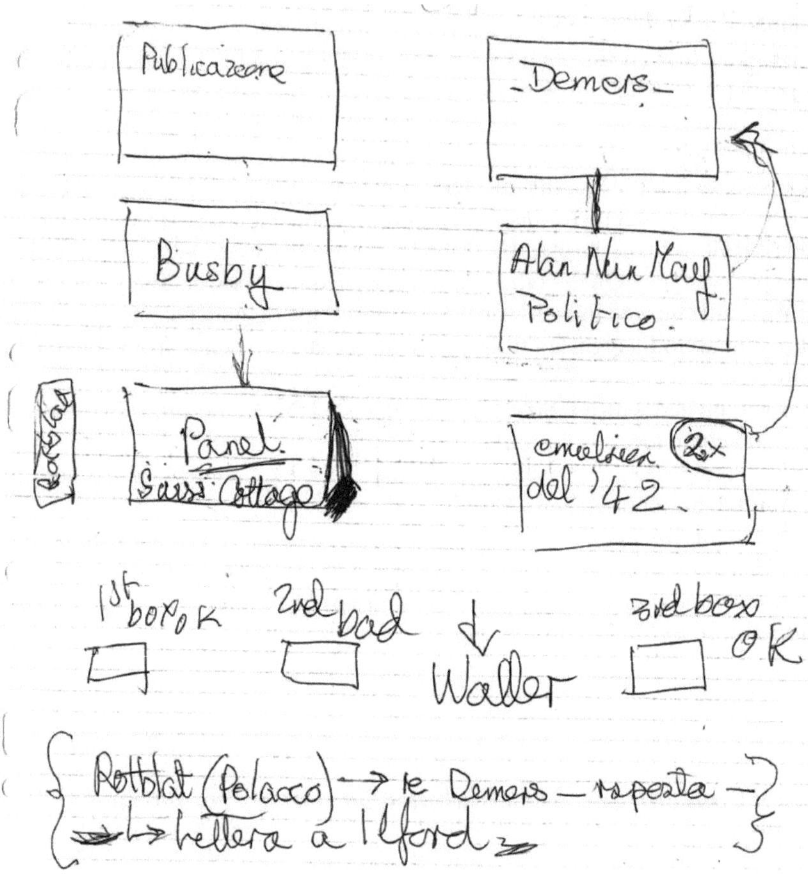

Fig. 3 Dilworth's block diagram of the topics on the concentrated emulsions. Courtesy of the BICF library of the University of Milan

Il me semble avoir expliqué comment les émulsions concentrées sont arrivées à Bristol. Il me semblait évident qu'avec ce saut en qualité la méthode pouvait être employée. Je sentais que ça justifiait mon sacrifice en abandonnant le Brésil et qu'à mes yeux j'étais pardonné pour n'avoir pas pris part à la guerre. Le saut technique aurait permis à l'Europe détruite d'avoir de "l'information en boîte".

Les intérêts de Powell étaient les mêmes que les miens. J'ai trouvé en lui une étrange paresse, il semblait que ces années d'effacement avaient diminué ses qualités d'initiative.

Pour moi la chose urgente était de publier possiblement sur Nature.

La situation était la suivante:

On a commencé à rédiger le travail pour la Revue des Instruments scientifiques. J'ai repris la vielle distribution des particules Alpha pour prouver graphiquement que la précision était meilleure que dans les vieilles plaques. J'ai rempli le papier avec des particuliers techniques, Powell a ajouté un dessin de la chambre a diffusion, il se proposait de bâtir pour l'employer avec les nouvelles plaques. On a ajouté les signatures de Livesey et Gibson et Chilton, on a envoyé le papier a Ilford pour qu'il le corrige.

Chilton a envoyé le papier en changeant l'ordre des nomes des gens qui devaient signer. Sur le papier original les signatures se suivait dans l'ordre Chilton, Gibson, Livesey, Occhialini et Powell.[113] Dans l'ordre de Chilton c'était Powell Occhialini et le reste venait après, je ne me souviens pas. ça m'a donné une joie extraordinaire parce qu'il y avait un témoignage extérieur que mon rôle n'avait pas été accessoire.

Au milieu de ces opérations, nous avons reçu l'interdiction de publier pour le fait qu'au Canada le résultat analogue avait été obtenu artisanalement par un physicien canadien Demers.

A commencé alors une controverse. Ces émulsions étaient un secret de guerre et il y avait dans l'atmosphère une suggestion vague que l'opération de Bristol avait été un plagia et que nous savons du travail Canadien (check Rotblat Connie) comme était ami de Powell, il était communiste, il travaillait au Canada, la complicité entre Powell et [Rotblat] aurait pu être à l'origine de l'interdiction de publier.

Le travail est finalement sorti et à ce moment je ne savais pas, Chilton avait été un représentant d'Ilford. La personne qui avait rendu le travail possible était Cecil Waller avec son brevet.

J'avais rencontré Waller personnellement quand il était venu pour voir les plaques. Je me sens encore coupable pour le fait que handsome comme il était, bien habillé, avec des mains si bien soignées avec sa ressemblance avec Mountbatten, j'avais pensé qu'il était un dirigent d'Ilford et non un du laboratoire. J'espère d'être pardonné en considération des efforts que j'ai faits 3 ans après pour rendre possible le G.S. et le K après que Bristol eut laissé tomber Ilford pour Kodak.

Au moment du veto de publication j'avais déjà alerté Blackett et j'avais passé un long weekend avec lui; il était très intéressé pour ce problème. Je lui ai conté tout, je lui ai porté des copies de photos, il a été le premier à me communiquer la difficulté de publication, il savait indépendamment sur "Demers work".

Le dimanche nous l'avons passé comme au vieux temps à Cambridge, à nous promener à travers le jardin de Londres. Je lui ai dit que j'avais l'impression que Powell était persécuté. Il m'a demandé s'il était communiste. J'ai répondu que non "to my best of knowledge" mais que cette question était importante pour le fait qu'elle prouvait comme je n'en avais jamais été sûr a membre du parti.

Une brève digression pour me conter que Lord Portal avait eu la même doute (He call me a traitor) et puis il m'a expliqué que le fait que Powell n'était pas un membre de la Royal Society ne signifiait pas mauvaise volonté; il y avait des personnes qui avaient interrompus leur activité pour la guerre et avait une priorité.

Nous sommes revenus à Ilford et je lui ai expliqué la manière avec laquelle les plaques étaient nés. J'étais témoin du fait que Powell n'avait pas montré d'enthousiasme pour ces plaques tant que j'avais dû les développer moi-même.

---

[113] Gibson does not appear among the authors of the paper published in the *Journal of Scientific Instruments* (C. F. Powell *et al.*, *J. Sci. Instrum.*, **23** (1946) 102–106).

Toute l'activité d'Ilford avait été volontaire et je proposais un contact avec Ilford. De la discussion je suis rentré avec la promesse qu'on aurait fait une commission spéciale capable de signer des contrats avec Ilford pour promouvoir la production de ces plaques et l'amélioration qui était possible. "Tu ne serais pas dans cette commission" m'a dit Blackett mais étant à Bristol tu seras dans la condition de me communiquer si quelqu'un va discriminer (muscle in). Ainsi est né le panel photographique dans un parc de Londres à une date que je ne peux pas signifier mais que Rochester connait très bien. Incidentellement je n'ai jamais eu l'opportunité de rectifier l'idée de Rochester que les émulsions concentrées avaient été un produit du panel, c'était l'opposé.

Blackett voyait dans ces émulsions un moyen de relancer l'Europe. Le soir il a ouvert une bouteille pour un toast à la possibilité que l'Italie puisse participer au boom.

Quelques mois après Powell a été nommé membre de la Royal Society.

Je suis rentré à Bristol et j'ai communiqué à Powell l'attitude de Blackett qui était d'accord avec la ligne, il fallait que Bristol se mette en condition de pouvoir acheter les moyens. Je lui ai garanti l'attitude amicale de Blackett, mais je n'ai pas parlé du Panel.

J'ai répété qu'il fallait augmenter le personnel pour pouvoir exploiter la nouvelle technique et qu'il fallait créer pour cela une légion étrangère. Il a accepté cette fois l'idée du moment que tous les gens valides étaient sous les armes. C'est ainsi qui s'est matérialisé l'offrande à Lattes, que j'avais laissé avec Camerini au Brésil, d'une fellowship de quelque type. Quand il est arrivé il a suggéré Camerini qui pouvait venir avec ses propres moyens. Ainsi est né la légion étrangère à laquelle s'ajoutait Goldschmidt-Clermont et Franzinetti, tous à part moi, Juifs.

Le 4eme étage de Royal Fort où nous habitions était appelé par Goldschmidt Tortilla Flat but Goldschmidt est venu après.

Rosbaud

Rosbaud [Dilworth's Manuscript]

Rosbaud

I first met him at a conference in Zürich in August 1933. He introduced himself as a sort of journalist. He wanted to talk about Italian politics. He was anti-nazi and we agreed it would be useful to have contact between the Italian antifascist and the German anti-nazis. The conversation was interrupted by a young Indian, Bhabha, who wanted to talk about showers.

Thirteen years later a mister to the Royal Fort asked me if I remembered him. I said yes, it was in Switzerland, Zürich, your name is something like Roteblat. It was Rosbaud. It was an exciting encounter, dantesque, Virgilio versus Sordello. We did not talk politics this time. I took him to the "telepanto" and presented him with a copy of a photo I took directly on transparent film. He asked to see the others and I explained I was doing them to send as a gift to my father whom I had not yet seen. Rosbaud suggested making an album of them, as he had done for Rochester and Butler's[114] cloud chamber photos. I took him at once to Powell, who was enthusiastic. It looked as through my poetic Saturday night work might become profitable.

I had not told Rosbaud the whole truth, some of the photos I intended to send to Germaine in Thailand as a proof that, in a certain type of photo I was a master as she was in another.

Rosbaud and I met after that for dinner in London. The first time, just the two of us. We talked about the situation in Germany. I wanted to invite Houtermans[115] to come to England and I was surprised by Rosbaud severe attitude versus Fritz. I did not ask him why, maybe I did not want to know. He gave me a list of German physicists who had been actively anti-nazi, who could be invited instead of Houtermans (that test I used later in a discussion with Heisenberg). The conversation ended with the phrase "as far as Houtermans" is concerned, leave him picking up the cigarette ends of the American Officers.[116]

(This is the story of a young man who liked two glorious years at the Cavendish. Twelve years later, after a world war, he found himself in another situation.)[117]

At the second dinner with Rosbaud there was also Peyrou.[118] Rosbaud introduced Peyrou as his spiritual son. I used that definition when I presented Peyrou for a place in the University of Bruxelles.

The Album of Rosbaud never came off . When Powell talked of it with Tyndall everybody seemed enthusiastic. I went ahead with my collection of photos, Powell made mosaics with his great manual dexterity. I found a simple way to make the mosaics, eliminating the hot iron by working on the wet parts. Mott apparently proposed to publish with the Clarendon Press of Oxford, to which he was scientific advisor. Powell agreed without consulting me. I knew too late, weeks later. I did not make it a question of vanity. I protested, I pointed out that Rosbaud was a hero of the German resistance (my argument would have been stronger had I known he was

---

[114] Clifford Charles Butler (1922–1999).

[115] Friedrich Houtermans (1903–1966).

[116] This phrase alludes to a story told me by Houtermans himself. He was giving a lecture course on physics to officers of the American occupation forces. There was at the time a shortage of cigarettes in Germany, a very serious problem for Fisl. The Americans of course had no problems, so after the end of his lecture he would go round the room, picking up the cigarette ends from the ashtrays and using the remaining tobacco to roll a few cigarettes. Typically, Fisl excogitated a rhythm to his lectures, alternating panels of slow, boring explanations during which the Americans would light up, with brief, staccato phrases full of important data during which they stubbed at their cigarettes to take furious notes. That why he got plenty of long stubs. [Handwritten footnote by Dilworth].

[117] The text in round brackets is written upside-down.

[118] Charles Peyrou (1918–2003).

the Griffon)[119] received in England as a recompense for his war work. The contract had been made, not by me, but as a common agreement between me and Powell. I never had the courage to meet and talk to Rosbaud about it. There was a sort of rough practice in what happened after (the book appeared as a signed "Powell and Occhialini" in spite of my protests to Mott).

Rosbaud so was one of those "brief encounters" in my life. I do not know what I will say when I'll meet him in Purgatory as the Communists say, whatever I can try to say it will be "even worse".

It is difficult to convince a person who has risked his life to conserve his dignity that one can submit to the paranoia of an old toothless professor for £29 a month. The Griffon will tell me I behaved like a mercenary.

Politics and Personalities

Politics and Personalities [Typescript]

Politics and Personalities

At this point I will try to explain the political background, both on a local and national scale, and the nature of the various personalities involved.

The majority of the University physicists were liberals - conservatives, the young ones also, but with a very interesting nucleus of communist sympathisers. There was no evident hostility between the two groups, but a foreigner could detect the tensions.

Tyndall, liberal-conservative, was bound to an almost maniacal defence of Powell. There was a list of people whom he considered the enemies of Powell. I discovered it by mentioning names and watching the reaction. This attitude, I understood from Powell, was due to the remorse that professors have toward their pupils when they realise that they have exploited them. For Tyndall I was an ally of Powell and that excused a possible leftist connection for some of the staff. The fact that I was a friend of Powell and of Blackett was enough to brand me as a communist.

The attitude of Mott is more difficult to define, and I have given up any attempt to judge him. Certainly he is an honest liberal who had, however, occasionally quasi-racist paroxysms.

The member of the staff I most respected was Fröhlich. I had a brotherly friendship with Fertel and Bates,[120] who came with me on my second caving trip to the Pyrenees; of their political attitude I never enquired. Another great friend was Burch.[121]

---

[119] The story of Rosbaud was told by Arnold Kramish in the book *The Griffin: Paul Rosbaud and the Nazi Atomic Bomb That Never Was* (Houghton Mifflin) 1987.

[120] Wilfred John Bates (1922–1995).

[121] Cecil Burch (1901–1983).

Frank[122] had a lot of charm. He was the only member of Mott's group I got to know well. His wife was a firm ally in my moments of difficulty. Frank introduced me to Jones[123] of Aberdeen and it was in talking with him that I realised that another 'War of the Roses' had been declared in England. I was bound to suffer from the animosity of Powell's political enemies who found in his lack of participation in war work a pretext for attacking him. The others, the friends of Powell, scared me for their lack of vision. Some of them considered me a reactionary, the others a Trotskyist.

There were many reasons for my deep and lasting affection for Blackett. They did not include, as some humourist suggested once, any homosexual attitude. From my side it was a total esteem; it included Pat, Nicky and Giovanna, (and the 'German' girls who came and went in his household). A strong component was a sense of gratitude which continued to grow as the years went by. As I dictate this, it occurs to me that Blackett also had gratitude for me, as shown by the respect he had for my opinion and my decisions, and the lack of sarcasm in his judgement of me. (With one exception, when he was at Birkbeck and my passion for Rony Gamow made me leave London to go to Cambridge. He thought I was doing that for Chadwick).

My attitude towards Powell was protective. I had met a very retiring person, prisoner of a marxist ideology, unsure of himself, like an invalid just leaving hospital. I watched him transform under my eyes.

Lattes one evening when he had had a glass too many, accused me publicly of being a coward because, he said, though I had shown I was capable of dying in defence of my friends I was incapable of defending my own interests. Maybe he was right, but how many people have the courage to ask back the money they have lent to a friend?

When I arrived in England from South America I was full of complexes and remorse for having betrayed my florentine anti-fascism, persecuted as I was by the fear that my father, a university professor, could be considered responsible by the fascists; full too of gratitude to England for having offered me a possibility for action, and so redemption. Then came the realisation that my sacrifice in leaving Brazil had been in vain. The Fuchs affair had transformed me into a security risk. On the other hand there was Powell, married to the sister of a German refugee I had known in 1933 in England, scorned by the Tories for his communism and his abstention from the war effort. I have never forgotten that Powell, and Bristol, accepted me when I was desperate.

Gratitude sometimes can be a bad counsellor.

[122] Frederick Charles Frank (1911–1998).

[123] Reginald Victor Jones (1911–1997).

"Quelques Considérations…" [Occhialini's Comments (Labeyrie Manuscript)]

27 dec. 1992

Quelques considérations pour expliquer la situation politique à Bristol et en Angleterre.
Rose vermeil et rose blanche
Vient en ordre après émulsions concentrées

27 dec. 1992

À un moment de cet exposé il faudra tâcher d'expliquer le déploiement des forces et des intrigues soient locales que sur échelle nationale. La majorité des universitaires physiciens était libérale, conservative. Les jeunes aussi mais avec un noyau très intéressant de sympathie communiste. Il n'y avait pas d'hostilité évidente entre les groupes mais un étranger pouvait détecter les tensions.

Tyndall libéral conservateur était lié à la défense de Powell d'une manière maniacale. Il y avait une liste des personnages qu'il considérait comme les ennemis de Powell, je l'ai découvert en introduisant le nom de ces personnages et en observant ses réactions. Cette liaison dépendait comme Powell m'ai fait comprendre dans une situation clé aux remords que les professeurs ont envers leurs élèves quand ils s'aperçoivent qu'ils les ont exploités pour Tyndall. J'étais un allié de Powell et ça excusait ma possible connexion avec la gauche pour une partie du staff.

Le fait que j'étais ami de Powell et de Blackett était suffisant pour me définir comme un communiste. L'attitude de Mott est plus difficile à définir et j'ai renoncé à passer un jugement sur lui parce que je serais injuste. Certainement l'homme est un libéral honnête qui pouvait connaître des moments de paroxysmes pratiquement raciste. L'élément que j'ai respecté le plus dans le staff a été certainement Frölich et j'ai eu une amitié fraternelle avec Festel et Bates qui m'ont accompagné dans mon deuxième voyage spéléologique aux Pyrénées. Je n'ai jamais investigué leur attitude politique. Grande amitié aussi pour Cr. Burch.

Editorial note: an exclamation mark was handwritten on the side of the following paragraph.

Frank était aussi une personnalité très attrayante, c'était le seul du groupe de Mott que j'ai fréquenté. Sa femme a toujours été dans les moments difficiles que j'ai connus un allié et une aide. Frank m'a fait connaître Jones of Aberdeen et ça a été en discutant avec lui que j'ai compris que la guerre des roses était déclenchée en Angleterre et que j'étais destiné à subir l'inimitié des ennemis politiques de Powell qui avaient trouvés dans le fait qu'il n'avait pas pris partie active à la guerre un prétexte pour l'accabler.

Les autres les amis de Powell me faisaient peur pour leur horizon limité. Ils me considéraient avec raison une partie comme réactionnaire et une partie comme trotskiste.

"Chère Connie..." [Occhialini's Comments (Labeyrie's Manuscript)]

31 dec. 1992

Chère Connie
c'est pour mettre en ordre une question qui m'a préoccupé pendant les derniers mois. À l'Accademia dei Lincei on a parlé en ma présence du violente attaque de Perkins contre Occhialini à un congrès d'Histoire de la Physique qui s'est tenu à Rome et auquel tu as participé.[124] À la discussion semble avoir participé Belloni.[125]

Je ne sais rien de plus mais la discussion devait regarder Bristol 1945. Je n'ai pas de document écrit sur cette période, j'ai seulement mon histoire à conter.

Avec ce que j'écris je veux ajouter à ma réputation d'imbécile cette de mythomane comme il s'est passé avec Benvenuto Cellini. Tu n'étais pas à Bristol dans cette période, mais dans nos conversations j'ai été très British dans mes silences.

Pour comprendre la situation il faut présenter le personnage Beppo. Arrivé d'Amérique du Sud plein de complexes et de remords pur avoir trahi son anti-fascisme florentin. Persécuté pour la peur que son père professeur universitaire soit tenu responsable par les fascistes et après les Allemands.

Sa gratitude envers l'Angleterre qui lui avait trouvé une possibilité d'action et de rédemption. Sa ligne haine—amour pour l'Italie qui l'avait forcé à sortir comme professeur italien après une tentative avortée d'aller combattre la guerre d'Espagne.

Sa réalisation que son sacrifice en abandonnant le Brésil avait été inutile parce que avec le suspect généré par l'affaire Fuchs il était devenu après son acceptation des Anglais un "security risk".

D'un autre côté il y avait Powell marié avec la soeur d'une exilée allemande que Beppo avait connu en 1933 en Angleterre. Le nom de Powell était execré dans l'ambiance Tory anglaise pour le suspect de son communisme et le reproche de ne pas avoir pris une partie active à la guerre.

La gratitude est une très mauvaise conseillère quelquefois. Je n'ai jamais oublié que Powell et Bristol m'avait accueilli dans un moment dans lequel j'étais au désespoir.

Le suicide de Rufini et de sa femme auquel j'avais confié les documents qui prouvaient que je n'avais pas été un espion payé par les Anglais avait rendu même possible mon arrestation par les nouvelles autorités soi-disant anti-fascistes de l'Italie.

Il existe dans le secrétariat des Affaires étrangères de Piazza Firenze une accusation explicite que je n'ai pas voulu voir parce que j'avais peur d'y trouver la signature d'un collègue.

L'arrivé de Jean François a interrompu la suite.

---

[124] De Maria Michelangelo, Grilli Mario, Sebastiani Fabio (Editors), *Proceedings of the International Conference on Restructuring of Physical Sciences In Europe And The United States 1945–1960.* Università "La Sapienza", Rome, Italy, 19–23 September 1988 (World Scientific Publishing, Singapore) 1989.

[125] Lanfranco Belloni (1944–2017).

J'ai trouvé ainsi à Bristol un Laboratoire composé d'une pièce plus un petit studio, 3 microscopes, et un Y4 monoculaire. Au 3eme étage une chambre noire, comme staff Heitler frère du théoricien et Guggenheimer. Le travail était impeccable mais les gens étaient décidés ou peut-être résignés à passer toute leur vie à étudier les butterflies.

Powell avait démontré que la précision dans la mesure des longueurs et des angles dans l'émulsion était comparable à celle de la Chambre de Wilson.

La recherche courait à Bristol comme un fleuve qui s'approche de l'embouchure.

Il y avait une dame très experte M.me Andrews. Dans le Bristol intellectuel que j'ai connu le clivage entre la gauche et la droite était ressenti.

Un ami Brésilien qui travaillait au Burden Institue et que j'avais recommandé à Blackett du Brésil m'avais complètement informé sur la situation comme elle était vue par les gauchistes. Powell était mal aimé par la droite, pratiquement persécuté avec la seule défense de son professeur et Directeur de Royal Fort. L'intensité de l'esprit paternel de défense de Tyndall envers Powell était maniacal.

Je suis arrivé à Bristol 15 jours après Pâques, envoyé par Blackett avec la mission de découvrir si la méthode des plaques pouvait s'appliquer aux rayons cosmiques. J'ai reçu pour ça un chèque sur la Midland Bank (£50?) (£80?). Il m'a assuré que cet argent était une somme mise à sa disposition par la Royal Society.

Dès mon arrivée j'ai obtenu les clés de Royal Fort ou je passais toute la journée et une partie de la nuit. J'avais le sens de l'urgence comme à Cambridge en 1932. Je refusais les invitations courtoises de familles et j'arrivais dans le pub à 9h.30 le soir pour sortir à 10 h.

Travail

Lecture continuelle des tous les tests photographiques, scanning d'une plaque de rayons cosmiques de la vielle exposition à la Jungfrau de Powell, Heitler, Fertel. Powell m'a chargé comme partie du mon apprentissage de mesurer les 2 longues lignes du Thorium.

La séparation que j'ai trouvée était meilleure que la sienne, incrédule il m'a donné une distribution avec proton et deutéron complètement inconnue par moi le résultats ont été 2 courbes qu'il m'a demandé le permission de substituer aux courbes déjà prêtes pour la Royal Society et pour un futur congrès. Il n'y avait aucun acknowledgement dans le travail publié mais ça m'a rendu heureux pour la 1ere fois après des années, j'étais utile. Quelques jours après Tyndall m'appelait pour m'offrir une place dans le Royal Fort à £25?, 27? Extase de mon côté, référence à la mission de Blackett pour les rayons cosmiques. Promesse que j'aurais eu l'aide et l'opportunité (je n'ai jamais pu l'avoir).

À ce moment à travers l'observation de la plaque de Jungfrau et aussi des autres plaques j'étais arrivé à des conclusions que j'ai communiqué à Powell. La technique du développement était trop primitive. Le développeur choisi appelé pompeusement XRay développer était un métal Hydroquinone très alcalin. Est-ce que le half tone pouvait être amélioré? Amélioration voulait dire densité sur les traces augmentées et ça pouvait être fait en chambre de Wilson augmentant la pression, ---- émulsions: augmenter la quantité d'argent ou en choisissant des grains plus sensibles (dimension accrue). J'ai demandé une proposition à Ilford sur cette ligne. Powell était sceptique, j'avais des difficultés pour ne pas l'offenser choisissant les mots les plus diplomatiques pour le convaincre que la méthode avait porté sur des particules de masse photonique et de vitesse réduite. Les mesotrons avaient besoin d'une technique plus sensible. Powell était satisfait avec le résultat spectaculaire mais difficile à interpréter obtenu avec les émulsions actuelles.

Il a énoncé le principe qu'il fallait un progrès exceptionnel dans la qualité de l'émulsion pour justifier les troubles de rebâtir une nouvelle "range energy relation".

Toutefois pour les intérêts que je portais sur la base de mon allégeance à Blackett il m'aurait avisé quand l'occasion avec une rencontre avec les gens de Ilford se serait passée.

———

31 dec. 92[126]

L'opportunité s'est présentée quelques semaines je ne sais pas quand mais après quand Chilton est venu. Nous avons eu une réunion à 4 Powell, Rosenblum, Chilton et Beppo dans la Bibliothèque.

Rosenblum était intéressé en structure fine de particules Alpha (sa découverte).

Chilton se souvenait de m'avoir rencontré comme un undergraduate.

J'ai exprimé avec grande sobriété ce que je demandais en disant que Manchester était intéressé dans ce développement et que j'étais sans expérience dans le champ des émulsions mais que le problème aurait demandé une Chambre de Wilson à pression supérieure à l'atmosphérique et que l'équivalent était une augmentation dans les grains de Bromure d'argent.

dans la pourcentage entre l'argent et la gélatine.[127]

Houtermans

Houtermans [Dilworth's Manuscript]

Houtermans

My contact with Houtermans occurred in 3 acts; the first, beginning in 1932 lasted for a year. He came from Germany, stayed [in] Cambridge and then went to London to work for the Gramophone Company at Hayes, Middlesex. It was very affectionate relationship, based on my unreserved admiration for his brilliance, his attitude, his knowledge of physics and the originality of his ideas, founded in German culture.

[126] The date was written by Dilworth.

[127] This last sentence was handwritten by Occhialini.

We separated in 1934 with my return to Italy and his departure for Kharkov due both to his dissatisfaction with his job in England and to his politics, very left for the time.

In 1938 on in one of my periodic returns from Brazil, I met Schnax[128] who told me of the tragedy of his imprisonment. I found among our common friends of the left the impression that it had been his brilliance that had ruined him, that he had talked too much. There was an uncomfortable sensation, the idea that his wit had brought him to commit indiscretions. Schnax denied all that. I did not realize how serious was the problem. Fisl[129] and Schnax had seen the problem as a choice between his image and that of the Soviet regime—and they had chosen that of the Soviet regime.

On the other hand, in Paris where I went to see Joliot, as Schnax had asked me, I found that Joliot and the other French physicists had an indulgent attitude towards the role of Fisl's temperament in his misadventures. Joliot told me he would do what he could but that the relations between the URSS and the French government were not so cordial as they had been.

At the end of the war having returned to England I heard of the exchange of Houtermans between the Russians and the Gestapo, and of Houtermans visit to the laboratory of Kharkov under the auspices of the German military (there was the meagre detail of his leave in uniform with a military cap). English communist friends kept an embarrassed silence, German visitors suggested delicately to leave him alone.

The third act was brief. In 1950 I came to London for a Royal Society meeting and there met Houtermans. We were both invited to dinner at George's house. Houtermans wanted to explain, so while the other guests went into the dining room, we remained alone in the studio to talk. We talked for about 4[130] hours. The guests left the dining room, and George, who had understood the situation, left us alone.

At the end of this conversation I was completely convinced. The years had not aged Fisl, but he was changed. I had the impression that we had met as adolescents but now he had become a man. He seemed to be prepared for this encounter. I asked few questions, he explained everything very clearly, his affection for URSS, his indignation for the injustice of the Government who had sent him back to Germany as a Nazi spy, the danger of his situation in Germany and his rescue by von Ardenne[131] and Heisenberg, his brotherly feeling for his Russian colleagues, in particular Leipunskii[132] and his desire to help them which led to his return to Kharkov to help people who might need it. This was the hard point of the discussion. Anti-nazi Germans I had encountered in previous years at Bristol had warned me that Russian colleagues asked about this episode had evaded the question. After this conversation, a dead friend resuscitated, had come into my life never more.

[128] This was Houtermans's wife's nickname.

[129] This was Houtermans's nickname.

[130] The "4" is circled out. The text "$2\frac{1}{2}$?" was written and circled out on its side.

[131] Manfred Ardenne (1907–1997). He was the director of a private research laboratory in a Berlin suburb from 1928 to 1945.

[132] Aleksandr Leipunskii (1903–1972).

This impression did not change in the years that followed. For me, his marital difficulties, his imprisonment were unimportant compared with the Kharkov affair.

Was the explanation he gave me, correct?

He covered the essential points, his arrest, his years in prison, his affection for the Russian people, his friend and his colleagues, his indignation that the government had exchanged him as a Nazi spy; his determination in prison to survive to convince his friends in Western Europe that he had not betrayed socialism, but that Stalinism was wicked. The important part was about his return to Kharkov with the German army. "I had to go there to see if there was something I could do to help my friends. It is not true that was wearing a Nazi cap." I looked at his broken shoes and ragged trousers ends and was full of compassion and of self-criticism. When I went into the studio with him I had thought it would have been a very short discussion and that we would never have met again afterwards, now I was ashamed of that intention. On the contrary, I invited him to Brussels to meet Picciotto and later arranged for him to take my place for some time during my absence on a Unesco mission to Brazil.

Now I have the impression that our Russian colleagues who were still very severe about the Kharkov still in the late '50 s have changed attitude. I had many discussions about it with German colleagues, in particular Crowther[133] (?)[134] and Gentner.[135] Gentner recognized with a smile that he had been in the same situation in Paris after the fall of France. The difference was that he was able to convince Joliot of his motives. Fisl did not have that chance.

During that evening I discovered that while we had parted in 1934 as adolescents, now I had found a man. Through his suffering, he had changed. He was still witty but with a keen judgement, a sense of generosity and an almost psychotic rebellion against injustice.

I was able to send him some boxes of nuclear emulsions for his work, and I was surprised to receive letters of thanks from his German colleagues to whom he had given some of them, telling them from whence they came. He was concerned about the financial difficulties of Mrs. Geiger and arranged a personal contributions to help her.

I have the impression that the problem of the Kharkov incident remained. I feel still the need, not to excuse but to understand him, and also to understand intelligent people who condemned him.

A very witty person is in an unstable equilibrium. He is as it were, a moralist and the intellectual does not forgive a moralist who errs. The wit, who may be called Triboulet or Oscar Wilde can be victim of his public when he commits an error [I already said that the man of 1950 was different from the of 1934.]

---

[133] James Crowther (1899–1983). He was English.

[134] The question mark is in Dilworth's manuscript.

[135] Wolfgang Gentner (1906–1980).

On the other hand although wit has a component of wisdom, when it expressed not in writing but verbally, it can be considered as denoting irresponsibility. One laughs freely at satires on others, not on oneself. I never heard anyone complaining at being hurt by Houthermans wit, but people with Communist sympathies seemed to have found it natural to suppose that a sense of humour tolerated in England should cause him difficulties in URSS. That has been denied by both Fisl and Schnax, who were so alone to the danger of too open speech that they talked of certain things to each other only in bed, under the sheets.

The Book

The Book [Dilworth's Manuscript]

The Book

Our ancestors at the dawn of the pre-history were troubled by the feeling that between God and the Archangels all was not well. Something strange was happening to the Book of Genesis, which had now appeared in a 2nd Edition.

The first edition had recounted in great detail the evolution of the world (Universe). The authors were the Archangels and God, named in alphabetical order. The second edition was attributed to God and then, in order, the Archangels.

The human depended heavily on the Archangels, who invented for them the useful gadgets like fire, the wheel, the lever and so on. They saw a lot of them. God was inaccessible, he was busy writing the 3rd edition of the book.

The authorship of the 3rd edition was given as, simply, "God and the Archangels", that of the following 4th, "God and Co-workers", and finally, "God and God's School".

By this time, the text of the book had undergone considerable transformation. There was no mention of the Archangels in the Genesis, but this did not seem strange to Men. The Book told the story of their appearance on Earth because they had become of central importance, the Archangels had become an unimportant historical debris.

Then God struck sure of the support of Man, He expelled the Archangels (all of them) and wrote the final, concise edition of the Book, destroying all previous editions.

In the final version, the first chapters of the book were replaced by a single phrase: "At the beginning God created Heaven and Earth", followed by 25 sentences which substituted the old text and reduced the Creation to its essentials. Since then all Creation sings the praises of God and God alone.

Les Livres [Occhialini's Comments (Labeyrie's Manuscript)]

4 janvier 1993

Les Livres

Les êtres primitifs qui habitaient la terre au commencement de la préhistoire se sont rendus compte que tout n'allait pas bien entre Dieu et les Archanges; le désaccord regardait le livre sur la genèse qui était sorti dans la 2eme édition.

La 1ere édition est très détaillée, elle décrivait la manière élaborée avec laquelle la terre avait évolué. Elle était signée par ordre alphabétique par les Archanges et Dieu.

La 2eme édition nommait encore les Archanges mais Dieu venait en 1ere position, les humains avaient besoin des Archanges qui inventaient pour eux des gadgets utiles comme la roue et le feu. Ils les voyaient souvent, Dieu était inaccessible parce qu'il était en train de rédiger la 3ere édition du Livre qui sortit avec les signatures de Dieu et les Archanges.

Dans l'édition suivante le Livre était signé God et coworkers et dans la suivante God and God's school (Houtermans) et successivement God et Adam. À ce moment le livre était en confusion.

L'éliminations des Archanges du texte avait rendu le Livre confus, en particulier la genèse était devenue très compliquée. Les humaines avaient compris que les Anges étaient perdants. Il n'y avait plus aucune référence aux Archanges. Les hommes ne voyaient pas de contradictions. Le Livre contenait leur apparition sur la Terre parce qu'ils étaient les êtres importants, les Archanges étaient des débris sans importance dans l'Histoire.

A ce moment God sûr de l'appui des hommes frappa avec grande décision, les Archanges furent expulsés (tous). Dieu édita la dernière édition du Livre raccourcie dans une forme élémentaire, les copies précédentes furent envoyées au pilon (macero).

Le Livre fut châtré de tous le premières chapitres qui furent substitués par une simple phrase: "Au commencement Dieu créa le Ciel et la Terre", suivaient 25 versets qui substituaient le vieux texte et réduisait la création à sa forme essentielle. De ce moment toute la nature et la création chantent les louanges de Dieu et de Dieu seul.

La forme originale était très bonne, et elle a été écrit perdue et retrouvée plusieurs fois. Houtermans ajouta "The School of God" au texte.

Je pense que l'histoire est encore bonne avec quelques changements elle pourrait être presque publiable. Peut-être une copie si j'avais le temps pourrait être envoyés à Perkins.

"Next February I Will Be ..."

"Next February I Will Be ..." [Dilworth's Manuscript]

Next February I will be the oldest member of my family. An historian friend of mine once said that parents should write down their story for their children, the truth, nothing bad but not the whole. Some privacy should be respected in default of my brothers. I will here try to do my duty not only by my daughter, but also by my nieces and nephews.

This common interest closes at a certain epoch. What happened afterward to my brothers concerns my daughter only.

The beginning comes from the stories Toshie (Charlotte Cont) Price, my mother, told me. George Dariell Dilworth, her husband, was a silent man. He was bent by two severe inhibitions. The first, an incurable apparently innate honesty, so strong as to override any personal interest. The second, according to Toshie, was a fear of his own physical strength, due to his having almost killed another man in a fist-fight in his youth.

[Sheets with Various Notes]

[Sheets with Various Notes] [Dilworth's Manuscript]

launching fields ([...] theirs of course) were often paid few out of the personal stipend /30.000 lire/al mese). But the rewards were high. First of all, the close community, few people and the chance of encountering the great ones of the previous epoch: Heisenberg, Pauli, Dirac, Bohr, Schrödinger. There they were, to be known personally. Then there were the new particles, thick and fast they came at last, more and more they came. All this because cosmic rays were free and nuclear emulsions cheap. These who had survived the second world war found themselves on a great bonanza—encouraged by the equation: particles = nuclear research = atomic bombs.

So the great accelerators were born. In Europe it meant CERN. There with the memory of the tragedy of war, the physicists imposed a pacifist strategy. In USA it was more different, but in the end, with a determined attitude against the cold war, they succeeded. American colleagues invited to their conferences those whom they knew would be suspected—and then made petitions and pressure and were then admitted.

By the mid-fifties the majority of the cosmic-ray physicists had converted to work with the being accelerators, capable of splitting many more new particles. A few, however, refused this [...] work. They took the alternative road, to the cosmological origin of cosmic rays.

One of this was Marcello Ceccarelli.

Editorial note: The following paragraph in square brackets was written upside-down.

[That Bohr had objections does not seem to have influenced the Nobel committee since months later Dirac was received the Nobel prize,]

1st January
Old box of plates 2xAg (Bock). Page in Powell's diary—suggested Rotblat suggestion

27 December
Paper–content–signatures–Demers problem.
Paper published—did not know Chilton was only representative.
Encounter with Waller—too handsome
"Au moment du veto de publication j'avais déjà alerté Blackett"
Took him photos.
He was the first to warn me about Demers problem
–Panel
–Légion étrangère"

25 April

It was only a month later when the "Security People interfered I knew (it was Waller).

I hid my emotion by the phrase "This time the SOB of Chelten had delivered the goods."

Lettice "camera microscopique Leitz"
Return—in Albion asked for photographs
—Busby—Leica—film—developer kit—print paper–(format postcard)—left Royal Fort at 2 am
—To lab at midday with postcards—bought enlarger a Leica lent by Powell's father—cut—"After the visit to Blackett"

2 weeks delay before Blackett had free weekend
—knew about Demers
—asked if Powell was communist
—Powell not in RS

Some days before—Waller in Bristol
–Panel

Two questions:
Page 3, 3rd paragraph.
"Heraults" what are they?

Page 5, 2nd paragraph
I do not understand Rutherford's thought. Did he expect there to be a negative neutron). Can you explain better?

Giuseppe Paolo Stanislao Occhialini (1907–1993) A Short Biography

Leonardo Gariboldi and Pasquale Tucci

Introduction

"The development of my studies is typical of those of many people of my generation. Those among us who started their research trying to understand the nature of cosmic rays had been induced to the hunting of new particles. Actually, for more than twenty years, the only source of new particles had been the cosmic radiation. In such a way, the positron, the π- and μ-mesons, the K-mesons, and the hyperons had been discovered. In 1954, when the Bevatron, the big accelerating machine, able to produce strange particles and the anti-proton, started to work, this time of grace of the cosmic rays found its end. At that time the physicists, who had an easy, adventurous, and less expensive way, found themselves in front of a choice. To go on with the particles, but with the accelerating machines, or to go back to cosmic rays and improve the research on their origin. There is who chose the first alternative—many people in Italy took the way to Frascati and to the CERN—a hard and very exciting way. Only a minority felt itself more bound to the old cosmic rays—some of them oscillated, hesitated. To those people (and I am one among them), the year 1957 was a strong draw. The launch of the first Sputnik offered the possibility, only dreamt in the thirties, to go out of our atmosphere, of our magnetosphere, to go and study the real primary radiation. Stratospheric balloons existed even before, that's true, and they went till a few grams of residual atmosphere, but this idea of being able to get rid of our whole curtain was fascinating. Here we are, the equerry of cosmic rays in space—but another transformation is even happening, the more we go on, the more

L. Gariboldi (✉)
Dipartimento di Fisica "Aldo Pontremoli",
Università degli Studi di Milano, Milano, Italy
e-mail: leonardo.gariboldi@unimi.it

P. Tucci
Università degli Studi di Milano (Retired), Milano, Italy

L. Gariboldi et al. (eds.), *The Scientific Legacy of Beppo Occhialini*,
https://doi.org/10.1007/978-3-031-37034-2_2

we notice that we are becoming close relatives of astrophysicists. From particles to astrophysics, it seems that this subject of cosmic rays be unable to keep a proper personality!".[1]

In these few lines, we can appreciate Giuseppe Paolo Stanislao Occhialini's feeling for his exciting scientific adventure in the researches on cosmic-ray physics and for their human background. Occhialini was a privileged actor whose adventurous scientific life went through three of the classical periods of cosmic-rays physics: the first and second particle periods, and the astrophysical one [1].[2]

His Father and the Arcetri Period

Giuseppe ("Beppo") P. S. Occhialini[3] (1907–1993) was born in Fossombrone (Pesaro Province, Italy) on December 5th, 1907, son of Raffaele Augusto Occhialini.[4] (1878–1951), a physicist, and Etra Grossi[5] His father played an important role in the years of Beppo's training, as the latter writes in his own autobiography:

> The milieu created by his father's personality, culture, and rigour, was fundamental for the choice of the course of his work and for his social and political formation.[6]

Augusto Occhialini studied at the University of Pisa where he got a degree in physics in 1903. A collaborator of Angelo Battelli until 1918, he directed the physics lab and taught physics and electrotechnics. He was the assistant of Antonio Garbasso in Florence (1918–1921), professor of physics in Sassari (1921–1924), Siena (1924–1928), and Genoa (1929–1951). Augusto Occhialini was known not only in Italy but also abroad, since he used to spend his holidays visiting the most important laboratories of physics abroad. His scientific production comprises works of experimental

---

[1] Hand-written note by Occhialini's wife. Undated, written after 1957. Original in Italian. Occhialini Papers 8, 3, 7.

[2] The classification advanced by Brown and Hoddeson (the historical table is on p. 7) is based on a division in five time intervals of the whole history of cosmic-rays physics: (a) the prehistory (up to 1911); (b) the age of discovery and exploration (1911–1930); (c) the first particle age (1930–1946); (d) the second particle age (1947–1953); e) the astrophysics age (from 1953 on).

[3] An enjoying fact concerns Beppo's proper name: "*The origin of his initials, G. P. S., supplies an amusing example of Occhialini's unorthodox approach. At the beginning of his career, he styled himself, as most Italians, with a single first name: Giuseppe. Upon joining Blackett, he opted for multiple initials. He explained that he added "P" for Peppino (a nickname for Giuseppe), and "S" for Sommerfeld, a pseudonym under which he had run as a sprinter in his student days (during an exam period). In actual fact, the Biographical Encyclopedia of Scientists lists, in addition to Giuseppe, the names Paolo and Stanislao*" [2]. A copy of his birth certificate, delivered in the '50s (having been the original documents destroyed during the war) attests his full name to be Giuseppe Paolo Stanislao, though it is right that he signed his first papers only with his first name.

[4] Biographical notes about Raffaele Augusto Occhialini can be found in [3].

[5] Biographical information about Giuseppe Occhialini can be found in [2, 4–9].

[6] Reference [9], quotation on p. 322.

physics: electromagnetism, gas physics, and atomic physics. Among his most known opera is the treatise "*Elettrotecnica*" (1921–1922) and "*La radioattività*" (1910) written with Chella and Battelli.

Augusto Occhialini's family followed him every time he moved to a new university. Beppo spent his youth first in Pisa, and then in Florence. He attended the Scientific High School in Florence, and he decided then to study physics at the University of Florence in 1926. The years spent at the Institute of Physics in Florence represented

> one of the three fundamental steps in his existence. The other two are the period at the Cavendish Laboratory in Cambridge and the period at the Wills Laboratory in Bristol.[7]

The course of physics in Florence started in 1924, and the director of the Institute of Physics[8] was Antonio Garbasso, a physicist who was also Senator of the Kingdom of Italy and mayor of Florence. The Institute was on a hill in Arcetri, not far from Galileo's villa.

> The graduation in physics had been instituted recently; the activity of the physics laboratory was supported by the faith of its animator, Garbasso. The view offered through the windows made forgive the meagre equipment, the lack of functionality of its conventual structure, and the difficulty of access.[9]

A considerable development of the activity of research done at the Institute of Physics in Arcetri started with the arrival of Bruno Rossi[10] in 1927 and Gilberto Bernardini in 1928. Enrico Persico, professor of Mechanics and Theoretical Physics, remained in Florence until 1930, when he transferred to the University of Turin.

In the Institute of Physics, under Garbasso's direction, they researched[11] optical and X spectroscopy, in particular the Lo Surdo-Stark effect. Of valuable importance for the cultural and technical training were the collective reading of the main scientific journals, and the Physical and Astrophysical Seminar that promoted the contact with both Italian and foreign scientists. The Physical and Astrophysical Seminar was founded by Giorgio Abetti and had Gilberto Bernardini as an ardent supporter. The spectroscopy naturally kept physicists in touch with astrophysicists.

Besides Occhialini, among the students of physics, there were Daria Bocciarelli, Attilio Colacevich, Lorenzo Emo Capodilista, Giulio Racah, and Guglielmo Righini. Occhialini graduated in physics with Rossi in Florence in 1929, with a dissertation on some research on cosmic rays inspired by his father who was familiar with Millikan's

[7] Reference [9], quotation on p. 322.

[8] Information about the course of physics in Florence, in the period took into account here, is in [10].

[9] Reference [9], quotation on p. 322.

[10] On Bruno Rossi, and his role in cosmic-ray physics in this period, see [11–14].

[11] On the general situation of physics in Italy of that time, see [15, 16].

theory.[12] After his graduation in physics, Occhialini remained at the Institute of Physics in Arcetri as volunteer assistant, then regular assistant (1932–1937).

The change of subject of research from spectroscopy to cosmic rays[13] gave birth to the Arcetri School,[14] that had to face the definition of a new, stimulating, and less expensive program of research. The solution was an article written by Walther Bothe and Werner Kolhörster [22] on the application of Bothe and Geiger's coincidences method to the study of cosmic rays. Bruno Rossi directed the research of the Arcetri group towards this new direction. To continue the studies begun by Bothe and Kolhörster, the Arcetri School produced some GM counters[15] to study their features. At this time, Rossi invented "Rossi's circuit" [23] that permitted to detect the simultaneous discharge of several counters (multiple coincidences), thus realising a decisive advance compared to Bothe's coincidence circuit.

At this time, Occhialini was one of Rossi's main assistants, with Daria Bocciarelli. The articles written in this period are but signed by Rossi alone, and in his memories Rossi does not highlight particular contributions of his assistants.

> Besides working with Rossi with counter-coincidence circuits, Occhialini was already interested in imaging particle detectors. These generate a photograph (or its electronic equivalent) of the particle's physical reality, rendered visible for the first time. At the end of the 1920s the most important instrument in this field was the Wilson chamber [...] Emilio Segré, a physics Nobel Prize winner, vividly remembers meeting Occhialini for the first time, perhaps in 1929, in Rome: '...two things worried him: how to avoid military service and how to make a one-meter diameter Wilson chamber ...an idea which was then considered megalomaniac and rather ridiculous...' .[16]

In 1931, the first kind of research Occhialini made was a particular application of Geiger-Müller counters to the measurement of the energy of the β-rays emitted by weak-radioactive sources. Although Bruno Rossi did not sign the paper [24], his role, together with Bocciarelli's one, was valuable.

In summer 1930, Rossi went to Bothe's laboratory to learn the best use of the GM counters. In Berlin, Rossi met several important physicists of that time, among them Patrick Blackett who was an expert in the use of the cloud chamber. Rossi was interested in the latter technique because of the results obtained with it by Skobeltsyn in Leningrad.

---

[12] "*Beppo Occhialini, still a student, was looking for a theme as argument of the Thesis for his Laurea. One day he told me this story: his father Augusto, Professor of Physics in Genova, used to spend the summer months in Germany for his research programs in Atomic Spectroscopy and coming back in fall 1927 reported about the experiments of Bothe and Kohlhorster on the Cosmic Radiation. Beppo proposed "Cosmic Rays" as theme for the thesis to his tutor and Bruno, after a reconnaisance tour in Germany, decided to make "Cosmic Rays as the main field of the research activity for his group in Florence*." Quotation from [17] on slide 7.

[13] On the early history of cosmic rays, see [18, 19], and the fundamental recollections in [20].

[14] On the Arcetri School, see [21].

[15] According to Occhialini, the Geiger-Müller counter was like the Colt in the Far West: a cheap instrument usable by everyone on one's way through a hard frontier.

[16] Reference [4], quotation on p. 334.

> Rossi thus asked Blackett to admit at the Cavendish one of his young collaborators, so that the latter could learn the cloud chamber technique to study cosmic rays.[17]

In summer 1931, the Arcetri group decided to send a physicist to Cambridge, to the Cavendish Laboratory, to learn from Blackett the cloud chamber technique to introduce it then back in Florence. Their choice fell on Bernardini who could not go because of the military service. Occhialini was sent instead of him, with a letter of introduction written by Bruno Rossi, and with a CNR scholarship for three months. The first English period started so, and it lasted not three months but three years.

The Cambridge Period (1931–1934)

When Occhialini went to Cambridge,[18] Ernest Rutherford directed the Cavendish Laboratory.[19] The research at the Cavendish mainly concerned nuclear physics and was still made in a relatively simple way. The Cavendish continued a tradition of research made producing visible phenomena. It was not the case that they used scintillators, photographic plates, and cloud chambers. Very less used at the Cavendish was instead the GM counter. The technique of cloud chamber[20] was developed for nuclear physics applications, after Rutherford's suggestion, first shortly by Shimizu and, then, by Blackett.[21]

The modernity of the Cavendish could only impress young Occhialini. He worked side by side with Blackett, and together they developed the controlled cloud chamber technique.[22]

> He brought with him the technique of the coincidence counting of cosmic rays developed by Rossi. The marrying of the counter technique with the cloud chamber was an obvious step.[23]

This marrying permitted a valuable progress in the obtaining of useful images: 76% of the photographs contained tracks of particles, and they could take one photograph every 2 min. In practice, the particles themselves started the controlled cloud chamber when they came into it and ionised. This fact aroused Blackett and Occhialini's comprehensible enthusiasm.

[17] Reference [7], quotation on p. 64.

[18] On the history of physics in Cambridge in the '30s, see [25] and Blackett's own reminiscences in [26].

[19] On Rutherford at the Cavendish, see [27].

[20] On the history of the use of the cloud chamber in physical researches, see the chapter "*Cloud Chambers: The Peculiar Genius of British Physics*" in [28], pp. 65–141.

[21] On Blackett's contributions to cosmic-ray physics with the cloud chamber, see [29].

[22] On the controlled cloud chamber, see [30].

[23] Reference [31], quotation on p. 18. Besides Lovell's work, biographical informations on Blackett can be found in [32].

> I can still see him, that Saturday morning when we first ran the chamber, bursting out of the dark room with four dripping photographic plates held high, and shouting for all the Cavendish to hear 'one on each, Beppe, one on each!'.[24]

The controlled cloud chamber was applied to the study of cosmic rays, the subject of research that had engaged Occhialini in Arcetri. In a work [34] of August 21st, 1932, published on *Nature*, Blackett and Occhialini proposed for the first time the results on their joined studies of the penetrating corpuscular radiation by means of the photographic technique applied to a Wilson cloud chamber triggered by a Rossi's coincidence circuit. The experimental disposal of the cloud chamber used by Blackett and Occhialini was similar to that proposed by Johnson, Fleisher, and Street [35].

The most important paper published by Blackett and Occhialini on their researches with the controlled cloud chamber was "*Some Photographs of the Tracks of Penetrating Radiation*" [36], communicated to the Royal Society by Rutherford on February 7th, 1933. The core of the paper was the study of cosmic-rays showers, with side, but absolutely not less important, items such as the positive electron (the positron)[25] and the non-ionising links.

Blackett and Occhialini interpreted the positron according to Dirac's theory as the anti-electron whose destiny was to annihilate with an electron producing one or more photons. The Meitner-Hupfner effect—that is the γ-rays anomalous absorption—was so interpreted as caused by the process of pair production, as well the re-emission of low energy γ-rays as caused by the process of annihilation of positrons and electrons. Positrons could also account for the apparent backwards trajectories of negative electrons from a neutron source.

In 1948, the Nobel Prize in Physics was awarded to Blackett for his contributions to the development of the Wilson method and his discoveries, made by this method, in nuclear physics and on cosmic radiation. It is commonly stated that Occhialini was quite too young—he was only twenty-five years old—when he invented the triggered cloud chamber in Cambridge with Blackett, but nomination to a Nobel Prize is actually independent of any consideration on the age of the nominees. In his Nobel Lecture [40], Blackett expressed himself outspokenly by mentioning several times Occhialini's fundamental contribution to their researches in Cambridge. In a letter to Occhialini's father, Blackett motivated his disappointment for having awarded the prize alone, without a common assignation to Occhialini himself:

> I am very happy and proud to have awarded this prize, but I would have been still happier if Beppo had been honoured at the same time. For it was certainly his arrival in Cambridge which stimulated my embarking on this field of cosmic rays which I have never left. And our work together in 1932–33 was a real collaboration of the happiest kind.[26]

Occhialini always admired Blackett with reverential attitude and never suffered from this exclusion. Blackett was indeed already a master of the Wilson cloud cham-

---

[24] Reference [33], quotation on p. 144.

[25] On the discovery of the positron see Hanson's classical studies [37], De Maria and Russo's historical reconstruction [38], and Roqué's analysis [39].

[26] Letter from P. M. S. Blackett (Manchester) to A. Occhialini (Genoa), November 21st 1948. Occhialini Papers 1, 1, 3.

ber, both from a theoretical and an experimental point of view, before Occhialini's arrival to Cambridge, and continued to develop that technique also after Occhialini's return to Florence. Blackett's nomination was based on his sixteen years long work with the cloud chamber.

An Intermezzo: Arcetri and São Paulo

Once back to Arcetri, in 1934, Occhialini found a situation that had worsened with respect to the one he had left in 1931. Garbasso died prematurely, Persico had moved to Turin, Rossi was professor in Padua. Above all, it was difficult to him to agree with the new political climate of fascist Italy, and to adapt himself to the new life and work conditions. He could not be indifferent to the war events in Ethiopia and Spain. In 1932, he had a deep crisis on conscience in taking an oath and enrolling as a member of the fascist party, to be allowed to continue his university career.

The Italian scientific world was well aware of the importance of the results obtained by the young Occhialini while in Cambridge. In 1934 he was awarded the Sella Prize of the Reale Accademia dei Lincei. It was but very hard to find a financial support to continue such kind of researches in Italy. Occhialini asked the Italian National Council of Researches (CNR) to finance the construction of a cloud chamber to study cosmic rays and neutrons. Guglielmo Marconi solicited this support too, but Occhialini got no answer at all.

Occhialini was in Lucca at the Military School in 1934–1935 for his military service, and taught physics at the Royal Institute of Arts (1935–1936) and at the Faculty of Architecture (1936–1937) both in Florence. He was then enrolled as professor at the Scientific High School in Macerata from 1937 on.

In June–July 1937, Occhialini was invited by Gleb Wataghin to join him in organising a school of physics in São Paulo.[27] Occhialini discussed with Dr. Parini, of the Head Office of Italians Abroad, about his transfer to the staff of Italian middle schools and his integration in the staff of Italian schools abroad. In August 1937, Occhialini left Italy to Brazil and was nominally admitted to the school "Dante Alighieri" in São Paulo. Actually, he became appointed professor of Experimental Physics at the University of São Paulo.

Due to the lack of local human resources, Brazil had to call from Europe many professors to build a new large university, the University of São Paulo, with the new Faculty of Philosophy, Sciences, and Letters.[28] The presence of Italian scientists at the University of São Paulo was supported by the Italian government, that paid their salary, because it was considered an activity of cultural and political mission in a Latin country with a consistent Italian immigration.

[27] On Gleb Wataghin in São Paulo, see [41].

[28] On the early history of physical sciences in Brazil, see [42–46].

> In a modest lab, populated by a group of Wataghin's young students, such as M. De Souza Santos,[29] M. Schönberg,[30] P. A. Pompeia, A. de Moraes, and, in a second time, U. Camerini and C. Lattes,[31] he [Occhialini] prepared an experiment to observe large cosmic rays showers with Wilson chambers and counters.[32]

They were interrupted during the Second World War, when most of the Brazilian physicists were engaged in the researches concerning the production of sonars able to detect German U-boots.

In March 1942, Brazil joined the nations fighting against Italy and Occhialini was recalled back to Italy because of the breaking of diplomatic relations. Many Italian professors came back to Italy but the English government refused to permit Occhialini the free passage. During this period, Occhialini escaped to the Agulhas Negras (or Itatiaya) mountains, between São Paulo and Rio de Janeiro, where he worked as an alpine guide. When Italy signed the armistice, on September 10th, 1943, he offered to fight with the United Nations, but the invasion of France delayed his arrival to England till the beginning of 1945.

In the meanwhile, Occhialini was again in touch with Brazilian researchers and worked in the Biophysics laboratory, directed by Carlos Chagas Jr.[33] There, he met a French-Canadian researcher, Charles Leblond, a pioneer of cell biology who performed physiology experiments with brain tissues that had absorbed radioactive compounds. The track left by the radioactive material, where it had been absorbed, was the phenomenon that attracted Occhialini's attention. It suggested to him a new way to study elementary particles. By the use of plates with thick emulsions, it could be possible to fix the tracks of the particles which, while penetrating into the plate, excited the grains of the exposed emulsion, and to study their physical properties.

The Bristol Period

Blackett, at that time at the Admiralty, called Occhialini from Brazil to work with the allied project to build the atomic bomb. As a first step, Blackett asked the Foreign Office, in February 1943, to look for information on Occhialini in Brazil, suggesting to get in touch with the British embassy in Brazil, and make clear the anti-fascist position of Occhialini himself.

The answer from the Embassy to the Foreign Office contained useful information on Occhialini's status and his willingness to collaborate with the Allied forces. The suggestion to let Occhialini come to England to work in the atomic research was kept secret, mostly because of the after-effects on his family in Italy.

---

29 On Marcelo Damy de Souza Santos, see his interview [47].

30 On Mario Schönberg, see his interviews [48, 49].

31 On Cesar Lattes, see [50–52].

32 Reference [9], quotation on p. 323.

33 On the Biophysics Laboratory in Rio de Janeiro, see [53].

Occhialini's position in 1944 was not at all easy. Brazilian scientists tried to make him stay in Brazil to go on with his researches in cosmic-rays physics. Similar proposals came from the Ohio University with a Rockefeller fellowship, but Occhialini preferred to wait for the safe-conduct to leave Brazil to England. He was but not admitted to join war research because of his nationality. Occhialini landed unconditionally at Cardiff on January 23rd, 1945, and spent a short time at the Department of Scientific and Industrial Research (DSIR) in London with Sir Edward Appleton, while waiting for an engagement in the Allied war efforts. This limbo state happened but to have no solution at all. After having spent a few days at the Research Laboratories of the General Electric Company in Wembley, because of the refusal of his employment from the Aliens War Service Department, Occhialini accepted the invitation of A. Tyndall from the Wills Laboratory.[34] He arrived in Bristol in September 1945 where he had been exempted by the Secretary of State from the Special Restrictions applicable to enemy aliens under the Aliens Order.

In the H. H. Wills Laboratories, they made use of nuclear emulsions,[35] experimentally produced by Ilford and Kodak, and they exposed them at high altitude to detect the disintegration due to cosmic-rays collisions against the atmosphere.

> With his stubborness and his intuition, Beppo started again from basics, working with the technicians from the Ilford photographic laboratories. He transformed their emulsion plates into a formidable instrument for detecting and studying elementary particles, about which almost everything remained to be understood. Occhialini immediately thought of cosmic rays, his first love in Arcetri when he worked with Rossi, as the ideal source of high-energy particles.[36]

They expected to find among the disintegration products the π-meson, a particle whose existence had been predicted by Hideki Yukawa, in 1935, as the particle responsible for the strong nuclear interaction.[37] Cecil Frank Powell[38] had been interested in the technique of nuclear emulsions from 1938 on, following a suggestion of Walter Heitler, and, after the end of WW II, was also a member of the emulsion panel chaired by Joseph Rotblat.

By referring to precedent works of other physicists, Powell, Occhialini, Livesey and Chilton wrote in 1946 a perfect description of the technique of photographic emulsions in nuclear and particle physics [62]. The tracks of the particles recorded in photographic emulsions were studied in order to obtain some physical quantities. By using some empirical relations, physicists were able to gain useful information on the energy and the velocity of the particle, leading thus to a determination of its mass.

---

[34] On the history of the Department of Physics in Bristol, see [54]. On the history of the Cosmic Ray School of Physics in Bristol, see [55].

[35] On the history of the use of nuclear emulsions in physical researches, see the chapter "*Nuclear Emulsions: The Anxiety of the Experimenter*" in [28], pp. 143–238.

[36] Reference [4], quotation on p. 336.

[37] On the discovery of the π-meson and the contemporary discovery of the V-particles, see the recollection of reminiscences and papers in [56]. See also [57, 58]. On the researches concerning particle physics in the '30 and '40s, see also the recollections in [59].

[38] Biographical information on Powell is in [60, 61].

In 1946, the main disadvantage of the photographic emulsions lay in the wide gaps in the succession of the grains forming a track left by a particle, preventing a correct determination of the true length of the tracks. It was thus very important to be able to produce new kinds of photographic emulsions, containing higher quantities of silver bromide.

The Bristol group with Cecil Waller of the Ilford invented a new kind of borax-loaded plates. It was an impressive step in the technique of nuclear emulsions. Lattes, at the University of São Paulo, had built in the meanwhile a cloud chamber to study slow mesons. After having received from Occhialini some photographs of protons and α-particles tracks, Lattes asked to join Powell's team to study the new Ilford emulsions. In winter 1946, thanks to Occhialini and Powell's help, but also to the lack of experimental physicists in minor British universities, Lattes obtained a scholarship and could really go to England.

During one of his holidays as a speleologist in the caves on the Pyrénées, Occhialini exposed about two dozens of C2 Ilford emulsions on the Pic-du-Midi. It was Occhialini, on the very night of his coming back to Bristol, who developed the emulsions after having recovered them from the one-month exposure. The difference between normal plates and borax-loaded ones was evident at once. Since the borax slowed down the image fading, the loaded emulsions were able to keep the record of a greater number of tracks, whereas in the normal plates the detection power considerably decreased after about a week.

> When they were recovered and developed in Bristol it was immediately apparent that a whole new world had been revealed. The track of a slow proton was so packed with developed grains that it appeared almost like a solid rod of silver, and the tiny volume of emulsion appeared under the microscope to be crowded with disintegrations produced by fast cosmic ray particles with much greater energies than any which could be generated artificially at the time. It was as if, suddenly, we had broken into a walled orchard, where protected trees had flourished and all kinds of exotic fruits had ripened in great profusion.[39]

Powell decided to concentrate the whole lab staff in the study of low energy normal events. Marietta Kurz (a member of "Cecil's beauty chorus") found the track of a meson (a π-meson) till its stopping point, and a second track, beginning from the end of the first one, of a second meson (actually: a muon), stopping in the same emulsion too. Within a few days, Irene Roberts found a similar case. These results were published on *Nature*, in May [63].

> The excitement of the discovery of the π-meson was intense and the occasion was such that Occhialini of undoubted rationalistic outlook could only express his feelings by going into the R.C. Cathedral to light a candle!.[40]

To confirm the discovery of the π-meson made on the Pic-du-Midi, the Bristol team decided that they had to get quickly other recordings of similar events to grant to their work the necessary scientific validity. Lattes found the indication of a meteorological station at high altitude on the Bolivian Andes, Chacaltaya station,

[39] Reference [60], quotation on p. 36.

[40] Reference [54], quotation on p. 38.

with an extremely advantageous geographical position granting a cosmic-rays flux 100,000 times greater than that on the Pic-du-Midi. The British government decided to finance Lattes' mission, in the conviction that the development of nuclear physics would have had political-military advantages. The first plate was developed after one month in La Paz, and Lattes found a complete track of the "double meson". All the other emulsions were developed and studied in Bristol and gave as a result about thirty tracks of "double mesons". The discovery doubtless represented an important step in the history of the comprehension of the structure of matter, and the results obtained were convincing in considering that the observed process was a fundamental one. The first meson (the π-meson) was thus identified with Yukawa's meson, and the second meson with Anderson's meson (the muon). It was furthermore postulated the typical decay reaction of the π-meson that, to grant momentum conservation, included a neutral small mass particle (later identified with the muonic neutrino).

The positive result of these researches was of course the outcome of the different contribute of various scientists and institutions. The official communication, apart from the published articles, was at the Conference on Cosmic Rays and Nuclear Physics held at the Institute of Advanced Studies in Dublin (July 5–12th, 1947). Powell, spokesman of the Bristol team, exposed their conclusions strengthening their scientific character, but the scepticism of a valuable part of physicists circles, more engaged in the analysis of Conversi-Pancini-Piccioni's experiment, required further open discussions of the argument and close examinations of the calculations to confirm their validity. The University of Bristol officially recognised Occhialini's important contribution to the researches he made in cosmic-rays physics and the development of the related technology with the award of the doctorate *honoris causa*.

The rapid development of the nuclear emulsions technique was strictly connected with a parallel development in the microscopic technique. Nuclear emulsions exposed on the Pic-du-Midi by the Bristol group, which collaborated with Cooke, Troughton and Simms, were studied by means of a reflecting microscope.[41] By the examination of the nuclear emulsions, Occhialini and Bates, using a reflecting microscope, reversed the normal microscopic procedure, inverted the plate, and observed through the backing glass plate. They were thus able to observe stars and heavy fragments, not visible by the ordinary technique [65].

The thickness of the emulsions always set a limit to the number of useful observed events. The aim of the researches in the technique of nuclear emulsions was to produce emulsions with a thickness comparable to the length of the tracks left by the particles to be recorded. A growing thickness of the emulsions led but in a first time to the impossibility to process the whole photographic material in the same way, and then to the impossibility to process in any way even an increasing part of the photographic material itself. Conventional nuclear emulsions were about 200 μm thick and could be processed in a non-uniform way: the surface of the emulsion (the "top") was more processed than the layer against the supporting material (the "bottom"). Grain density was thus observed to be different in tracks left by particles with the same ionising power, as a function of the depth in the emulsion.

[41] On the reflecting microscope, see [64].

Occhialini, with his future wife Constance Dilworth and Ron Payne, were able to invent the "Temperature Development" processing method [66–68] which led them to the production of a nearly uniform development of 300 μm thick emulsions, with a variation in grain density of less than 10% between the "top" and the "bottom". New nuclear emulsions and their processing were thus studied by Occhialini with the help of Waller of Ilford, and of Berg, Berriman, Herz and Stevens of Kodak. The problem to be solved was the permeation of the whole emulsion with the developer before the developer could act, that is the separation of the processes of permeation and development by stopping the permeation before the development. Occhialini and his group preferred to separate the two processes by acting on the soaking temperature and making use of the lower temperature coefficient of the rate of permeation with respect to the one of the rate of development.

The role played by Occhialini in the discovery of the π-meson was as important as that one he played in Cambridge.[42] In this case too, Occhialini did not win the 1950 Nobel Prize in Physics awarded to Powell alone for his development of the photographic method of studying nuclear processes and his discoveries regarding mesons made with this method. The denied acknowledgement was less self-evident than in Blackett's case. Powell never quoted Occhialini in his Nobel Lecture [70]. Hypothetical political motivations have been advanced to explain Occhialini's exclusion to the prize:

> Blackett and Powell separately won the Nobel Prize for their work on elementary particles. Both awards were made in difficult, Cold War years, and Occhialini had never made a secret of his political ideas.[43]

Occhialini's father, in a letter to his son, on November 11th, 1950, wrote his very hard comment:

> I saw a telegram to be sent to Powell signed by the Istituto di Fisica and I stopped it. In that signature, mine own was implicit; they did not ask me for it, and I would not have signed it because that would have meant a hypocrisy. Your only signature is worth in a general case; you can allow it or not depending on what you feel. Sending a telegram that implies your signature and not ours is, according to me, a double hypocrisy, that would lower you in this moment. My suggestion is to send your own telegram, as generous as your not repayed work. Only so you will pass for a sacrificed in front of an imposer, and you will be over the miseries.[44]

These very hard words are a father's word, but it is a matter of fact that Occhialini suffered quite a lot from the exclusion of him from the Nobel Prize awarded to Powell.

---

[42] "*Here we had two of the greatest physicists of their time, experimenters who were destined to become Nobel Prize winners and men who had the ability to inspire the countless researchers who were attracted to work with them. It would be hard to find British cosmic-rayites who had not worked in either of their laboratories, and indeed many of the European (not forgetting Eastern European) cosmic-ray physicists had received part at least of their training in Manchester or Bristol.*" Reference [69], quotation on p. 27.

[43] Reference [4], quotation on p. 340.

[44] Letter from Augusto Occhialini (Genoa) to Giuseppe Occhialini (Brussels) (1950). Occhialini Papers 5, 1, 2.

He maybe suffered more from what seemed to be the lack of an official recognition of his own role by Powell himself.

In the same year when Powell was awarded the Nobel Prize, an important award of Occhialini's contribution to cosmic-rays physics and the relative technology came with the Charles Vernon Boys Prize of the Physical Society of London. Powell showed a particular happiness for this award to Occhialini. This fact might contribute to see from a different point of view Powell's accusing silences on the role played by Occhialini in his researches in Bristol.

> We were very glad to hear that the Physical Society of London has awarded you the Vernon Boys Prize for 1950. We were particularly glad that this award was made by an English Society and therefore represents some small act of appreciation in recognition of the contributions you have made to physics and in particular to English physics. I hope it will remind you also of the happy and exciting time we spent in this laboratory from 1945–48.[45]

Powell, not only brought pressure to bear on the Physical Society, but suggested to award Occhialini the Nobel Prize. Actually, he wrote to Wolfgang Pauli in order to convince him to put Occhialini's name forward to the Nobel Committee in Physics. Pauli, who was a great friend of Occhialini and had admired his scientific work since the years the latter spent in Cambridge with Blackett, was happily willing to support his nomination.

A New Laboratory in Brussels and His Return to Italy

Occhialini left Powell's group in Bristol and went to Brussels to work at the Centre de Physique Nucléaire of the Université Libre de Bruxelles. He worked continuously in Brussels from 1948 to 1950. Thereafter, he was appointed professor at the University of Genoa from 1950 to 1952, and at the University of Milan from 1952 on. While teaching and making research in Italy,[46] Occhialini continued to collaborate with the Centre de Physique Nucléaire where he spent a lot of time every year until his sabbatical year at the MIT in 1959. The Brussels and Genoa/Milan groups in the '50s can be considered a single one group of research under the scientific leadership of Occhialini.

Occhialini was called to Brussels, together with Constance Dilworth,[47] by Max Cosyns, a friend of his speleological adventures, to give birth to a new laboratory where they could study nuclear emulsions. They also printed a new journal, the "*Bulletin du Centre de Physique Nucléaire de l'Université Libre de Bruxelles*". Yves Goldschmidt-Clermont, who had worked in Bristol with Occhialini, joined them too.

---

[45] Letter from Cecil Powell (Bristol) to Giuseppe Occhialini (Brussels), May 12th, 1950. Powell Papers. Bristol University Special Collections DM 1947/E.303.

[46] On the state of physical researches in Italy soon after WW II, see [71].

[47] On biographical information on Constance Charlotte Dilworth, see [72, 73].

In the Brussels laboratory they went on with the researches on the new NT2 and NT4 Kodak plates by Berriman, and G5 Ilford plates by Waller. The emulsions group, under the scientific leadership of Occhialini, became soon one of the most important groups of research on nuclear emulsions, after the one in Bristol.

While in Brussels, and then in Genoa and Milan, Occhialini played a fundamental role in the development of several groups of research in nuclear emulsions in Italy and in the birth of a European network of groups devoted to the studies of cosmic radiation. Brussels was the school where to learn the emulsion technique for Bonetti and Scarsi from Genoa, Merlin from Padua, Cortini from Rome, Levi-Setti from Pavia. The new way of life in Brussels, and then in Genoa and Milan, was mainly like the old way of life in Bristol, full of hard work on cosmic-ray physics and the development of the technique of nuclear emulsions.

A very sad fact shattered the laboratory in Brussels in 1952: the so-called winch affair. On August 14th, 1952, during a speleological exploration of the cave of Pierre-Saint-Martin, Marcel Loubens, a member of a speleological group formed by Occhialini, Cosyns, Labeyrie, Casteret and Tazieff, died while climbing to the earth's surface, since the hook connecting his cable to the winch broke and he fell for about forty meters and crashed on the rocks below. He was in his last agony two days long, assisted by Occhialini, Labeyrie and Tazieff. Occhialini was able to tie a particular kind of knot used in sailing to join the cable with the winch. The speleologists outside the cave could so pick them up very slowly. Unfortunately Loubens' body was salvaged only two years later.

The very deep crisis in the Brussels group, due to the irrimediable dissidence between Occhialini and Cosyns, lasted through 1953. Baudoux was appointed new director of the group in May 1952. The winch affair caused but a tremendous internal division. Occhialini's scientific collaboration with Cosyns came so to its end. Occhialini went on to work with the Brussels emulsion group, while Cosyns had to leave Belgium and went to Paris. The Brussels group was able to get over the worst, but

> the internal division and later loss of staff seriously compromised the scientific activity and led to the loss of that preeminence which the group had till then maintained.[48]

Occhialini succeeded in making it again a part of the European network of laboratories that were flying stacks of plates on balloons at high altitude. Occhialini asked Powell to let the Brussels group join the collaboration organising flights of nuclear emulsions.

Occhialini was Professor of Physics at the Institute of Physics of the University of Genoa from 1949 to 1951. He had namely won a competition for the chair of physics in Cagliari, but he eventually had the one left by his father because of the latter's age. With the help of Alberto Bonetti, he was able to give birth to a small group of research in Genoa too.

---

[48] Reference [74], quotation on p. 737.

During 1952, Occhialini was also at the Brazilian Centre of Physical Researches in Rio de Janeiro. His travel was supported by the UNESCO and had the aim to help Lattes and Camerini in the organisation of the group of research in Rio de Janeiro and its researches made in the laboratory on Chacaltaya. The new laboratory in South America should have worked following the same lines used by the European network: the use of nuclear emulsions as detecting device of cosmic rays, exposition of plates at high altitudes, flights of stacks of plates on balloons.

Occhialini arrived in Rio de Janeiro on February 8th, 1952, and was pleased to find the Brazilian Centre of Physical Researches better than any optimistic prevision. The moral among the young researcher was very high, the financial position was quite good, and the technical positions were also good due to the establishment of a workshop with eight technicians, a glass laboratory and a vacuum pump able to mass produce very good GM counters. During his stay for a period of about six months, Occhialini was able, with the precious help of Lattes and Camerini, to install the electronic department which had the task to design and produce all the coincidence apparatuses to be sent to Chacaltaya. In the Chacaltaya laboratory, a controlled Wilson chamber was built and tested, and a small group of people started to work on nuclear emulsion.

In the short time spent working both in Genoa and Brussels, Occhialini and his team developed further methods to obtain better images with nuclear emulsions, the wire method of loading emulsions [75] and the use of cylindrically-shaped emulsions [76]. Emulsions produced with the Wire Method gave interesting results in the study of cosmic-rays phenomena, with a previous analysis of the possible distortions of the tracks introduced by the wires themselves.

During the development of nuclear emulsions, one the difficulties met with was the fact that the emulsions became more and more opaque. One of the main sources of this opacity of the emulsions was the deposit of colloidal silver in the gelatine. This problem was particularly annoying with thick emulsions, from 100 μm on, affecting in such a way the new thick emulsions used with the "temperature development" technique. Occhialini, Dilworth and Eric Samuel found a simple method to avoid the deposit of colloidal silver [77].

Another use of variations of temperature was made by Occhialini's group to clear nuclear emulsions of the tracks left by cosmic radiation (or other radiation sources) in them before their use. Improvements in the production of more sensitive nuclear emulsions and in the technique concerning their development and examination permitted the study of relativistic phenomena of particle physics and the observations of the β-decay of the μ-meson, not to mention the possibility of new analysis on already known phenomena. In particular, further analysis on disintegration stars put in evidence the existence of kinds of particles not yet studied, such as highly charged particles.

An Actor in the Long Search for Strange Particles

Occhialini was Professor of Physics at the Institute of Physics of the University of Milan[49] from 1951 on, where he moved after his father's death, sharing his time with the emulsions laboratory in Brussels too. The coming of Occhialini to Milan was supported by Giovanni Polvani and Piero Caldirola, and must be seen within the frame of the birth of the INFN (National Institute of Nuclear Physics).

The scientific production in the early '50s concerned both the study on the development and use of nuclear emulsions, and the studies on new kinds of particles.

> The effort of the organizing and scientific direction of the group of research of Brussels and Genoa and then of Milan found an answer in the scientific production which, in part, inserted itself on an international level: mesons and pions capture in the elements of the nuclear emulsion; measurements of very high energy particles with the method of multiple scattering and with the curvature in emulsion; individuation of the decay schemes of the positive hyperon, and contributions to the study of the hyperfragments.[50]

Maybe the most important characteristic of Occhialini's studies on cosmic rays was the participation of his groups to international cooperations organising flights of balloons carrying stacks of nuclear emulsion plates at high altitude. The prelude to the European collaborations was, in 1947–1948, the launch of balloons to expose nuclear plates at cosmic radiations at 30 km altitude. This launch was organised by Powell's group in Bristol with the collaboration of the University of Padua. The further development of the Bristol activities with the Italian groups saw the engagement of the INFN Sections of Milan, Padua, and Rome. The first great expedition, involving thirteen groups, was, in 1952, the launch from the Italian bases of Naples and Cagliari, and the following recovery of stacks of plates (with a glass support) after the landing on sea of the balloons. Besides the importance of the tracks recorded in the plates themselves, the study of the flight of the balloons permitted to get useful information on the velocity and direction of wind at high altitude.

In June-July 1953, eighteen groups of research launched twenty-five balloons from the Elmas airport in Sardinia, and exposed more than one thousand stripped emulsions, corresponding to a volume of 9.3 dm^3, seven hours long at an altitude between 25 and 30 km. All emulsions were processed in Bristol, Padua, and Rome; in October 1953, a meeting was held in Bern in order to distribute the processed plates among the different groups. The first results were discussed in an international congress held in Padua in April 1954. Further results were the subject of the second course of the International School of Varenna, in summer 1954.

The third significant experience was the launch of the so-called G-Stack from Novi Ligure, in October 1954. The G-Stack was a single stack of emulsions with a volume of about 15 dm^3. The choice to launch a single "giant" stack came from the aim to study in the most advantageous way part of the recorded tracks in their whole length in order to obtain precise values of their energy and decay modes. The

---

[49] On the history of scientific and physical studies in Milan, see [78–80].

[50] Reference [9], quotation on p. 324.

most important result of the G-Stack was the determination of the equality of the values of the masses of the then supposed different K-mesons, and the statement that the different decay modes were alternative decay processes of a same particle. The solution of the related θ-τ puzzle was, in such a way, a first step to the discovery of the non-conservation of parity in weak interactions.

After the G-Stack, the results on elementary particles obtained by means of accelerating machines soon outnumbered the ones found in cosmic rays.[51] In the second half of the '50s, Occhialini thus went on with his studies on elementary particles by exposing the nuclear emulsion plates to beams of particles produced by accelerating machines at the CERN or elsewhere.

The conferences held by cosmic-rays physicists in the early '50s were of the utmost importance in the history of physics. The 1953 International Cosmic Rays Conference held in Bagnères-de-Bigorre was an epoch-making event where "order emerges from chaos".[52] The discussion on the plethora of K-meson decay modes, thanks to "Rossi argument",[53] led to the conclusion that many different events actually corresponded to different decay modes of one kind of particle.

The results of the plates impressed in the 1953 launches from Sardinia discussed in April 1954 at the above-mentioned "International Congress on Heavy Unstable Particles and High Energy Events in Cosmic Rays held in Padua". The main subject of the Padua congress was a problem that arose from the analysis of the tracks left by the τ-particle, or $K\pi_3$, that is a K-meson decaying into three coplanar charged π-mesons. The problem was known as the θ-τ puzzle. Another problem discussed at the Padua Congress was the new $K\mu_2$-decay mode (the "Camus"), suggested by the French group of the École Polytechnique of Paris. To solve both the θ-τ puzzle and the $K\mu_2$ problem, the groups of Bristol, Milan, and Padua decided to undertake the G-Stack flight.

The results of the G-Stack experiment[54] were the main subject of the "International Conference on Elementary Particles" held in Pisa in 1955. Seven different decay modes (τ, τ', $K\pi_2$, $K\pi_3$, $K\mu_2$, $K\mu_3$, Ke_3) were definitely assigned to only one particle, the K^+-meson. The θ^0-decay mode corresponded to the K^0-meson. The θ-τ puzzle continued to be unsolved since the spin-parity of the τ-decay was confirmed to be 0^-. Strange particles led to the definition by Murray Gell-Mann of the strangeness S, a new quantum number. The θ-τ puzzle was solved in 1957, when cosmic rays were

---

[51] The G-Stack flight itself had been prepared in a very short time in order to precede the results that would have been with the accelerating machines soon after. "*Leprince Ringuet expressed, with an elegant metaphor, the attitude to this threat. Rather than to 'retire to the country and wait six months for Brookhaven to give the answers', the community would continue to work in the field in the hope that the higher energy components of the cosmic rays would still reserve some surprises. It was in this sort of climate that the G-Stack collaboration was born. It was a last minute attempt to beat the machines.*" Reference [74], quotation on p. 739.

[52] See the paragraph on Bagnères-de-Bigorre in [81].

[53] "*I would like to take the point of view that two particles are equal until they are proven different.*" Reference [81], quotation on p. 8.

[54] For the results of the G-Stack experiment see [82, 83].

no more the source of K-mesons, but accelerating machines, the Berkeley Bevatron in a first time, were used instead.

Researches by means of nuclear emulsions had an improvement in Milan thanks to the application of an already known principle to microscope technique. The afforded problem concerned the possibility to make small relative shifts between the observed object and the ocular micrometer. The solution consisted in the measurement of the rotation around axis perpendicular to the optical axis of one or more leaves with plane and parallel faces, introduced along the optical canal of the microscope. This method was already applied in optical devices, such as the Klausen's blade micrometer, and permitted to shift the image of the object in the plane of the micrometer, and to measure its shift. A device of this kind, a MS2 Koristka microscope, was made for Occhialini and his group by Dr. Cantù of Koristka.

A second important European collaboration followed a few time after the G-Stack Collaboration: the K^--Collaboration. The new aim was a thorough study of the interactions and decay of K^--mesons. For the publication of the three classical papers on the K^--Collaboration [84–86], Occhialini's group in Milan and Brussels was more and more involved in the studies on K^--mesons, both from cosmic radiation and artificially produced. During this period, Occhialini stopped to sign his contribution to published articles.

The Astrophysical Period

We can consider the second half of the '50s as a "transition period" in Occhialini's scientific career. The K^--Collaboration was not only a renounce to cosmic rays as the primary source of elementary particles, but was also the last great experiment made with stacks of nuclear emulsions. Even if Occhialini played an important role both in the coordination of the K^--Collaboration and in the further development of new kinds of microscopes to scan nuclear emulsions, the importance of this last technique was decreasing.

> In this transition period Beppo Occhialini continues to be a drawing leader; he coordinates the collaboration among the groups on a European scale, he continues the experimentation and the technical development on new kinds of high precision microscopes, he is engaged at close quarters in the analysis of the events due to the capture and interaction of the K^-. In this connection I remember, in Milan, the classification of the "hyperfragments" widespread among the microscopists, following the increasing difficulty of interpretation: "Normal", "G.O.K." (God only Knows), "D.O.K." (Devil only Knows) and "B.O.K" (Beppo only Knows).[55]

The solution was offered by the launch of the Sputnik in September 1957. Bruno Rossi, at the Massachusetts Institute of Technology (MIT), began at once a series of researches in space physics on the interplanetary plasma and cosmic γ-rays. Occhialini, accompanied by Constance Dilworth, decided to spend a sabbatical year

[55] Reference [5], quotation on pp. 614–615.

at the MIT as Visiting Professor to rise, one back to Milan, a group of cosmic and space physics. The Milan group of nuclear emulsions was converted to the new space adventure by means of spark chambers on balloon and satellite,[56] in a strict collaboration with the French group in Saclay, while the cloud and bubble chamber group was sent to work at the CERN.

> The University of Milan group, led by Occhialini, was increasing in both the number of its staff and the importance of its research projects. [...] It was an exciting time; everything was still to be discovered. Were there, for instance, neutrons reaching the Earth from the Sun and influencing our atmosphere? Although most of the primary cosmic radiation had been found to be made up of protons, were there also electrons among them? What about positrons? And what of the never forgotten γ-rays (of which Millikan wrote in the 1920s) —were there any at all, coming from the Sun, or from our Galaxy, or from deep space, like the X-rays that Rossi and Giacconi had just discovered? The Milan group attacked all these problems, with Beppo's enthusiasm and energy acting as a motor.[57]

The Milan group was one of the main characters of Italian space physics, together with the ones in Bologna and Rome. Occhialini was, with Castagnoli and Puppi, the actor of the passage of the Italian groups of cosmic physics from the INFN to the CNR in 1966–67 with the constitution of the Italian Group of Cosmic Physics (GIFCO).[58] The Milan group evolved into the Laboratory of Cosmic Physics and Related Technologies (IFCTR), led by Occhialini till 1974, with two dozen researchers and technicians.

Occhialini was one of the founding fathers of the European space physics,[59] as well as Pierre Auger, Robert Boyd, Marcel Golay, Bengt Hultqvist, Reimar Lüst, Harrie Massey, Bernard Peters, Pol Swings, Hendrik van de Hulst. After the first steps of the European Preparatory Committee for Space Research (COPERS—Comité Preparatoire Europeen pour la Recherche Spatiale) in Paris, Occhialini was one among the most important members of the Council and of the Scientific and Technical Committee of the new-founded European Space Research Organisation (ESRO). He was Chairman of the COS-Group (Advice Committee for Cosmic Rays Physics) and member of the restricted Launching Program Advisory Committee (LPAC) devoted to choose and define European space missions that were organised following the

---

[56] "*Beppo is certainly not a home cooking man: he discerns the potentiality of the "spark chamber" as a detector fit for balloons and satellites and begins a collaboration with Saclay which masters the spark chamber technology and that is also bring to pass a partial change, from particles to space.*" Quotation from [5] on p. 616.

[57] Reference [4], quotation on p. 338.

[58] The GIFCO was born with the constitution of four laboratiories: the ITESRE in Bologna, the IFSI in Frascati, the IFCTR in Milan, and the CosmoGeofisica in Turin, with two divisions in Florence and Palermo. The Palermo division will be the fifth laboratory in 1981, the IFCAI. After the event, this well-intended operation might have suffered of some deficiencies in the coordination of the groups: "*The CNR responded to the initiative, with the agility of an asthmatic pachyderm, by mistaking (I'm quoting Occhialini's words) the timing typical of archaeological excavations in Pompei, for that necessary to the reaction to a fast countdown for a rocket on the launching pad.*" Quotation from [5] on p. 617.

[59] On the history of space physics in Europe, see [87, 88].

"Street-car" Principle: each mission was a cluster of experiments proposed by the various scientific communities.

Among the different space missions organised by Occhialini too, we can remember the HEOS A1, the TD1, the HEOS A2, and the COS-B.

The HEOS A1 mini-satellite was launched from the Eastern Test Range, Florida, on a Thos Delta DSV3-E launcher, on December 5th, 1968. HEOS A1 penetrated interplanetary space to about 33 earth radii, and was re-entered the atmosphere on October 28th, 1975. There were devices to make several experiments on-board: electric fields, magnetic fields, high-energy cosmic rays anisotropy, low-energy solar protons, solar wind, flux and spectrum of cosmic rays, and primary cosmic-ray electrons. Milan and Saclay had developed the instrument that measured the primary cosmic-ray electrons.

The TD 1 mini-satellite was launched from the Western Test Range, California, on a Delta-N launcher, on March 12th, 1972. TD 1 had not been active since May 4th, 1974, and decayed on January 9th, 1980. Seven experiments were carried on-board: multicolour celestial scanning in the UV and IR, UV stellar spectrometry, spectrometry of primary charged particles, spectrometry of celestial X-rays, solar γ-rays, celestial γ-rays.[60] Milan developed the instrument for solar γ-rays, while that for celestial γ-rays had been developed by Milan, Munich ("Monaco" in Italian) and Saclay ("MIMOSA", the name of a flower). As for the studies on γ-rays.

> The TD-1 experiment was not successful due to high background and the failure of the vidicon system. Its impact was two-fold: it was the beginning of the later Caravane-COS-B collaboration with Leiden and ESTEC as additional partners, and it cleared the way for the acceptance of the wire spark chamber in Europe.[61]

The HEOS A2 mini-satellite was launched from the Western Test Range, California, on a Thor Delta launcher, on January 31st, 1972. It was the first vehicle to penetrate into the area of the neutral point at the border of the Earth's magnetic field and that of interplanetary space. Seven experiments were carried on-board: magnetic field measurements, plasma measurement, solar v.l.f. observation, particle-counter telescope, high-energy electrons, solar wind measurement, and micrometeorite detector.

The COS-B satellite was launched on a Delta 2913 launcher, on August 8th, 1975. COS-B failed on April 26th, 1981. This satellite was the product of the Caravane Collaboration, formed by the Laboratory for Space Research (Leiden), the CNR Institute of Cosmic Physics and Informatics (Palermo), the CNR Laboratory of Cosmic Physics and Related Technologies (Milan), the Max-Planck Institute for Extraterrestrial Physics (Garching), the CEN Service of Physical Electronics (Saclay), the ESRO Scientific Laboratory (ESLAB) (Nordwijk). COS-B permitted to draw the first detailed γ-map of the Galaxy and to have a first catalogue of discrete sources in the range of a few 100 MeV.

---

[60] On the history of gamma-ray astronomy, see [89].

[61] Reference [89], quotation on p. 45.

> With COS-B, as well as other ESRO instruments, Beppo showed how he could work constructively on a supernational scale while maintaining his unique Italian character. COS-B was launched in 1975 and soon became the first great European success in high-energy astrophysics.[62]

In 1975, ESRO and ELDO, an industrial organisation projecting European launchers, merged giving birth to the European Space Agency (ESA). After the launch of COS-B, the new, more bureaucratic, way to manage the ESA was the beginning of the end of Occhialini's scientific activity. COS-B was Occhialini's last scientific success, even if he did not play a central role throughout the seven years the mission lasted. Besides the valuable scientific results obtained with COS-B, we have to note that the project permitted the Milan group to learn how to build second-generation space instruments.

During the '80s, Occhialini spent more and more time in Marcialla, a hamlet of Certaldo, Tuscany, not far from Arcetri. The choice between two proposals of an X-ray astronomy satellite made Occhialini emerge from his retreat. He supported, in a committee with Bruno Rossi[63] too, the SAX satellite, that was then renamed Beppo-SAX after him.

In 1979 Occhialini and Uhlenbeck were awarded together the prestigious Wolf Prize, the last of a consistent series of awards (1934 Sella Prize, 1935 Vallauri Prize, 1949 Einaudi Prize, 1951 Charles Vernon Boys Prize, 1955 Feltrinelli Prize). He was doctor *honoris causa* of the universities of Brussels (1949) and Bristol (1959), member of several scientific academies in Italy and abroad, and foreign fellow of the Royal Society.

Occhialini, Bruno Rossi, and Bruno Pontecorvo died in the same period. Occhialini, in particular, died in Paris on December 30th, 1993. Italy lost three among the main characters of its scientific history in the twentieth century. It was but a loss for the whole scientific world too.

Some Personal Notes

Occhialini was an eminent scientist. He made research with great attention and enthusiasm, a feeling that he transmitted to his students and collaborators. He was considered to be a legendary figure because of his discoveries and the vicissitudes of his life. Occhialini was surely an experimental physicist and not a theoretician according to the usual meaning of these categories. He had a deep sense of nature's laws and was more interested in how nature actually works than in the mathematical laws describing natural phenomena. Nevertheless, he was in wonderful relation with some theoretical physicists, such as Wolfgang Pauli.

Many physicists in Europe and America consider themselves disciples of Occhialini's school of physics. Occhialini was certainly a great teacher but not in

[62] Reference [4], quotation on p. 339.

[63] After Bruno Rossi was named the NASA X-Ray Explorer.

a professional sense. When he felt sure enough to give a lesson, it was really an event. His few lectures were exemplary, more suitable for a graduation *ad honorem* than for a normal lesson to a physics class. As a teacher in the laboratory, Occhialini was instead peerless. He tried to teach, first of all, the experimental sense and the interest in the ways nature works. He wanted also to teach his enthusiasm to make a difficult research successfully and in a neat way. Another equally direct way to teach Occhialini's participation to the drafting of a scientific paper. He discussed the text with pedantry and insight so that his students could learn how to write a paper.

Occhialini was not only a great scientist but also a sincere humanist.

> Beppo had many friends because he felt well among the people. He was sincerely interested in every human being he met, of whatever condition, and treated everybody on an equal footing. He was unselfish, kind and tender; but he reacted, even harshly, to abuses and did not forgive falsity. He did not like to speak to the public, but he talked with pleasure. He was a storyteller, and he hold his listeners spellbound when he talked about his discoveries in physics, his speleological explorations, the scientists and artists he met, the beauty of Tuscany and Umbria landscape, his adventures during his travels.[64]

Occhialini had a very deep cultural life. He was so endowed of a refined sensibility to be able to quote the finest passages in Shakespeare's works by heart. His own thought was quite always original and stimulating and was the very proof of his tremendous literary and artistic culture. At the bottom of his culture there was always the human being in his multiform aspects. He was very fond of sport: he was a skilled alpinist and speleologist,[65] a fan of soccer, and an enthusiast motorcyclist.

Both in Brussels and in Milan, Occhialini was at the heart of an informal humanistic cenacle. In the Brussels laboratory, coffee-breaks were often interrupted by conversations on subjects not concerning physics: psychoanalysis, literature, music, and so on. Occhialini was considered a model from this point of view. In Milan, he and his wife organised once a week, particularly in winter, an informal meeting with an important physicist, and hosted friends, acquaintances, and students. These evenings spent in his home are among the best memories of the people who knew him personally.

References

1. Brown L. M., Hoddeson L. (Editors), *The Birth of Particle Physics* (Cambridge University Press, Cambridge) 1983.
2. Telegdi V. "Giuseppe Occhialini. 5 December 1907–30 December 1993", *Proceedings of the American Philosophical Society* **146** (2002) 218–222.
3. Emanuelli G. *Scienziati fisico-matematici marchigiani: 1846-1951: Luigi Donati, Oreste Murani, Temistocle Calzecchi Onesti, Angelo Battelli, Ferdinando Lori, Raffaello Augusto Occhialini* (Steu, Urbino) 1964.

[64] Reference [8], quotation on pp. 238–239.

[65] Occhialini was one of the discoverers of a large cave in the French Pyrénées, the Pierre St. Martin Cave, 1342 m deep, which will be for a long time the deepest cave the world over.

4. BIGNAMI G. F. "Giuseppe Paolo Stanislao Occhialini. 5 December 1907–30 December 1993", *Biographical Memoirs of the Fellows of the Royal Society of London* **48** (2002) 331–340.
5. SCARSI L. "Giuseppe Occhialini: Il secondo periodo italiano (1950-1993)", *Il Nuovo Cimento C* **20** (1997) 613–618.
6. POLVANI G. "Per i venti anni dal ritorno in Italia di Giuseppe Occhialini", *Rendiconti del Seminario Matematico e Fisico di Milano* **39** (1969) 11–16.
7. RUSSO A. "Vita di uno sperimentatore", *Sapere* **4** (1996) 62–69.
8. TAGLIAFERRI G. "Giuseppe Occhialini", *Rendiconti dell'Istituto Lombardo Accademia di Scienze e Lettere* **128** (1994) 231–240.
9. OCCHIALINI G. P. S. "Occhialini, Giuseppe", in *Scienziati e Tecnologi Contemporanei*, vol. II (Mondadori, Milan) 1974, pp. 322–324.
10. MANDÒ M. *Notizie sugli studi di Fisica (1859–1949)*, in *Storia dell'Ateneo Fiorentino. Contributi di studio* (Parretti, Firenze) 1986.
11. CLARK G. W. "Bruno Benedetto Rossi. April 13, 1905–November 21, 1993", *Biographical Memoirs of the National Academy of Sciences* **75** (1998) 310–341.
12. ROSSI B. B. "Early days in cosmic rays", *Physics today*, vol. **34**, No. 10 (1981) 34–41.
13. BUSTAMANTE M. C. "Bruno Rossi au début des années trente: une étape décisive dans la physique des rayons cosmiques", *Archives internationals d'histoire des sciences* **44** (1994) 92–115.
14. DE MARIA M., MALIZIA G. and Russo A. "La nascita della fisica dei raggi cosmici in Italia e la scoperta dell'effetto Est-Ovest", *Giornale di Fisica* **33** (1992) 207–228.
15. GALDABINI S. and Giuliani G. "Physics in Italy between 1900 and 1940: The universities, physicists, funds, and research", *Historical Studies in the Physical Sciences* **19** (1988) 115–136.
16. RUSSO A. "Science and industry in Italy between the two world wars", *Historical Studies in the Physical Sciences* **16** (1986) 230–281.
17. SCARSI L. *Cosmic Rays in Florence. Bruno Rossi and the "Group of Arcetri"* in *ECRS Florence 30 August–3 September 2004*.
18. XU Q. and Brown L. M. "The early history of cosmic ray research", *American Journal of Physics* **55** (1987) 23-33.
19. SWANN W. F. G. "The History of Cosmic Rays", *American Journal of Physics* **29** (1961) 811–816.
20. SEKIDO Y. and Elliot H. (Editors) *Early History of Cosmic Ray Studies. Personal Reminiscences with Old Photographs* (Reidel, Dordrecht) 1985.
21. ROSSI B. "Arcetri, 1928-1932", in *Early History of Cosmic Ray Studies. Personal Reminiscences with Old Photographs*, edited by SEKIDO Y. and ELLIOT H. (Reidel, Dordrecht) 1985, p. 53.
22. BOTHE W. W. G. and Kolhörster W. "Das Wesen der Höhenstrahlung", *Zeitschrift für Physik* **56** (1929) 751–777. BOTHE W. W. G. and KOLHÖRSTER W., "Die Natur der Höhenstrahlung", *Die Naturwissenschaften* **17** (1929) 271–273.
23. ROSSI B. B. "Method of Registering Multiple Simultaneous Impulses of Several Geiger's Counters", *Nature* **125** (1930) 636.
24. OCCHIALINI G. P. S. "Uno spettrografo magnetico per raggi β emessi da sostanze debolmente radioattive", *Rendiconti della Reale Accademia dei Lincei* **14** (1931) 103–107.
25. HENDRY J. (Editor), *Cambridge Physics in the Thirties* (Adam Hilger Ltd, Bristol) 1984.
26. BLACKETT P. M. S. "The Old Days of the Cavendish", *Rivista del Nuovo Cimento*, Special Issue, **1** (1969) xxxii–xxxix.
27. OLIPHANT M. *Rutherford. Recollections of the Cambridge Days* (Elsevier, Amsterdam) 1972.
28. GALISON P. *Image and Logic. A Material Culture of Microphysics* (The University of Chicago Press, Chicago) 1997.
29. BUSTAMANTE M. C. "Blackett's Experimental Researches on the Energy of Cosmic Rays", *Archives internationales d'histoire des sciences* **47**, No. 138 (1997) 108–141.
30. BLACKETT P. M. S. "On the Technique of the Counter Controlled Cloud Chamber", *Proceedings of the Royal Society of London* **146** (1934) 281–299.

31. LOVELL B. "Patrick Maynard Stuart Blackett, Baron Blackett, of Chelsea", *Biographical Memoirs of Fellows of the Royal Society of London* **21** (1975) 1–115.
32. NYE M. J. *Blackett. Physics, War and Politics in the Twentieth Century* (Harvard University Press, Cambridge) 2004.
33. OCCHIALINI G. P. S. contribution to the "Memorial Meeting for Lord Blackett, O.M., C.H., F.R.S. at the Royal Society on 31 October 1974", *Notes and Records of the Royal Society of London* **29** (1975) 144–146.
34. BLACKETT P. M. S. and Occhialini G. P. S. "Photography of Penetrating Corpuscular Radiation", *Nature* **130** (1932) 363.
35. JOHNSON T. H., FLEISHER W. JR. and Street J. C. "A Cloud Expansion Chamber for Automatically Photographing the Tracks of Corpuscular Cosmic Rays", *The Physical Review* **40** (1932) 1048.
36. BLACKETT P. M. S. and OCCHIALINI G. P. S. "Some Photographs of the Tracks of Penetrating Radiation", *Proceedings of the Royal Society of London Ser. A* **139** (1933) 699–727.
37. HANSON N. R. *The Concept of the Positron* (Cambridge University Press, Cambridge) 1963; "Discovering the Positron", *The British Journal for the Philosophy of Science* **12**, No. 47 (1961) 194–214, **12**, No. 48 (1962) 299–313.
38. DE MARIA M. and Russo A. "The Discovery of the Positron", *Rivista di Storia della Scienza* **2** (1985) 237–286.
39. ROQUÉ X. "The Manufacture of the Positron", *Studies in the History and Philosophy of Modern Physics* **28** (1997) 73–129.
40. BLACKETT P. M. S. "Cloud Chamber Researches in Nuclear Physics and Cosmic Radiation. Nobel Lecture, December 13, 1948" in *Nobel Lecture, Physics 1942–1962* (Elsevier, New York) 1964, pp. 97–119.
41. PASSOS VIDEIRA A. A. and Bustamante M. C. "Gleb Wataghin en la Universidad de São Paulo: un momento culminante de la ciencia brasileña", *Quipu* **10**, No. 3 (1993) 263–284.
42. SCHWARTZMAN S. "Struggling to be Born: The Scientific Community in Brazil", *Minerva* **16** (1978) 545–580.
43. SCHWARTZMAN S. "Science and Higher Education in Brazil: an Historical View", *Woodrow Wilson International Center of Scholars, Latin American Program, Working Papers* **8** (1979).
44. DE SOUZA SANTOS M. D. "Os precursores da física no Brasil" in *Centros e Museus de Ciência visões e experiências* edited by CRESTANA S., GOLDMAN DE CASTRO M., and PEREIRA G. (Saraiva, São Paulo) 1998, pp. 40–49.
45. BUSTAMANTE M. C. and Videira A. A. P. "Bernhard Gross y la física de los rayos cósmicos en el Brasil", *Quipu* **8** (1991) 325–347.
46. MASCARENHAS S. "Bernhard Gross and his Contribution to Physics in Brazil", *Brazilian Journal of Physics* **29** (1999) 217–219.
47. SCHOBER J. and Belisário R. "Marcelo Damy de Souza Santos", *Ciência e Cultura* **55** (2003) 10–12.
48. HAMBURGER A. I. "Mario Schenberg (1914-1990)" in *Cientistas do Brasil - Depoimentos* (Sociedade Brasileira par o Progresso da Ciência, São Paulo) 1998, pp. 89–101.
49. FRANKEN T. and Costa C. "Entrevista com o Professor Mario ù Schenberg" (FGV/CPDOC-Historia Oral, Rio de Janeiro) 1991.
50. NUSSENZVEIG M. *et al.*, "Cesar Lattes. Modéstia, ciência e sabedoria", *Ciência hoje* **19** (1995) 10–22.
51. CARUSO F., MARQUES A. and Troper A. (Editors), *Cesar Lattes, a descoberta do méson π e outras histórias* (CBPF - MCT, Rio de Janeiro) 1999.
52. BELLANDI J. and Pemmaraju A. (Editors), *Topics on Cosmic Rays. 60th anniversary of C. M. G. Lattes* (Editora da Unicamp, Campinas) 1984.
53. MARIANI M. C. "O Instituto de Biofísica da UFRJ" in *Universidades e Instituições Científicas no Rio de Janeiro* edited by SCHWARTZMAN S. (CNPq, Brasília) 1982, pp. 196–208.
54. TYNDALL A. M. *A History of the Department of Physics in Bristol. 1876-1948. With Personal Reminiscences* (1956) in *Histories of Physics in Bristol* edited by CHAMBERS R. G. and HART M., 2005: http://www.phy.bris.ac.uk/history.html.

55. LOCK W. O. "Origins and Early Days of the Bristol School of Cosmic-Ray Physics", *European Journal of Physics* **11** (1990) 193–202.
56. FOSTER B. and Fowler P. H. (Editors), *Forty Years of Particles Physics. Proceedings of the International Conference to Celebrate the* 40th *Anniversary of the Discoveries of the* π*- and V-Particles, held at the University of Bristol, 22–24 July 1987* (Adam Hilger, Bristol) 1988.
57. RIBEIRO DE ANDRADE A. M. "The Socio-Historical Construction of π-Meson" (Museo de Astronomia e Ciências Afins, Rio de Janeiro) 1997.
58. RIBEIRO DE ANDRADE A. M. *Físicos, Mésons e Política: a dinâmica da ciência na sociedade* (Hucitec/MAST, São Paulo) 1999.
59. BROWN L. M. and Hoddeson L. (Editors), *The Birth of Particles Physics* (Cambridge University Press, Cambridge) 1983.
60. POWELL C. F. "Fragments of Autobiography" in *Selected Papers of Cecil Frank Powell* edited by BURHOP E. H. S., LOCK W. O. and MENON M. G. K. (North Holland, Amsterdam) 1972, p. 17.
61. FRANK F. C. and Perkins D. H. "Powell, Cecil Frank", *Biographical Memoirs of Fellows of the Royal Society* **17** (1971) 541–563.
62. POWELL C. F., OCCHIALINI G. P. S., LIVESEY D. L. and Chilton L. V. "A New Photographic Emulsion for the Detection of Fast Charged Particles", *Journal of Scientific Instruments* **23** (1946) 102–106.
63. LATTES C. G. M., MUIRHEAD H., OCCHIALINI G. P. S. and Powell C. F. "Processes Involving Charged Mesons", *Nature* **159** (1947) 694–697.
64. BURCH C. R. "Reflecting Microscope", *Proceedings of the Physical Society* **59** (1947) 41–46.
65. BATES W. J. and Occhialini G. P. S. "Applications of the Reflecting Microscope to the Nuclear Plates Technique", *Nature* **161** (1948) 473.
66. DILWORTH C. C., OCCHIALINI G. P. S. and Payne R. M. "Processing Thick Emulsions for Nuclear Research", *Nature* **162** (1948) 102–103.
67. DILWORTH C. C., OCCHIALINI G. P. S. and Vermaesen L. "On Processing Nuclear Emulsions. Part 1. Concerning Temperature Development", *Bulletin du Centre de Physique Nucléaire de l'Université Libre de Bruxelles* **13a** (1950).
68. BONETTI A., DILWORTH C. C. and Occhialini G. P. S. "On Processing Nuclear Emulsions. Part II. After Development Techniques", *Bulletin du Centre de Physique Nucléaire de l'Université Libre de Bruxelles* **13b** (1951).
69. WOLFENDALE A. W. "History of British Contributions to the Astrophysical Aspects of Cosmic Rays and Gamma Rays since the Second World War", *Quarterly Journal of the Royal Astronomical Society* **29** (1988) 27–37.
70. POWELL C. F. *The Cosmic Radiation. Nobel Lecture, December 11, 1950* in *Nobel Lectures, Physics 1942-1962* (Elsevier, New York) 1964, pp. 144–157.
71. AMALDI E. "Gli anni della ricostruzione", *Giornale di Fisica* **20** (1979) 186–225.
72. GARIBOLDI L. "Constance Charlotte Dilworth", *Il Nuovo Saggiatore* **20**, No. 3-4 (2004) 16–21.
73. GARIBOLDI L. "A Note in Memory of Constance Charlotte Dilworth Occhialini", *Boletim da Sociedade Brasileira da Física* **4** (2004) 1–10.
74. BELLONI L. and Dilworth C. C. "From Little to Big. A Story of a European Postwar Collaboration with Nuclear Emulsions" in *The Restructuring of Physical Sciences in Europe and the United States 1945-1960* edited by DE MARIA M., GRILLI M., and SEBASTIANI F. (World Scientific, Singapore) 1989, pp. 732–744.
75. MEULEMANS G., OCCHIALINI G. P. S. and Vincent A. M. "The Wire Method of Loading Nuclear Emulsions", *Il Nuovo Cimento* **8** (1951) 341–344.
76. BONETTI A. and Occhialini G. P. S. "Cylindrical Emulsions", *Il Nuovo Cimento* **8** (1951) 725–727.
77. DILWORTH C. C., OCCHIALINI G. P. S. and Samuel E. "Eclaircissement des plaques photographiques nucléaires", *Bulletin du Centre de Physique Nucléaire de l'Université Libre de Bruxelles* **2** (1948).
78. TAGLIAFERRI G. "Le scienze esatte all'Università di Milano" in *Storia di Milano*, vol. **18** (Istituto della Enciclopedia Italiana, Roma) 1995, pp. 659–677.

79. BELLONI L. "Giovanni Polvani e l'Istituto di Milano", *Il Nuovo Saggiatore* **4**, No. 3 (1988) 35–49.
80. TAGLIAFERRI G. "Post War Restarting of Cosmic Ray Research at Milan University" in *The Restructuring of Physical Sciences in Europe and the United States 1945-1960* edited by DE MARIA M., GRILLI M. and SEBASTIANI F. (World Scientific, Singapore) 1989, pp. 716–731.
81. BALDO CEOLIN M. "The Discreet Charm of the Nuclear Emulsion Era", *Annual Review of Nuclear and Particle Sciences* **52** (2001) 1–21.
82. DAVIES J. H. *et al.*, "On the Masses and Modes of Decay of Heavy Mesons Produced by Cosmic Radiation", *Il Nuovo Cimento* **2** (1955) 1065-1102.
83. DAVIES J. H. *et al.*, "Observations on Heavy Mesons Secondaries", *Suppl. Nuovo Cimento,* **6**, Ser. X, No. 2 (1956) 398–424.
84. BHOMWIK B. *et al.*, "The Interaction and Decay of K^- Mesons in Photographic Emulsion. Part I. General Characteristics of K^--Interactions and Analysis of Events in which a Charged π-Meson is Emitted", *Il Nuovo Cimento* **13** (1959) 690–729.
85. BHOMWIK B. *et al.*, "The Interaction and Decay of K^- Mesons in Photographic Emulsion Nuclei. Part II. The Emission of Hyperons from K^- at Rest", *Il Nuovo Cimento* **14** (1959) 315–364.
86. EVANS D. *et al.*, "The Interaction and Decay of K^- Mesons in Photographic Emulsion. Part III", *Il Nuovo Cimento* **15** (1960) 873–898.
87. KRIGE J. and Russo A. *A History of the European Space Agency 1958–1987. Volume I. The Story of ESRO and ELDO 1958–1973* (ESA Publications Division, Dordrecht) 2000.
88. KRIGE J., RUSSO A. and Sebesta L. *A History of the European Space Agency 1958–1987. Volume II. The Story of ESA 1973–1987* (ESA Publications Division, Dordrecht) 2000.
89. PINKAU K. "The Early Days of Gamma-Ray Astronomy", *Astronomy and Astrophysics Supplement Series* **120** (1996) 43–47.

The Occhialini-Dilworth Archives

Etra Occhialini and Pasquale Tucci

The Occhialini-Dilworth Archives are formed essentially of two blocks of papers: the first one coming from the Physics Department of the University of Milan and the second one coming from the country house of Occhialini and Dilworth in Marcialla, a small village near Certaldo and Florence.

The papers coming from the Physics Department were in the room occupied by Giuseppe (Beppo) Occhialini until his retirement in 1983. Afterwards the room was given to Giorgio Sironi who took care of Occhialini's papers and arranged them in two large metal cupboard in order to avoid their dispersion. When Costance (Connie) Dilworth retired in 1985 also her papers were moved to Sironi's room.

In 1997 Occhialini's and Dilworth's papers, were transferred, by the good offices of Sironi and Tucci, and the assent of Dilworth, to the Section of the History of Physics of the "Istituto di Fisica Generale Applicata" of the University of Milan, where were deposited other archives of Milanese physicists: Polvani, Succi, Tagliaferri.

Less linear the course of Marcialla's papers. When Beppo Occhialini and Connie Dilworth moved from Milan to their country house in Tuscany, they brought with them everything they had collected during their life—furniture, objects, records, photographs, some paintings, and above all, books, papers and notes. As well as pieces of other lives, Beppo's parents'—Augusto and Etra Occhialini.

Some years ago (after Beppo's death) Connie had ordered—in a chronological sequence—Augusto's papers, separating the personal from the scientific ones.

Etra Occhialini passed away on July 3, 2019. Etra has done a great service to physicists and historians by allowing his parents' vast archive to be made available to the scholarly community.

E. Occhialini
Milano, Italy

P. Tucci (✉)
Prof. of History of Physics (retired in 2013), Università degli Studi di Milano, Milano, Italy
e-mail: ptucci@icloud.com; pasquale1043@gmail.com

L. Gariboldi et al. (eds.), *The Scientific Legacy of Beppo Occhialini*,
https://doi.org/10.1007/978-3-031-37034-2_3

Occhialini's and Dilworth's daughter Etra decided that the papers of her grandfather Augusto Occhialini could be handed over and kept in the institution where he was active. Her choice was therefore to give them to the Arcetri Department of Physics and Astronomy a place that had been in her opinion meaningful for Augusto Occhialini's scientific story.

Different was the situation of Beppo's and Connie's papers. Connie was never able to put order in Beppo's and her own papers. (She once told Etra she didn't feel "ready".) When Connie died, in 2004, Etra found herself with a great quantity of documents—meaning by "documents" not only writings but also photographs, notes, objects—in an apparent (in all senses) disorder, much in the Occhialini style, that she had no capacity to select and order. Not being a physicist—a non regretted decision taken very early—Etra couldn't tell what was relevant and what not from a scientific point of view among those documents.

So she got in touch with Giorgio Sironi and Pasquale Tucci, and asked them to help her. Etra had in fact decided that the best place where to place her parents' records of a lifetime of common work were the university research groups, now splitted between the University of Milan and the University of Milano Bicocca, which continue Beppo and Connie activities. So in summer 2005, Giorgio Sironi from the University of Milano Bicocca and Pasquale Tucci from the University of Milan went to Tuscany and together with Etra gathered and transferred to Milan all there was that could become part in the future of the "Occhialini-Dilworth" Archives. By common agreement, all the collected materials were deposited at the Section of History of Physics of the "Istituto di Fisica Generale Applicata" of the University of Milan, where were deposited the papers coming from the Department of Physics. It would have been, in fact, extremely difficult to create distinct archives separating Beppo's and Connie's private life and scientific activity.

So far under the name "papers" we have included all the records, hand-written and not, left by Beppo and Connie. And that is not casual: it would be, in fact, completely undue to name them "archives" if with this name we mean a set of papers linked by some logic. All that we have until now collected does not seem to have always had an order.

Etra was aware of the amount of work it would take to really build up the archives. It is certainly difficult to give a so-called rational order to the life and work of two persons who were (and were believed to be) extremely rational, but also "peculiar" to the point of appearing "extravagant" when compared to most people. Which is what made them somewhat special. For those who have known Beppo at work it is difficult to imagine him filing his correspondence or his notes.

Agnese Mandrino gave a first order to the papers coming from the Physics Department. She tried to keep connected the papers who seemed to be a programmed archival unit, although sometimes the connection among them could seem rather weak: a little folder in which they were inserted, an envelope with some handwriting and so on. She was aware that what seemed an embryo of classification could be instead the casual result of manipulation suffered by the papers.

Anyway nucleuses of papers which could undoubtedly be considered related by some deliberately given connection, have not been disassembled, as imposed by the archival principles that Agnese Mandrino has always followed closely and meticulously. Papers which seemed completely unlinked have been collected in archival series which follow the main steps of Occhialini's scientific life. At the end of her work the archivist has drawn up an inventory of the papers coming from the Department of physics.

Several problems have arisen from the papers coming from Marcialla: some of them are analogous to those encountered for the material coming from the Physics Department. Other problems are instead completely new. The prevailing view is to insert the papers coming from Marcialla in the archival series created for the papers coming from the Physics Department, if their content allows this kind of insertion. As for the Marcialla papers that are connected with the extant archival series, new series will be created trying to save, however, that little but precious work made by Connie to give an order to some of the papers.

The distinction between private papers of Beppo and Connie and documents dealing with their research activity is very weak and, sometimes, it simply does not exist. This is why the documents in the ordered part of the Archives are open to scholars on request and with the limitations imposed by the delicacy of the topics discussed in some documents, as correctly reported and stressed by the archivist to the responsible of the Archives. The same procedure will be followed with the papers coming from Marcialla the border between private and public being, in this case, even narrower.

Agnese Mandrino, is head of the Library and the Historical Archives of the Brera Astronomical Observatory in Milan. The inventory is available to scholars who consult the Occhialini-Dilworth Archives at the Library of Biology, Computer Science, Chemistry, Physics (BICF) of the University of Milan.

In the meantime, other Occhialini's and Dilworth's papers have come to light. A transcription of them can be found in this volume.

Constance Charlotte Dilworth

Leonardo Gariboldi

Streatham (London) February 5th 1924—Florence May 16th 2004.[1]

A little more than ten years after the death of the great and unforgettable Giuseppe ("Beppo") Paolo Stanislao Occhialini (1907–1993), his wife, Constance ("Connie") Charlotte Dilworth, also an illustrious star of the adventurous story of the "cosmiciens" after World War II, left us. The great scientific discoveries, the charisma and personality of Beppo meant that, rightly, he appeared unquestionably as the leader of the close-knit couple—more generally, of the research groups in which they worked, since the days of Brussels in the distant 1948 – but that wasn't enough to relegate Connie to the shadows, hidden by the figure of her husband. Dilworth's contribution to their research always proved to be essential—in the technique of nuclear emulsions, in the study of cosmic rays, in high energy physics, and in astrophysics—above all when, over the years, Occhialini was increasingly involved in institutional activities.

Constance Charlotte Dilworth was born at Streatham Manor, Leigham Avenue, County London, on 5 February 1924, to George Darwell Dilworth, a merchant, and Charlotte Cant Price. She completed her high school studies at King's College (University of London) from 1941 to 1944, obtaining the B.Sc. in 1944 and the M.Sc. in 1945.

Her research activity began, already as a student, at the Admiralty Research Laboratory in 1943–1945, to continue with postgraduate studies as a researcher at the H. H. Wills Laboratory in Bristol, under the direction of Cecil Frank Powell (1903–1969). Dilworth's first researches concerned solid state physics, in particular the study of the effects on the electric current of the insulating layers on the contact surfaces between

[1] This paper was first published in Italian in: *Il Nuovo Saggiatore* 2004, **20** (3–4): 16–21.

L. Gariboldi (✉)
Leonardo Gariboldi, Dipartimento Di Fisica "Aldo Pontremoli", Università Degli Studi Di Milano, Milan, Italy
e-mail: leonardo.gariboldi@unimi.it

L. Gariboldi et al. (eds.), *The Scientific Legacy of Beppo Occhialini*,
https://doi.org/10.1007/978-3-031-37034-2_4

semiconductor crystals, according to Schottky and Wilson's theories [1]. However, Dilworth's studies were soon turned to research on nuclear emulsions.

In 1938, Powell began his studies on the development of techniques for recording the traces left by elementary particles in photographic emulsions: from the analysis of the traces it was possible to obtain some information on the particles, which would have provided a substantial contribution to the development of nuclear physics studies. The big step forward came with the arrival in Bristol of Occhialini, who had been in England for a few months after the years spent in Brazil, a country from which Beppo had his old collaborators, Cesare Lattes and Ugo Camerini, called to fill the numerical shortages of the Wills Lab. Beppo's collaboration with Waller of Ilford Ltd. allowed the development of the nuclear emulsion technique with a high concentration of silver halides which replaced the previous half-tone emulsions. The technical studies made it possible to obtain emulsions, the nuclear research emulsions, of ever increasing thickness, up to a limit of 2 mm, even if it was then preferred to use the thickness of 600 μm, considered the most compatible with the optical properties. Two fundamental articles on the treatment and uniform development of emulsions date from 1948, which Dilworth wrote with Occhialini, Ron Payne (in Bristol), and Eric Samuel (in Brussels) [2].

In the summer of 1946, Beppo took a few dozen plates to 2867 m on the Pic-du-Midi, in the French Pyrenees, a place he frequented on his speleological holidays with a Belgian physicist, Max Cosyns (1906–1998), who shared his passion for caving. The plates were examined in London and the difference between the traces left by the mesons (at the time, only the μ mesons were known experimentally) and those left by the protons clearly appeared. However, the most important result was the discovery, in the spring of 1947, of the trace of a meson (the π meson), which decayed into another meson (the already known μ meson). The new π meson was identified as the nuclear force meson predicted by Yukawa's theory. The exposure of other plates at 5600 m in Chacaltaya, in the Bolivian Andes, allowed to obtain other traces of the two mesons, confirming the announced result. The analysis of the plates made it possible to determine both the mass and the average lifetime of the π meson which, by decaying in the upper atmosphere, gave rise to a flow of penetrating μ mesons observable at sea level.

A letter from Dilworth to Beppo, dated January 23, 1950, gives us a bit of the atmosphere of those years: "*Dear Beppo, I got back from Bagnères* [a place in the Pyrenees that has marked the history of cosmic ray physics] *yesterday. I had great good luck on the journey back. From Bagnères to Tarbes: no heating on the train. Tarbes to Paris: I had a sleeper 2nd class but the other sleeper was not occupied, so I opened the window and got into bed quick, read the thermometer from time to time, it was always 3–4 °C. Paris to Bruxelles: the train was heated (20° in the compartment), but there was only one other in the compartment and when I explained the problem he gallantly opened the window and sat in his overcoat. This made 10 °C in the compartment. In addition I had the plates sitting on top of a biscuit box full of snow, at least it was full when I left Bagnères and was still half full at Bruxelles. So the plates did not get much fading. They are now in the icebox. I was not completely satisfied that we had got everything out of the plates so we are making two more trials today. The last trials were pretty good anyway, so as soon as we have these dry we can develop the plates. We are trying with a bit higher pH.*"

In May 1948, Occhialini's friendship with Max Cosyns made Occhialini and Dilworth move as free researchers to the Centre du Physique Nucléaire of the Université Libre in Brussels, where Yves Goldschmidt-Clermont, one of the future leaders of the Experimental Physics Division at CERN, had returned, also from Bristol. Thus a research group was formed in Brussels, which continued the study of nuclear emulsions, under the scientific direction of Occhialini, especially of the new NT2 and NT4 nuclear plates by Berrimann (Kodak) and of the G5 by Waller (Ilford) sensitive to the minimum ionization.

On October 4, 1950, Dilworth and Occhialini were married in Ixelles, a union that was then gladdened, the following year, by the birth in Uccle of their daughter, Etra Mary Giovanna Occhialini, known to *cosmiciens* with the nickname of Conniepooh. In the years spent in Brussels, Dilworth's bond with Italian physics grew stronger and stronger. While Occhialini's research activity in the Brussels laboratory was shared with that at the University of Genoa, of which he became a professor in 1949, and, from 1952, with that at the University of Milan, numerous young Italian physicists flocked to Brussels and contributed, upon their return, to the development of research into nuclear emulsions in their universities of origin. We recall, among others, the names of Alberto Bonetti, Michelangelo Merlin, Giulio Cortini, Riccardo Levi-Setti, Giovanna Tomasini and Livio Scarsi. A series of papers that Dilworth wrote on the development of the nuclear emulsion technique together with Occhialini and their collaborators dates back to these years [3].

Among Dilworth's scientific research connected with her scientific period in Brussels there are numerous studies, promoted by Occhialini as scientific director: the Auger effect in the μ mesons capture by the nuclei present in the emulsions, i.e. the emission of slow electrons (Auger electrons) emitted in transitions without emission of radiation in the atomic field associated with the μ meson capture (with Occhialini, Cosyns, Schönberg and Page) [4]; the study of the measurable magnetic deflection of long tracks of charged particles in electron-sensitive nuclear emulsions exposed at various altitudes in magnetic fields of 34,000 gauss to apply the results to high altitude cosmic rays (with Goldschmidt-Clermont, Goldsack and Levy) [5]; research on the nature of cosmic rays and their showers [6].

The creation—in particular in Milan with Occhialini—of a research group dedicated to high energy physics became possible through the establishment in Italy of the National Institute of Nuclear Physics, which allowed the transfer of Dilworth from Brussels to the University of Milan in 1954 as an INFN researcher (1954–1960), of Bonetti and Scarsi from Genoa, and of Levi-Setti from Pavia to Milan, to continue the research school begun in Brussels [7].

The 1950s saw the evolution of large national and European cooperations which allowed the organization of high-altitude flights of emulsion stacks on balloon which, after their recovery, were developed and analyzed by various laboratories leading to the discovery of new particles, including the first "strange" particles [8]. The decision to take part in this new challenge was taken, in Italy, during the International Physics Congress in Como in 1949. The mode of interaction between cosmic rays and the geomagnetic field favoured experiments conducted at low latitudes, so that, from 1952, they began a series of balloon flights from Italy (Naples, Cagliari, the Po Valley) in collaboration with various Italian (Milan, Genoa, Rome, Padua, Turin) and European (Brussels, Bristol, Bern,[2] Göttingen) universities and also with CERN, carrying increasingly large volumes of emulsions: the 1952 flights brought to high altitude just over 1 dm^3 of emulsions to pass to more than 9 dm^3 of emulsions (equal to 37 kg) in the summer of 1953, to reach the famous G-Stack Collaboration of Bristol-Milan-Padua (where G stands for giant) in October 1954 which transported as many as 15 dm^3 (63 kg) of stripped emulsions (without glass support, therefore the particles only passed through a continuum of emulsion layers), for the study of the K meson [9].

Dilworth found herself fully involved in the study of phenomena connected with the capture and disintegration of K mesons, both by cosmic rays and by the Bevatron of the Lawrence Berkeley National Laboratory and by accelerating machines at CERN (the K-Stack), the studies on the τ-θ puzzle (one of the first proofs of parity non conservation) and on the Kμ mesons (the "Camus") problem, the studies on hyperfragments, and the identification of the decay schemes of the Σ hyperon. All these researches, which saw Dilworth as protagonist of the Milan group alongside Occhialini, yielded a substantial number of articles published throughout the decade, above all in *Il Nuovo Cimento* [10]. A particular contribution by Dilworth and Bruno Rossi [11], which stimulated the birth of the G-Stack collaboration, were the observations on three different types of decay (whose abundance strongly depended on the detection technique used) of the K mesons:

$$K\pi_2 \rightarrow \pi + \pi^0$$

equally abundant in emulsions and in the cloud chamber;

$$K\mu_2 \rightarrow \mu + \nu$$

in the cloud chamber and with a mass similar to that of the meson τ;

$$K\mu_3 \rightarrow \mu + ?^0 + ?^0$$

abundant in emulsions, but very rare in the cloud chamber.

We also recall the important human role played by Dilworth alongside Occhialini, as is well expressed by the words of Cormac O'Ceallaigh: "*I will never forget the fever and excitement associated with the effort. Occhialini was in ultimate charge and strode up and down the scene like an avenging Jehovah or thundering Jove, cursing and swearing* [...] *He was just as quick to apologise when the dust had settled. He was ably supported by his wife Connie, another example of a great husband and wife team*" [12].

---

[2] The University of Bern conferred an honorary doctorate on Constance Dilworth in 1954.

In the mid-1950s, a period of uncertainty for the *cosmiciens* thus came. The development of large accelerating machines allowed particle physicists to produce and study particles at gradually increasing energies more comfortably than by organizing balloon flights. The congress of Bagnères-de-Bigorre in 1953, or that of Siena in 1955 (where the results of the G-Stack were exhibited), can be considered as the final event, the "*swan song*" to quote the words of Livio Scarsi, "*for nuclear emulsions as a winning technique for the study of particles: the lack of "timing" and possibility of command constitute a fundamental handicap for exposure to accelerators*" [13]. In fact, the techniques for detecting cosmic rays, such as nuclear emulsions, were used for the study of artificially produced particles substantially until the K-Stack of 1956. While a large part of *cosmiciens* enthusiastically converted to the new situation, for others this period was later considered the end of an era. In hindsight, it was only a sort of interregnum: the successful launch of Sputnik on October 4 1957 was interpreted by the Milan group as the beginning of the relaunch of the study of cosmic rays. In these years of transition, however, an Emulsions Committee was set up at CERN, which proved to be useful in paving the way for other specialized committees. In Dilworth's own words: "*Perhaps the most important legacy of the G-Stack was its demonstration of the feasibility of large inter-group collaborations, thus ensuring the exploitation by the European community of the CERN facilities. The collaborating groups passed soon after to a similar effort with stacks of emulsions exposed to the Berkeley accelerators (the K collaboration). In due course many of these emulsion groups transferred their experience of collaboration and their troupes of scanners to the examination of films from bubble chambers operated at the CERN accelerators, a task which required similar methods of patient exploration and suppression of individual creativity in all but a few senior members*" [14]. Furthermore, the advent of accelerator machines had meanwhile led the Milan group to take an interest in other techniques, such as the bubble chamber and the spark chamber.

The new opening towards cosmic rays began for Dilworth and Occhialini with the invitation of Bruno Rossi to the Massachusetts Institute of Technology, for the 1959–1960 academic year. Dilworth went there as a staff member to work in the Laboratory for Nuclear Science on a program of meteorological and geophysical research on cosmic rays, using rockets and satellites. As a result of this sabbatical, Dilworth published, with Rossi et al., an article on instrumentation for interplanetary plasma research [15].

Upon their return to Milan, Dilworth and Occhialini created a group dedicated to cosmic physics research, a group which, in 1968, took the form of a "Laboratory of Cosmic Physics and Related Technologies", a member of GIFCO (Italian Group of Cosmic Physics). From 1964, Occhialini and Dilworth were involved in the programs [16] of the newly founded ESRO (European Space Research Organization), in particular Connie will be chairperson of the Space Committees. The first collaboration took place with the French group of Jacques Labeyrie of the Center d'Études Atomiques of Saclay (Paris), with the flight of spark chambers (the instrument considered optimal by Occhialini) on balloons for the study of the primary electronic component in cosmic rays [17]. The success of these first studies encouraged the Milan-Saclay collaboration to continue with the launch of instruments, both on balloons and satellites, for the study of different components of cosmic rays: electrons, atmospheric and terrestrial albedo [18] neutrons. In 1968 ESA launched the HEOS-A satellite into orbit for the study of interplanetary magnetic fields and the solar wind, a satellite that was fully operational for 16 months and lost functionality in the early 1970s. HEOS-A was followed by the launch, in 1972, of the TD-1 satellite (Milan-Monaco-Saclay), so named after the Thor-Delta launch base, considered the most ambitious ESRO project (Mimosa project), with on board instrumentation for the study of UV, X, γ rays, heavy nuclei of cosmic rays, and solar radiation [19].

These "first generation" satellites were followed by the design of the ambitious ESRO COS-B project, of the Caravane collaboration (Milan, Munich, Leiden, Saclay, Palermo, Noordwijk), the high energy satellite of ESA (European Space Agency, founded in 1975), with the Gamma Ray Telescope on board, launched in 1975 and which provided the first complete γ map of the Galaxy. In the meantime, the EXOSAT satellite [20], launched in 1983, was being designed for the observation of active galactic nuclei, stellar coronas, cataclysmic variables, white dwarfs, X binaries, galaxy clusters and supernovae remains. Dilworth was always engaged, both in the planning activities and in the study of the results obtained from these satellites, throughout the 70s which saw her producing papers on cosmic rays [21], on solar electrons [22], and on X and γ astronomy [23].

Infrared astronomy also attracted the attention of the Milan group. In particular, Dilworth was one of the coordinators of the construction of the national observatory for infrared astronomy of the Gornergrat, active since 1979, near Zermatt, at 3100 m, equipped with a telescope of the Cassegrain type (the TIRGO). These are Dilworth's last research activities [24] before her definitive retirement, with Occhialini, in July 1985 to the Marcialla hamlet of Certaldo, in Tuscany, where Occhialini had been spending long periods for some time.

In addition to her research activity, Dilworth carried out an intense teaching activity at the University of Milan: from 1957 to 1959, she held a position at the School of Specialization in Atomic and Molecular Physics, from 1960 to 1964, she was appointed professor, extraordinary professor of Radioactivity (since 1964) and full professor of the same subject since 1967, and subsequently full professor of Advanced Physics.

Dilworth's honours include: an honorary degree from the University of Bern in 1954 and the B.A. of the University of London in 1981. Constance Dilworth was a member of the National Institute of Nuclear Physics, of the Italian Physical Society, of the American Geophysical Union, of the Royal Astronomical Society, of the Lombard Institute of Science and Letters.

Constance Charlotte Dilworth (1924–2004)

References

1. C. Dilworth "The Influence of Surface Films on the Electrical Behaviour of Contacts", *Proc. Phys. Soc.*, **60** (1948) 315.
2. C. Dilworth, G. Occhialini, R. Payne "Processing Thick Emulsions for Nuclear Research", *Nature*, **162** (1948) 102; C. Dilworth, G. Occhialini, E. Samuel "Eclaircissement des plaques photographiques nucléaires", *Bull. Centre Phys. Nucl. Univ. Libre Bruxelles* n° 2, August 1948.
3. Among the others: M. Cosyns, C. Dilworth, G. Occhialini "Obturateur thermique pour plaques nucléaires", *Bull. Centre Phys. Nucl. Univ. Libre Bruxelles* n° 6, January 1949; C. Dilworth, G. Occhialini, L. Vermaesen "On Processing Nuclear Emulsions", *Bull. Centre Phys. Nucl. Univ. Libre Bruxelles* n° 13a, February 1950; A. Bonetti, C. Dilworth, G. Occhialini "On Processing Nuclear Emulsions. Part II. After Development Techniques", *Bull. Centre Phys. Nucl. Univ. Libre Bruxelles* n° 13b, March 1951.
4. M. Cosyns *et al.* "The Decay and Capture of \mu-Mesons in Photographic Emulsions", *Proc. Phys. Soc. A*, **62** (1949) 801.
5. C. Dilworth *et al.* "The magnetic deflection of fast charged particles from photographic emulsion", *Philos. Mag.*, **41** (1950) 1032.
6. A. Bonetti, C. Dilworth "Heavy splinters in cosmic ray stars", *Philos. Mag.*, **40** (1949) 585; M. Cosyns *et al.* "Double Stars with Relativistic Particles from Cosmic Rays", *Nature*, **164** (1949) 129; M. Demeur, C. Dilworth, M. Schönberg "A probable example of multiple productions of mesons", *Nuovo Cimento*, **9** (1952) 92; C. Dilworth, S.J. Goldsack "The spectra and sign distributions of particles from cosmic ray stars at 2860 metres altitude", *Nuovo Cimento*, **10** (1953) 926; C. Dilworth, S.J. Goldsack, T.F. Hoang, L. Scarsi "Observations dans les emulsions photographiques de jets de grande et faible ouverture", *Comptes Rendus*, **236** (1953) 1551; C. Dilworth, S.J. Goldsack, D. Hirschberg "Determination of the mass of slow particles by the constant sagitta method", *Nuovo Cimento*, **11** (1954) 113; C. Dilworth, S.J. Goldsack, T.F.

Hoang, L. Scarsi "Remarks on recent analysis of cosmic ray jets", *Nuovo Cimento*, **11** (1954) 4.

7. A. Bonetti, C. Dilworth, M. Ladu, G. Occhialini "Misure in lastre nucleari", *Rendic. Accad. Naz. Lincei*, **17** (1954) 311; C. Dilworth "The Constant Sagitta Method", *Suppl. Nuovo Cimento*, **2** (1954) 203.
8. See the chapter on the discovery of strange particles, in: A. Russo *Le reti dei Fisici* (La Goliardica Pavese, Pavia) 2000.
9. For an autobiographical reconstruction of these events, see: L. Belloni, C. Dilworth "From Little to Big. A Story of a European Postwar Collaboration with Nuclear Emulsions" in: M. De Maria, M. Grilli, F. Sebastiani (eds.) *The Restructuring of Physical Sciences in Europe and the United States 1945–1960* (World Scientific) 1988, p. 732.
10. We recall, in addition to the various reports of the committees on K mesons, some of the main articles which also saw Dilworth as author: C. Dilworth, G. Occhialini, L. Scarsi "Heavy Mesons", *Ann. Rev. Nucl. Sci.*, **4** (1954) 271; M. Di Corato, C. Dilworth, L. Scarsi "An example of a τ^+ and of a $\tau K_e{}^+$", *Suppl. Nuovo Cimento*, **12** (1954) 202; J.H. Davies *et al.* "On the Masses and Modes of Decay of Heavy Mesons Produced by Cosmic Radiation", *Nuovo Cimento*, **2** (1955) 1063; G.L. Bacchella *et al.*, "On the K-value of the Tau Decay", *Nuovo Cimento*, **4** (1956) 1529; G.L. Bacchella *et al.* "Heavy Mesons and Hyperons Giving Rise to Capture Stars", *Suppl. Nuovo Cimento*, **4** (1956) 465; G-Stack Collaboration "Observations on Heavy Mesons Secondaries", *Suppl. Nuovo Cimento*, **1** (1956) 398; M. Baldo-Ceolin *et al.* "Nuclear Interactions of Neutral K-mesons of Long Lifetime", *Nuovo Cimento*, **6** (1957) 130; G.L. Bacchella *et al.* "Étude de l'absorption des mésons K^- au repos dans l'émulsion nucléaire", *Nuovo Cimento*, **8** (1958) 10; M. Baldo-Ceolin *et al.* "Stimulated Decay of the Hyperon in Light Hyperfragments", *Nuovo Cimento*, **7** (1958) 3; M.C. Amerighi *et al.* "On the Observation of π-mesons Emitted in the Interaction in Emulsion of K-mesons", *Nuovo Cimento*, **12** (1959), 91; B. Bhomwik *et al.* "The Interaction and Decay of K^- Mesons in Photographic Emulsion. Part I. General Characteristics of K-Interactions and Analysis of Events in which a Charged π-Meson is Emitted", *Nuovo Cimento*, **13** (1959) 690; B. Bhomwik *et al.* "The Interaction of K^- Mesons with Photographic Emulsion Nuclei. Part II. The Emission of Hyperons from K-interactions at Rest", *Nuovo Cimento*, **14** (1959) 315; D. Evans *et al.* "The Interaction and Decay of K-Mesons in Photographic Emulsion. Part III", *Nuovo Cimento*, **15** (1960) 873; B.D. Jones *et al.* "On the Observation of Fast Σ-Hyperons Emitted from the Interactions of K^--Mesons with Emulsion Nuclei", *Nuovo Cimento*, **19** (1961) 1077.
11. C. Dilworth, B. Rossi "Comparison of Results on K-particles Disintegration at Rest in Cloud Chambers and Photoemulsions", *Suppl. Nuovo Cimento*, **2** (1955) 238.
12. C. O'Ceallaigh "A Contribution to the history", *Comptes Rendus du Colloque International sur l'Histoire de la Physique des Particules, Paris, 21–23 juillet 1982*, p. 185.
13. L. Scarsi "Giuseppe Occhialini: Il secondo periodo italiano (1950-1993)", *Nuovo Cimento C*, **20** (1997) 613.
14. L. Belloni, C. Dilworth op. cit. ref. (9).
15. H. Bridge, C. Dilworth, B. Rossi, F. Scherb, E. Lyon "An Instrument for the Investigation of Interplanetary Plasma", *J. Geophys. Res.*, **65** (1960) 3053.
16. H. Bondi, J. Ostner "ESRO and its programme", *J. Phys. D: Appl. Phys.*, **4** (1970) 1.
17. B. Agrinier *et al.* "Investigation of Primary Electrons at 46° Geomagnetic Latitude; Composition, Origin, and Prehistory", *Proceedings of the 8th International Cosmic Rays Conference*, **3** (1963) 167.
18. G. Boella *et al.* "Analyse des gerbes créés dans une chambres à étincelles à plaque de plomb par des électrons monocinétiques de moment compris entre 0,5 et 1,5 BeV/c", *Compte rendu colloque intern. Électr. Nucl.* OCDE-Paris 1963; C. Dilworth, A. Fedrighini, V. Pelosi, L. Tallone "Analysis of GeV/c Proton Interactions. I. Elastic Scattering", *Proc. Siena Intern. Conf.*, **1** (1963) 591; G. Boella *et al.* "Neutron albedo flux at 45° geomagnetic latitude", *Proceedings of the 8th International Cosmic Rays Conference*, **2** (1963) 26; B. Agrinier *et al.* "Flux of Primary Cosmic Ray Electrons of Rigidity Above 4.5 BV", *Phys. Rev. Lett.*, **13** (1964) 377; B. Agrinier *et al.* "The flux and East-West asymmetry of primary electrons at a geomagnetic rigidity cut-off

of about 4.5 GV", *Proceedings of the 9th International Cosmic Rays Conference, London, 1965*, Vol. **1** (The Institute of Physics and The Physical Society, London) 1966, p. 331; G. Boella *et al.* "Measurement of the Cosmic Ray Neutron Flux at 4.6 Billion Volts Geomagnetic Cut-off Rigidity", *J. Geophys. Res.*, **70** (1965) 1019; G. Boella *et al.* "The atmospheric and leakage flux of neutrons produced in the atmosphere by cosmic ray interactions", *Earth & Plan. Sci. Lett.*, **4** (1968) 393; C. Dilworth, L. Scarsi, G. Sironi, A. Russo "East-West Asymmetry and Charge Sign Ratio of Primary Cosmic-Ray Electrons at 8.3 GV Rigidity Cut-Off", *Lett. Nuovo Cimento*, **1** (1969) 1.
19. C. Dilworth, G. Occhialini "Satelliti scientifici europei per astrofisica stellare", *S&T 73. Annuario della EST*, **6** (1973) 127.
20. C. Dilworth "Introduzione storica al progetto EXOSAT (HELOS)", *Mem. Soc. Astr. It.*, **44** (1973) 481.
21. C. Dilworth *et al.* "Composition and Spectra of Primary Cosmic-Ray Electrons and Nuclei above 10^{10} eV: Discussion", *Philos. Trans. R. Soc. London, Ser. A*, **277** (1975) 360.
22. C. Bland *et al.* "High Energy Solar Electrons Intercorrelated Satellite Observations", *Proceedings of the 3rd ESLAB/ESRIN Symposium Noordwijk, September 16–19, 1969*, Vol. **19** (Kulwer, Dordrecht) 1970; Y. Koechlin *et al.* "High energy electrons in solar flares", *Proceedings of the 11th International Conference on Cosmic Rays, Budapest, 25 Aug. - 4 Sept. 1969 – Acta Physica*, **2** (1970) 663; C. Bland *et al.* "Observations on electrons in the energy range 10–300 MeV", *Proceedings of the 11th International Cosmic Rays Conference, Budapest, 25 Aug. - 4 Sept. 1969 – Acta Physica*, **1** (1970) 195; C. Dilworth *et al.* "High Energy Electrons Detected during Solar Flares", *Solar Physics*, **23** (1972) 487.
23. C. Dilworth, L. Maraschi, C. Reina "On the Luminosity Function of Galactic X-ray Sources", *Astron. Astrophys.*, **28** (1972) 71; C. Dilworth, L. Maraschi, G. Perola "Galactic Positrons and γ-rays, and the Cosmic-ray Mean Path-length in the Galaxy", *Astron. Astrophys.*, **33** (1974) 43; C. Dilworth "The Interstellar Medium in X- and Gamma Ray Astronomy", *Proc. NATO ASI, Schliersee, Germany, April 2–13, 1973*, (1974) 69; C. Dilworth, L. Maraschi, G. Perola, E. Tanzi "A spherical model for the transient X-ray source A0620–00", *Mon. Not. R. Astron. Soc.*, **181** (1977) 339.
24. B. Cabli *et al.* "Preliminary results of measurements of atmospheric emission and absorption at infrared wave-length at the Gornergrat", *Mem. Soc. Astr. It.*, **47** (1976) 101; O. Citterio, C. Dilworth, N. Iucci "New Italian infrared telescope", *Sky & Tel.*, **62** (1981) 17.

Giuseppe Occhialini

Translated by Leonardo Gariboldi

Fossombrone (Pesaro), December 5, 1907.[1]

Experimental physicist, he made important contributions to the physics of cosmic rays and elementary particles by means of techniques and tools developed or perfected by him.

Following his father Augusto in his first university locations, he spent his childhood years first in Pisa and then in Florence. The environment created by his father's personality, culture and rigor was fundamental for the choice of his work addresses and for his moral and political formation.

He attended his secondary studies at the Liceo Scientifico in Florence and his university studies at the same location, where he graduated in physics in 1929, becoming a voluntary assistant and, subsequently, a permanent assistant at the Physics Institute in Arcetri directed by Antonio Garbasso.

The formative period in Arcetri represents, from 1927 to 1931, one of the three important stages in his existence; the other two are the periods at the Cavendish Laboratory in Cambridge and the period at the Wills Laboratory in Bristol.

The Physics degree was newly established; the activity of the physics laboratory was supported by the faith of its animator Garbasso. The view offered by the windows made up for the poor equipment, the lack of functionality of its convent structure and the difficulty of access.

The presence of E. Persico, professor of rational mechanics and theoretical physics, and the arrival in Florence, fresh from their degrees, of G. Bernardini from Pisa and B. B. Rossi from Bologna, as assistants, made possible the formation of a

[1] This paper was first published in Italian as an autobiographic entry in: *EST. Scienziati e Tecnologi Contemporanei*, vol. II (Milano: Mondadori, 1974): 322–324.

L. Gariboldi (✉)
Dipartimento Di Fisica "Aldo Pontremoli", Università Degli Studi Di Milano, Milan, Italy
e-mail: leonardo.gariboldi@unimi.it

L. Gariboldi et al. (eds.), *The Scientific Legacy of Beppo Occhialini*,
https://doi.org/10.1007/978-3-031-37034-2_5

group of enthusiastic young physicists. The laboratory's interest shifted from spectroscopy towards nuclear physics and cosmic rays. Thus was born around 1927–28 the school of Arcetri.

The inclusion of the laboratory in the new line of research was provoked by the publication of W. Bothe and H. Kohlhörster's work on the application of Bothe and Geiger's method of coincidences to the study of cosmic rays. Rossi grasped the consequences of this event and invented the multiple recording method, which today bears his name and with which he began the study of cosmic rays. A period of improvement at Bothe's allowed him to import into Florence a set of very important techniques for those times. The possibility was thus opened up of undertaking significant research even by an institute without great resources, because the source of the radiation came from the sky and the devices, the Geiger and Müller tube counters, were of simple construction.

On the other hand, two initiatives contributed to the cultural and technical formation of the group: the Physics and Astrophysics Seminar and the collegial reading of magazines. The first, founded by G. Abetti and which found its driving force in Bernardini, favouring contact with both Italian and foreign scientists, inserted the Faculty of Sciences of the University of Florence into the European environment. The second, promoted and directed by E. Persico, engaged all the members of the small community, which in addition to Bernardini and Rossi brought together D. Bocciarelli, A. Colacevich, L. Emo, G. Racah and G. Righini, to keep up to date of the contents of the magazines of a very rich library.

In 1931, following the assignment of a grant from the CNR [National Council of Researches], Occhialini was sent by the Institute to Cambridge to the Cavendish Laboratory, with the aim of learning the technique of Wilson's chambers from P. M. S. Blackett in order to introduce it in Florence.

The arrival in Cambridge made Occhialini discover a new world. The Cavendish was in fact a laboratory that was ahead of its time due to its modernity: scientists from many countries flocked around E. Rutherford (1871–1937) and J. Chadwick and alongside traditional researchers of purely academic training there were more applicative elements: among these were the specialists in electrotechnics and the 'tube makers' as electronic physicists were called at that time, who anticipated the structures that would ensure the subsequent developments of physics.

Among the ensemble of personalities that populated the laboratory, Blackett's stood out. A complete and classy physicist, he possesses a lively and critical political and social conscience, with a profoundly human sense of justice and an intellectual passion not limited to the field of physics. Blackett, who together with his father Augusto exercised the greatest influence on Occhialini, is one of those rare people to whom the quality of greatness can be attributed without hesitation.

Occhialini's stay in England was scheduled for three months; instead it extended for three years. During this period Blackett and Occhialini combined Wilson's chamber technique with that of coincidence counters. This led to the construction and use of the controlled Wilson chamber.

In February 1933 they published the paper with which, through the observation of the showers of positive and negative electrons produced by cosmic rays, the existence of C. D. Anderson's positive electron was confirmed and its nature was interpreted by linking it to P. A. M. Dirac's recent relativistic theory of the electron. They further proposed that the recently discovered phenomenon of anomalous absorption of high-energy γ rays was related to the creation of positive and negative electron pairs and that the scattered radiation associated with this effect was produced by their subsequent annihilation. Today the latter concept has entered the general cultural heritage at the middle school level. In February 1933 it was being accepted with some reservations not only by some theorists but also by some experimentalists. However, in the space of a few months the evidence of the production by γ rays of the new particles in the laboratory was provided by various groups, among others by Blackett, Chadwick and Occhialini. With this activity the first long stay in England ended.

Upon his return to Florence in 1934 he found the laboratory profoundly changed: Garbasso had died prematurely, Persico had moved, Rossi had become a professor in Padua. Furthermore, the political atmosphere, already difficult, had become unbearable.

In 1937 he received an invitation from G. Wataghin, the promoter of physics in that country, to move to Brazil. Occhialini settled there for several years, initially working in the Physics Institute of the Faculty of Science in São Paulo. In a modest laboratory, populated by a group of young students of Wataghin including M. De Sousa Santos, M. Schömberg, P. A. Pompeia, A. de Moraes and later U. Camerini and C. Lattes, he prepared an experience for the observation of large cosmic ray showers with Wilson chambers and counters, which was interrupted by the war. At the declaration of war, unable to take an open position for fear of reprisals on his father in Italy, he was excluded from the University of São Paulo and withdrew to isolation in the Itatiaya group of mountains.

After the September 1943 armistice, while awaiting a return to Europe, he was housed in the Biophysics Laboratory of the Faculty of Medicine of Rio de Janeiro, directed by C. Chagas.

He moved to England at the end of 1944; after an unsuccessful attempt to enter war research he went, at Blackett's suggestion, to the Wills Laboratory in Bristol, where Powell with little means had highlighted the power of the photographic method, showing how in the field of nuclear physics the possibilities of detecting charged particles and measuring energy by measuring the length of traces in emulsion were comparable and even superior to those offered by the Wilson chamber.

Occhialini arrived in Bristol with the intention of staying there for a few weeks, to learn this technique and to examine the possibility of applying it to research on cosmic rays.

Although the laboratory's resources and the state of the art with Ilford half-tone plates were not adequate at the time for research on cosmic rays, he decided that there was ample room for improvement and thus accepted the invitation of A. M. Tyndall, director of the laboratory, to make his stay permanent. So Occhialini and Powell, both excluded from the war effort, one an expatriate and the other frustrated by a lack of professional recognition, set to work, establishing a close and affectionate bond.

In the following months, improving the detection techniques and through the invention of concentrated plates carried out together with Powell and with Waller of Ilford, he showed how the nuclear emulsion was capable of great versatility, efficiency and precision. An expansion of laboratory funds followed, with a consequent increase in microscopes and microscope technicians. The lack of English physicists (still engaged in war research) led him to invite Camerini and Lattes from Brazil.

In the spring of 1946 the ground was ripe for research on cosmic rays, but the entire laboratory was still tied to the application of the new device to low-energy nuclear physics until the results of the first exposure to cosmic rays, carried out with the help of his friend Max Cosyns who, during a speleological campaign in the Low Pyrenees in August 1946, made it possible to transport a group of Ilford C2 plates to the Pic-du-Midi.

The wealth of information contained in these plates convinced Powell to move practically all the microscopist observers to their study. With these plates Lattes, Muirhead, Occhialini and Powell demonstrated the existence of the π meson and its decay into the μ meson.

In 1948 Cosyns invited Occhialini to the Free University of Brussels, where the young Goldschmidt-Clermont had returned from Bristol; and together with him he invited C. Dilworth (later C. Dilworth Occhialini) with whom in recent months in Bristol Occhialini had perfected the development-fixing treatment technique of the new emulsions so as to allow the use of emulsions up to 2 mm thick, and had begun the study of the new Kodak emulsions.

The scientific activity of the Brussels laboratory was concentrated on these techniques brought from Bristol and on the use of the new NT2 and NT4 nuclear plates by Berriman from Kodak, and later on of the G5 by Waller from Ilford, sensitive to the minimum ionization. Thus a group was formed in Brussels for the study of cosmic rays with the technique of nuclear emulsions of which Occhialini exercised the scientific direction. The group became a centre of attraction for physicists interested in the new techniques, including theoretical physicists such as M. Schömberg and G. Molière, experimental physicists such as G. F. Houtermans, together with a group of young people including A. Bonetti, G. Cortini, R Levi-Setti, M. Merlin, L. Scarsi, G. Tomasini, who came to learn the new techniques.

In 1950 Occhialini won a competition for a professorship and was called to Genoa, to the professorship vacated by his father due to age limits. While continuing his activity in Brussels, he formed, with the help of Bonetti, a small research group in Genoa. After his father's death in 1952 he moved to Milan where there was a thriving research business with Wilson chambers. After a six-month interlude as a UNESCO appointee to collaborate with Lattes and Camerini in organizing the Physical Research Centre of Rio de Janeiro and its activity in Chacaltaya, the formation of a cosmic ray research group began in Milan with the technique of nuclear emulsions. The activity of this group expanded rapidly thanks to the creation in Italy of the National Institute of Nuclear Physics. In 1954 his wife and daughter Etra joined him in Milan.

The effort of the organizational and scientific management of the research groups of Brussels and Genoa and then of Milan found an answer in the scientific production which was partly placed at an international level: capture of mesons and pions in the elements of the nuclear emulsion; measurements of very high energy particles with the multiple scattering method and with emulsion curvature; identification of the decay patterns of the positive hyperon and contributions to the study of hyperfragments. A part of these results was obtained through participation in international collaborations. The study of elementary particles through cosmic rays ended in 1954 with the exposure, at high altitude, of the G-Stack, promoted by Merlin and Powell, which allowed the study of the systematics of the decays of the K^+ meson. In that same year, the advent of large accelerating machines brought experimental interest back to the laboratory level and prompted the development of new techniques.

After a period of analysis of artificial mesons which, in the context of a European collaboration, highlighted the merits and limits of the use of emulsions in the study of interactions with complex nuclei in which both neutral and charged unstable particles were emitted, the group led by Occhialini shifted their activity towards the bubble chamber technique, participating in the expositions at the CERN beams.

The Occhialini spouses resumed their activity in the field of cosmic rays in 1960, when Rossi invited them to the Massachusetts Institute of Technology for a year to learn the techniques of space research. Thus, upon their return to Milan, an intense participation in the programs of the European Space Research Organization (ESRO) developed.

Occhialini was awarded the Sella, Vallauri, C. V. Boys, Einaudi, Feltrinelli prizes and an honorary doctorate from the Universities of Bristol and Brussels.

Giuseppe Occhialini

EST entries to refer to: Cosmic radiation; Nuclear plates; Meson; Cloud chamber; Elementary particles.

Also, in S&T 73: Dilworth-Occhialini C., Occhialini G., *Satelliti scientifici europei per astrofisica stellare.*

G. Occhialini (left), with G. Polvani, C. Salvetti and P. Caldirola on the occasion of a degree exam session held at the University of Milan in 1957. He trained in Arcetri, at the Cavendish Laboratory in Cambridge and at Wills Laboratory in Bristol (courtesy of I. Polvani)

The Arcetri School of Physics

Alberto Bonetti and Massimo Mazzoni

Introduction

The years between the first and the second World War (broadly from 1920 to 1940) are remarkable in Italy for the achievements attained in physical research. This was because of rather peculiar circumstances which made the Physical Institutes of the Universities of Rome and Florence the centre of advanced research and of formation of research leaders. Both groups originated through the dedication and the vision of enlightened men, Orso Mario Corbino in Rome and Antonio Garbasso in Florence, both good physicists open to the extraordinary discoveries of the years before and after the first World War, both sincere patriots willing to give their country a sound and up-to-date scientific culture.

The making and performance of the Group of Rome received wide attention and recognition in years due to the personality of Enrico Fermi in spite of the death of Corbino in 1937. The Group of Florence did not receive the same recognition, presumably because of the early death of Garbasso in 1933 and the quick dispersal of its members thereafter. Both groups were heavily hit by the racist campaign sparked off by fascism and culminated in the shameful racial laws of 1938.

In fact, the two groups were not formed in a desert. The tradition of scientific research in Italy gave remarkable results in time after Galileo, with such names as Torricelli, Spallanzani, Volta, Lagrange, Avogadro. But the Restoration after the French Revolution and the Napoleonic adventure, not only reinstated, in the first half of 19th century, the previous situation of political fragmentation in several small traditionalist principalities, among which the State of the Holy Seat, but also revived a

A. Bonetti (✉)
Università di Firenze, Florence, Italy
e-mail: albertob1920@libero.it

M. Mazzoni
Dipartimento di Astronomia e Scienza dello Spazio, Università di Firenze, Florence, Italy
e-mail: mazzoni@arcetri.astro.it

L. Gariboldi et al. (eds.), *The Scientific Legacy of Beppo Occhialini*,
https://doi.org/10.1007/978-3-031-37034-2_6

reactionary attitude towards the heritage of Enlightenment. Fragmentation and conformity did not affect the development of mathematical research, mainly because it did not need financial investments, but were detrimental to the development of "natural" sciences of dimension and openness comparable with contemporary European, i.e. English, French and German, research and capable of becoming an active part of it. Restricting the interest to Physics, the Galilean tradition was maintained almost only in the sense of a careful experimentalism with good measurements often based on original instruments, more devoted to the discovery and description of peculiar "effects" than to the assessment of theoretical developments. Achievements recognized at the European level were those attained by Leopoldo Nobili (the thermocouple), Macedonio Melloni (the infrared radiation), Ottaviano F. Mossotti (the structure of dielectrics), Giovanbattista Amici (the immersion objective), Father Angelo Secchi (stellar spectroscopy and the dawn of Astrophysics). No less important the contributions to "applied" Physics, such as the telephone of Antonio Meucci, the dynamo of Antonio Pacinotti, the rotary magnetic field of Galileo Ferraris. It is a sign of the weakness of scientific and technical consciousness, as well as of economic structure, that neither the inventors nor the dawning Italian industry took a direct advantage of such results. On the other hand, by now we are well into the second half of the 19th century, after the independence wars and the political unification (1849–1870); now it is the stage of the greatest efforts for the integration and modernisation of the country and for the recognition of Italy at an international level. Many scientists, in particular mathematicians and physicists, were very active politically, taking part also in military actions and getting involved in governmental duties. It is worthwhile mentioning Carlo Matteucci (1811–1868), active in Pisa, interested in Florence, a good physicist founder with chemist Raffaele Piria of the journal *Il Nuovo Cimento*. He was Minister of Public Education in 1862–63, just after the proclamation of the almost unified Italian Kingdom, and made a first attempt for a structural reform and modernisation of the system of Italian Universities, too many and generally too weak as a consequence of the already pointed out localism. The Physical Institutes in particular were generally understaffed and poorly equipped also because of the prevailing petty humanistic culture of the ruling class, to the detriment of a more open attitude in consonance with the rest of European culture.

Physical Research in Italy From the End of the 19th Century to the Outbreak of First World War

It was a time of exceptional flourishing of Physics all over Europe, and also in the United States, from electromagnetism, spectroscopy and statistical mechanics to radioactivity, relativity and atomic structure. At the same time mathematical research was going on along the path initiated in the 18th and 19th century, contributing heavily to the building of what was to become the modern Theoretical Physics. The Italian mathematicians were well on the front line of this path, with Enrico Betti and

Luigi Bianchi in Pisa, Giuseppe Peano in Turin, Gregorio Ricci Curbastro in Padua, Tullio Levi Civita e Vito Volterra in Rome. Not equally impressive the contributions of the contemporary Italian physicists, still bound to the experimentalist attitude inherited from their predecessors of the 19th century. While important results were obtained, among others, in electromagnetism and later in spectroscopy, some of them refused stubbornly Einsteinian relativity in spite of the position of their mathematical colleagues, and only a few caught the importance of the Rutherford-Bohr atomic model. The development of research and teaching in Physics from the end of the 19th century to the outbreak of the first world war, can be outlined through the story of four Universities, Bologna, Rome, Pisa and Florence.

Bologna

Bologna was dominated by the personality of Augusto Righi (1850–1920), perhaps the most prominent Italian physicist before the 1st World War. He is better known for his elegant experiments in the wake of H. R. Hertz, proving the identity of electromagnetic oscillations of any frequency and light, but his ingenuity and thoroughness were present in all the subjects he treated, including the methodological approach to physical research [1]. This gave him recognition at the European level and power in improving the facilities of his Physical Institute. However he remained in doubt about relativity, characteristically lamenting the lack of a sound "laboratory" experimental basis.[1] Perhaps only his death in 1920 prevented Righi from elaborating the successes of the new theoretical (and experimental!) Physics. His equally doubtful successor Quirino Majorana (1871–1957), a good experimentalist in the old tradition, to the end of his life made use of the good equipment of the laboratory to carry out carefully designed experiments aimed at falsifying the results of Michelson and Morley. Of course, those experiments kept confirming the constancy of the velocity of light irrespective of the frame of reference.

As a matter of fact, the Physical Institute of Bologna did not contribute to the formation of the schools of Rome and Florence (with the exception of Bruno Rossi, but this occurred through the initiative of Rita Brunetti, of Pisan and Florentine origin).

Rome

Modern Physics in Rome begins with Pietro Blaserna (1836–1918). Born in Friuli under Austro-Hungarian administration, he completed his education in Physics at the

[1] Even after the Eddington's expedition, in 1920, he wrote to the French physicist Violle: "Apres la brillante confirmation que l'éclypse de Mai a donné à la théorie d'Einstein, il est juste que des epreuves sûres en faveur soient fournies même par les expériences de laboratoire" [1].

University of Wien and then in Paris with H.-V. Regnault. Called by Carlo Matteucci in 1862 (just one year after the first step of the unification of Italy) as teacher at the Museo di Fisica e Scienza Naturale of Florence, he became in 1863 Chair Professor of Physics at the University of Palermo. In 1872 he was called by the newly established University of Rome to join E. Keller in the establishment of a “Scuola Pratica di Fisica” in recognition of his contributions to electromagnetic induction and to the dynamic theory of gases. Interested in Terrestrial Physics, he was president from 1879 to 1907 of the Consiglio di Meteorologia e Geodinamica. In 1881 he founded the Physical Institute of the University in via Panisperna and there he called (1908) Orso Mario Corbino (1876–1937), who was then Chair Professor in Messina after spending several years in Palermo with remarkable achievements in various fields (magneto-optics and the Macaluso-Corbino effect). In Rome Corbino continued his successful scientific career (photoelasticity and the effect of Volterra distortions; specific heat in high-temperature metals; improvements in X-ray generators), also with the collaboration of young Giulio Cesare Trabacchi (1884–1959) who was to become director of the Physics Laboratory at the Istituto Superiore di Sanità in 1922 (Trabacchi had an important part in the development of Nuclear Physics in Rome).

Like several other prominent colleagues, Corbino was deeply involved in the First World War. The war caused a violent stirring of emotions, being perceived by most Italians as the way to the completion of national unification; also, it stimulated initiatives to overcome the weakness of the economic and industrial structure of the country, disclosed by the war needs. The very first outcome was the 1915 Committee, made mainly of industrials and scientists. A further initiative (see later) was the creation of an Office for Research and Inventions, attached to the Under-Secretariat for Weapons and Ammunitions (Ufficio Invenzioni e Ricerca, UIR): this government support made the difference, because, while the first Committee gave scarce results, the Office, directed by mathematician and physicist Vito Volterra, was the first step toward the foundation (1923) of the National Research Council (CNR). Corbino with other colleagues had an important part in it, as well as in other bodies created with the aim of developing the interaction of the scientific and industrial world also beyond the needs of the war [2]. Corbino initiated an intense public life in which he displayed both at the governmental and at industrial level his technical preparation, his leadership and his broadmindedness. He became senator shortly after the war and briefly Minister of Education, the first scientist after Matteucci. He was Minister of Economy for a few months at the beginning of the fascist government in a particularly troubled political period. He was able to keep his authoritative stand in the industrial and scientific world without becoming a member of the fascist party. The public life did not prevent him from continuing his scientific activity and above all from building what was to become the Group and School of Physics of via Panisperna.

Pisa

Pisa is peculiar because of the existence of the Scuola Normale Superiore next to a University of old tradition (13th century). Founded by Napoleon in 1810 as the core of his "Italian program" of reform of knowledge, the Scuola Normale followed the model of its twin Ecole Normale Superieure in Paris, with the same scope of formation of high-level secondary-school teachers; in fact, since its origin it became a school intended for the preparation of a selected cultural elite. Teachers at the SNS were specially appointed lecturers, often from the University of Pisa. This was the case in the first thirty years of the 20th century, when directors were the mathematicians Ulisse Dini and Luigi Bianchi, while Physics teachers were Angelo Battelli and Vito Volterra.[2] Several personalities who appear in the following were "normalisti".

As for the University, Angelo Battelli (1862–1916) succeeded in 1893 to Riccardo Felici (1819–1902, renovator after Carlo Matteucci of the Studio di Pisa and well known for his contribution to the interpretation of electromagnetic induction). Battelli, originating from le Marche, "laureato"[3] in Turin, 1884, was briefly Chair Professor in Cagliari and Padua, showing from the beginning his taste and ability in rigorous experimentalism (thermal properties of vapours, Peltier effect, thermoelectricity). In Pisa, he founded the Italian Physical Society, revived successfully the journal *Il Nuovo Cimento* and rebuilt and re-equipped the Physical Institute, receiving and stimulating a number of researchers and pupils, whom he would involve directly in the design and running of experiments. He was against specialisation, perhaps to the detriment of coherence in his projects, but he was ready to open his mind and his activity to the more recent results (gas discharge, cathode rays, X-rays, radioactivity), which were leading to the experimental and theoretical approach to the structure of matter beyond the limits of the chemical atom. This interest of Battelli, and also the method, is well presented in the treaty on Radioactivity (1909), written with his pupils and co-workers R. A. Occhialini (see below) and S. Chella and translated (1910) in German and French.[4]

Battelli had little time left for elaborating the next results of Rutherford (1911) and Bohr (1913) on the planetary atom: a fatal illness in the last years of his life brought him to a premature death in January 1916, only 54. Furthermore, in those years (as in

[2] With the reform of Giovanni Gentile, 1928, the Scuola added explicitly the further mission of promoting the scientific and literary national culture, with special postgraduate courses open to graduated from all over Italy and since 2002 from all over the world [3].

[3] At that time the Italian "laurea" was based traditionally on a (minimum) 4-year curriculum (this is the case for Physics and Mathematics) and on a written thesis preferably on an original subject. In the period discussed here the theses in Physics were typically experimental. The Italian words "laurea" and "laureato" will be used in the following.

[4] See Gamba [4]. Together with a meticulous review of the experimental results updated to 1909, none of them original, but many of which carefully replicated in the laboratory, the book contains a critical presentation of the current dubious models of atomic structure in the light of an electric theory of matter—a remarkable behaviour for a researcher strictly devoted to the empirical basis of Physics. Notably, no reference is made to the work of Einstein, presumably because of a supposed absence of an experimental basis.

previous years) he was deeply engaged in public life, becoming repeatedly a Member of Parliament with particular interest in the school system. Like so many Italian fellow-scientists in the 19th century, Battelli was an active patriot: a few days after the engagement of Italy in the first World War, he wrote to a leading newspaper an open letter, July 11, 1915, urging the Government to take immediate steps for the formation of a body, a "Scientists Section", where scientists would put their expertise in the selection and production of ideas and inventions useful for the war effort (similar bodies were already active in Germany, France and Great Britain) [5]. A "National Committee of Inventions" was promptly formed with the active participation of pain stricken Battelli and others. Alongside with other similar initiatives, this led to the creation in 1917 of the Ufficio Invenzioni e Ricerca, mentioned in the above.

Luigi Puccianti (1875–1952), born in Pisa, succeeded Battelli in the direction of the Institute in 1917. An enthusiastic pupil of his, "laureato" in 1898 with a thesis on the absorption of near-infrared light in a large sample of organic liquids: this was actually a first observation of vibrational spectra of molecules. From 1900 to 1915 Puccianti was in Florence, first in the position of assistant and "aiuto" (aide) to Antonio Ròiti at the Istituto di Studi Superiori (see later), then keeping his activities in that Institute while acting, with a better salary, as professor of Physics at the Istituto Superiore di Magistero Femminile. Chair Professor in 1915, first briefly in Genoa and Turin, after two years he was back in Pisa to the end of his life. His scientific contributions are in electromagnetism and, more importantly, in spectroscopy, where he shares with Antonio Garbasso the merit of the rebirth of spectroscopy in Italian Physics (see later). He was a good and dedicated teacher and had the chance of being the director of a well organised institute with a good mathematical school nearby.[5] Differently from Battelli, Corbino and Garbasso, Puccianti did not engage in political and administrative life.

Florence

For centuries, Florence did not have a University, although there was intermittently a Studio opened since the 14th century [8]. This was conceived as a place for "natural" (according to the meaning of the time) investigations and was housed from the end of the 16th century in the Uffizi as "Gabinetto delle Matematiche". During the House of Lorraine grand duchy, this became in 1775 the Museo di Fisica e Storia Naturale, well equipped with instruments and collections and housed nearby Palazzo Pitti. In spite of the label Museum, it was intended to be also a laboratory for Physics experiments. A few years later, a small astronomical observatory (later known as la Specola) and a room for meteorological measurements were added. It is worth to stress that

[5] Perhaps Puccianti was not active enough to fill the gap between the traditional experimentalist culture and the culture of the "new" Physics which was stimulating the interest of eager young people. He was however generous and broadminded, to the point of asking the still student Enrico Fermi "to teach him something" of the new Physics "which he might still learn": see [6, 7].

shortly after the death of its first director, Felice Fontana (1730–1805), a very early professorial chair in Astronomy was instituted, during the period of the Napoleonic domination of Florence. In 1859 the provisional government, installed in Tuscany after the expulsion of the Lorenese family, gave rise to the Istituto di Studi Superiori Pratici e di Perfezionamento according to the plans of Carlo Matteucci for a kind of super-university concentrating high-level competence and adequate financial means, with emphasis on observational and experimental activities. The Museo di Storia Naturale became a part and the basis of the "Sezione di Scienze Fisiche e Naturali" of the Istituto, strongly oriented towards experimental research and education. The local rivalries and the financial difficulties of the newly born Italian State frustrated the project, not so much in the chemical and naturalistic section as in the physical section, in spite of the support of Matteucci. The situation worsened after his death (1868). However, the meteorological and geophysical observatory kept on working [9] under the Direction of, among others, Antonio Ròiti, Antonino Lo Surdo and then Antonio Garbasso (see later). In 1876 the Istituto was made "equivalent" to Italian universities, with the possibility of offering "laurea" theses of experimental kind but without the structure of a regular faculty (and classed B, i.e. mainly supported by local financial contributions, a condition whose consequences were felt in years, discouraging teachers at Chair Professor level to remain for long). In spite of that, in 1880 Antonio Ròiti accepted the offer of a chair professorship.[6] Since he was a respected scientist, he obtained quickly an "aiuto" and an assistant, and increased and updated the equipment of the Physical Institute, still named Sezione di Fisica. In spite of the absence of a regular "corso di laurea" and of the obsolescence of the seat, Ròiti was able to attract in Florence several eager young elements in the position of "aiuti" and assistants: among them Luigi Pasqualini, Luigi Puccianti, Antonino Lo Surdo, naming only the ones who are directly involved in the story of Florence Physics. When Ròiti retired in 1913, keeping for himself only the position of co-director of *Il Nuovo Cimento*, his place was taken by Antonio Garbasso.

Garbasso (1871–1933) was a remarkable mix of a naturalist scientist and a humanist: born in Piedmont, he changed into an enthusiastic Florentine, extending his patriotism towards united Italy into a kind of worship for the adopted Tuscan homeland. To a large extent, this was due to that deeply appreciated "natural and positive" approach to reality as distinctive of the "flower of the Latin culture, namely the Tuscan thought" [10]. "Laureato" in Turin, 1892, with a good physicist and teacher, Andrea Naccari, he completed his scientific preparation in Bonn with Hertz and in Berlin with Helmholtz and initiated his interesting scientific activities working on the optical properties of electromagnetic waves. After teaching appointments in Turin,

[6] With an honourable record as companion of Garibaldi in the 1866 war against the Austrian Empire, Antonio Ròiti (1843–1921), native of Ferrara, got the laurea in Pisa (1868) with Felici and was also a "normalista". He taught at secondary school level (at the time a non-diminutive position for many young scientists!) in Leghorn and Florence and, 1878, was Chair Professor in Palermo with a good recognition as a careful experimenter, gaining him an authoritative membership in the International Commission for Electric Standards. In Florence he took a particular commitment in teaching, producing a successful text, *Elementi di Fisica*, comparable with recognised texts at the European level.

in Pisa with Battelli (working with him on X-rays) and again in Turin, in 1902 won two competitions, one for Mathematical Physics in Pisa, the other for Experimental Physics in Genoa. This latter was his choice and there he remained for 10 years, continuing his research in electrodynamics and spectroscopy (in full development at the time in Europe, and about which he published a treatise in German). His research and teaching method was based characteristically on the association of the mathematical treatment of problems and the accurate experimental verification of results[7] but he did not refrain from proposing "analogic" models, as in the case of his "electromagnetic model" of atomic structure intended to explain line spectra.[8] Apart the limits of his model, Garbasso was ready to appreciate the new field of quantum spectroscopy, at the time when he was appointed professor of Experimental Physics at the Istituto Superiore of Florence in 1913. The staff he inherited from Ròiti comprised Puccianti, already active in Florence part-time since 1905, and the "aiuto" Antonino Lo Surdo.[9] Garbasso quickly set Lo Surdo to investigate spectroscopically the Doppler effect in the light emitted by the positive "retrograde rays" discovered, 1886, by Goldstein near the cathode of a discharge tube.[10] With an original design of the discharge tube Lo Surdo, in summer 1913, rediscovered in more efficient conditions the effect found in the same months by Stark. While a not interesting dispute followed about the priority, in which also Corbino was involved, Garbasso was able within 1913 to propose a first theory of the effect based on the Bohr model which had appeared a few months before. While his calculations contained an error pointed out to him by Bohr himself, Garbasso can be correctly considered the initiator in Italy of the use of the Bohr model along the first steps of quantum mechanics.[11]

[7] See Manlio Mandò, [8] page 599 and following. An amusing statement to the benefit of his students was: "Mathematics is very important for a physicist, almost as much as mercury".

[8] Garbasso expressed his conception of models with the following words: "Any theory in its essence is a model, better, is a description of a model ... the only connection between nature and model, in the most favourable case, is that the laws which describe the variations of corresponding quantities are the same in both systems ... so, a theory can be true without containing anything of the real". An interesting treatment of the impact of Garbasso and Puccianti in the development of spectroscopy in Italy is found in [11] and references therein.

[9] Antonino Lo Surdo (1880–1949), born in Siracusa, a good experimentalist with interests in terrestrial physics and spectroscopy. He became "aiuto" of Antonio Ròiti in 1908 and was also appointed director of the Meteorological Observatory at la Specola two years later. Lo Surdo moved to Rome in 1918 and became "aiuto" of Corbino, obtaining in 1919 the chair of Fisica Superiore. In 1937 Lo Surdo became director of the Physical Institute after the death of Corbino. He founded and directed to the end of his life the Istituto Nazionale di Geofisica of CNR.

[10] See [12] and references therein. This article gives a vivid picture of the results obtained in a few weeks confirming and completing the first observations, also Puccianti taking part in them.

[11] A touching presentation of the scientific and human figure of Garbasso is due to Rita Brunetti [13]. A comment showing the interest and the limitations of the scientific attitude of Garbasso is found in [14]. See also note ([8]).

The Hill of Science

The Sciences of Light: Astronomy

To provide a complete picture of the scientific environment in Arcetri, two institutions must be mentioned: the Arcetri Astronomical Observatory and the Laboratory that, in successive steps, became the National Institute of Optics. Besides the possibility of new overlapping fields of research, the benefits for the growing Florentine Physics were the international collaboration scenario (Astronomy) and a special care for applied Physics (Optics). The two institutions are close to the Physical Institute, although built before and after it, in a period just longer than half a century. The first was the Observatory.

It is well known that Galileo Galilei spent the last years of his life confined in the Villa del Gioiello, a country house at the Pian de' Giullari on the hill of Arcetri, a few kilometres from the centre of Florence. There he carried out his last heavenly observations and wrote fundamental Physics works. Accordingly, that hill seemed to be the most suitable place, when, in the second half of the 19th century, a new site for the Astronomical Observatory was sought. Still positioned downtown, it was by then incompatible with some aspects of the post-unification developing city, first of all with the street lighting. The decision was for Arcetri, at walking distance from Galileo's historical house. The new Observatory was inaugurated in October 1872.

Unfortunately the Astronomy research suffered from the same restrictions affecting Physics: first of all inadequate teams, in the present case two or three people, usually a director and an assistant. As for the scientific activity, the main fields beyond eclipses were "terrestrial" phenomena, like, e.g., northern lights or meteoric showers, and of course the hunt for comets. The name of Giovan Battista Donati (1828–1873) is associated to many celestial bodies, but he died only a few months after the inauguration of the Observatory.[12] His "aiuto" became the new director, but he too died some six months later: then it was pretty hard to find a replacement. Eventually Giovanni Virginio Schiaparelli (1835–1910), director of the Observatory of Brera (Milan) and world famous discoverer of the so-called Martian canals, had his German assistant, Wilhelm Tempel (1821–1889), appointed by Florence. The fame of Tempel too is based on the observation of comets and quite a few were named after him, but he was not a real astronomer. From our modern point of view, he was slightly more than an amateur, and actually he was only an assistant never in charge of the direction. However he had a very valuable ability: in those days when the photographic emulsions were not fast enough he was a gifted drawer, really skilled and accurate. His hand-painted plates were a good tool for the sky studies, and moreover nice to look at. Those plates yielded him the Royal Award of the Accademia dei Lincei in 1879. But a lithographer does not open research lines and for a while, after his death in 1889, the Observatory was neglected: twenty years elapsed since the opening and almost never there was a director.

---

[12] A short history of the Arcetri Astrophysical Observatory can be found in [15].

Only in 1893 the professor of Astronomy Antonio Abetti (1846–1928) came from Padua. He had to make a great effort in the restoration and maintenance of instruments. Thanks to him the already existing "Officina Galileo", specialized in Fine Mechanics and Optics, underwent a strong development. Abetti was known as a bright scientist: his scientific career already comprised work done at the Astronomy Institute in Berlin and an expedition in India, in 1874, to observe the transit of Venus before the Sun. His figure was recognised at the international level and even the great American astronomer George Ellery Hale visited him and his Observatory. Hale was looking for a European support for a new scientific journal, *The Astrophysical Journal*. His visit initiated a long-lasting collaboration on Solar Physics with a significant role in Arcetri's astronomical research. Abetti was mainly an observational astronomer but he understood soon the need of a deeper integration between Astronomy and Physics, the so-called "New Astronomy" or Astrophysics. Donati himself, after Father Angelo Secchi, carried out investigations on the spectral classification of stars. Aware of this evolution, Abetti favoured the plans of Antonio Garbasso to transfer the Physical Institute from the decaying seat in the centre of Florence to a new building on the same hill of Arcetri, close to the Observatory (see later). The emphasis on the international quality of research and the need of evolution from Astronomy to Astrophysics were the remarkable features of the scientific policy of Antonio Abetti. Following his steps, the son Giorgio Abetti (1882–1982) carried out studies and collaborations abroad, mainly in German universities. Back to Italy, he obtained a position at the Collegio Romano in Rome and in 1913–1914 he took part to an engaging multi-scientific expedition in the Himalayas. In 1917 he went to the USA as a member of a military mission organized by the just founded Italian UIR, the already mentioned Research and Development Board of the Department of War. In 1921 he was again at the Arcetri Observatory to become, shortly after, its director. Since the beginning, his scientific production was noticeable and most of it concerned astronomical spectroscopy. In the same year he succeeded in changing the Observatory's denomination to Astrophysical Observatory, as recommended by Garbasso [10]. The Faculty of Science introduced the teaching of Astrophysics, beside Astronomy, already after the end of the World War. The time was ripe for this new approach to heavenly phenomena. In fact the first Italian Astrophysical Observatory was established in Catania, as early as the end of the 19th century, along with the first Chair of Astrophysics [16]. This was a model for Arcetri, but while Catania was unable to develop an Astrophysics school, this succeeded in Florence. Indeed a few years later, in 1925, a Solar Tower was built on the hill to study high-resolution solar spectra. It has to be stressed that it was the first Solar Tower in Europe and the third in the world, after the ones already built by G. E. Hale. Actually, both Hale's scientific and financial help were instrumental for the design and the realization of the Arcetri Tower [17]. The contemporaneous establishment of the "corso di laurea" in Physics (see later) stimulated the formation of a school of Astronomy which produced several of the directors of Italian observatories after having been students or junior astronomers in Arcetri (see Table 4 later on).

This was the favourable scientific environment found by the bright students and teachers gathered around Garbasso. In those very years Giorgio Abetti devised the

Seminar on Astronomy, Physics, and Mathematics according to the model of the seminars in Anglo-Saxon universities. Both Italian and foreign physicist and astronomers were happy to present their ideas and results: lectures were held among others by Hall, Bethe, Persico, Fermi, but also by younger physicists as Rossi and Bernardini. More important, students were encouraged to attend and contribute to the lectures.

The Sciences of Light: Optics

With the expansion of the "Istituto di Studi Superiori", new research buildings were supposed to be built in the proximity of the Observatory. In addition to the new seat of the Institute of Physics, almost completed during the war, there was also a minor building, halfway between the Physical Institute and the Astronomical Observatory. This building was supposed to house the Chair of Terrestrial Physics which should inherit the activity of la Specola, and was meant also for meteorological measurements through balloon-borne instruments. So, in the words of Garbasso, one would join in the same area "Physics of Earth and Physics of Heaven, the most Tuscan ones among the Tuscan Sciences" [10]. Instead, the scope of the new building changed very soon. The director should have been Antonino Lo Surdo, director since 1910 of the old Meteorological Observatory. The idea was to keep Lo Surdo in Florence. But in 1918 Lo Surdo joined the Physical Institute in Rome and the building remained deserted. Nine years later, it became the seat of the National Optics Institute (INO).[13] That was the last step of a project stemming from the needs of the "Great War": in fact, as soon as the conflict began, scientific and industrial Italy had to face with a complete dependence on foreign countries for products based on Optics. Pointing systems, periscopes, binoculars, all these were imported mainly from Germany, but then Germany had become the enemy and among other restrictions a block on import was applied. All of a sudden, Italy realized that optical goods were not only for peace times. Not by chance, "Industrial Mobilization" was the specific aim of the UIR (see Sects. 6.2 and 6.3.1): it was decided to support the birth in Florence of a Laboratory of Applied Optics and Fine Mechanics,[14] following an original idea of Garbasso. Behind this undertaking, there was, of course, a strong military concern together with the will of some Italian enterprises, interested also in civil production. In fact, the real proponent of the whole project was the physicist Luigi Pasqualini (1888–1999), a former assistant of Ròiti, inventor, skilled technician first and then director of a workshop specialised in precision mechanics, the "Officina Galileo". Moreover, he could rely on the great experience gained in the Italian Navy as "electric" technician, in charge of the Torpedoes Laboratory, close to La Spezia. He was well aware of the Italian deficiency in Optics, which extended to the technique of

[13] For an analysis of the complex phases of such evolution and its links with the Florentine political and industrial environment see [18].

[14] As for the governmental side of science in that period see [19], in particular the Appendix with the report of Lo Surdo on a meeting held in Palazzo Vecchio to establish the Laboratory.

optical glass, and was strongly motivated in the development of a national industrial production of high-quality instrumentation; furthermore he was convinced that this required the formation of specialised technicians with scientific background. The Laboratory had to be the first step.

To carry out his plan, Pasqualini was able to involve other industries, local politicians and of course Garbasso (details on the role of Garbasso and the evolution of the project at the end of the war are found in the next section). But as soon as the Florentine project was officially approved, in September 1918, and even before the inauguration of the Laboratory, the war was over. As a consequence, the Optical emergency was over and the industrial interest decreased. At variance with the original intentions, Garbasso chose as director a university lecturer, his "aiuto", Raffaele Augusto Occhialini,[15] former "aiuto" of Battelli in Pisa. Occhialini started working in rather unfavourable conditions also because of the transfer of the whole Physical Institute and of the attached Laboratory to the new seat in Arcetri. He succeeded in publishing the first few issues of the journal *Rivista di Ottica e Meccanica di Precisione*, one of the statutory obligations of the laboratory, containing among other things his study on "moiré" interference fringes and their use in optical and mechanical applications. Unfortunately he was not aware that a rather complete study of the subject had been carried out by Augusto Righi about 30 years before (and had fallen in oblivion!). Frustrated Occhialini abandoned the subject. On the other hand he was on the verge of leaving Florence after winning a competition for a professorship.

The work on "moiré" interference fringes was picked up by the young Vasco Ronchi,[16] who had been appointed by Garbasso (1920) assistant to the (empty) chair of Fisica Terrestre under recommendation of Occhialini. Ronchi was for years the only scientist engaged in the activity of the Laboratory, mostly to determine the technical features of lenses on behalf of the Astronomical Observatory (the Amici's objectives!) and the Officine Galileo. Very soon he obtained (just by chance, as he was proud to say) an important result, that is a new method, based on moiré fringes, to verify smoothness and quality of an optical surface. This easy yet powerful tool is still

[15] Raffaele Augusto Occhialini (1878–1951): "marchigiano" like Battelli, born in Fossombrone, educated in Pesaro (see [20]), student of Battelli and also "normalista" in Pisa 1898, "laureato" in 1903, was his assistant and "aiuto" till the death of Battelli. Briefly in the same positions with Puccianti, 1916–1917. After the war (see text) "aiuto" of Garbasso in Florence, Chair Professor in 1921 in Sassari, in 1924 in Siena and from 1929 in Genoa. With good connections in Germany and the United States, he was an excellent teacher, and notable for his works on radioactivity, gas-discharge, spectroscopy, electrotechnics. He produced also a booklet on relativity of popular character.

[16] Vasco Ronchi, (1897–1988): student in Pisa and "normalista" from 1915, recalled for military service in 1917, back to Pisa in 1919, succeeded in completing the exams and graduating in that very year with the encouragement of Puccianti. Introduced to Garbasso by Occhialini, he was appointed assistant in the Institute of Physics in Florence from 1920 and, when Occhialini left for his chair in Sassari, he took responsibility of the Laboratorio di Ottica e Meccanica di Precisione. In the following years he succeeded in transforming that initiative, which had badly suffered in the aftermath of the war, in the Istituto Nazionale di Ottica with a notable stand in the Florentine and national scientific and technical panorama. He was instrumental also for the foundation of the Associazione Nazionale di Ottica. In his initiatives Ronchi had the support of Garbasso until the latter's death.

nowadays called the "Ronchi test". Thanks to it, Optical techniques gained an official recognition. From then on Ronchi spent all his efforts to revive the original project of the Laboratory. He kept the contacts with Pasqualini on the one hand and with the military ambient on the other, in particular with the "Istituto Geografico Militare", which had its seat in Florence and was obviously interested in optical devices. The person instrumental in the development of the Laboratory along the lines hoped by the still young Ronchi, was Gen. Nicola Vacchelli, responsible of IGM. On the other hand, with the advent of the fascist government, the policy towards military expenditures and towards the support of the related optical and mechanical industry changed. Pasqualini and Vacchelli joined Ronchi in promoting the renovation of the Laboratory with an extended program which included explicitly formation courses intended for civil and military (not only Italian) high-level technicians. A first step was the transfer of the instrumentation of the Laboratory from the inadequate rooms in the Physical Institute to the still empty pavilion which should have housed Fisica Terrestre. The second step was the acquisition of Gino Giotti, an optical expert working at the Merate Astronomical Observatory, who became an excellent co-worker of Ronchi and was also involved in the administrative management. The third step was the foundation of the Associazione Ottica Italiana, in view of promoting the coordination of the interests of the industries involved. The aim was to favour the diffusion of optical culture according to the original idea of forming skilled shop foremen.

At this point it was possible to transform the Laboratory in the Istituto Nazionale di Ottica under the direction of Vasco Ronchi. The inauguration took place in 1928 and the small pavilion was recycled in the seat of a kind of advanced vocational school, with room and some equipment for applied research. In time, the increasing activity led to the expansion of the primitive construction into the present building. Thanks to a strong governmental support and to the determined character of his director the INO underwent a fast growth and reached significant objectives [21] favouring the practice rather than the theory, with a feeling for the evolving civil and cultural needs. Although the part of the program aiming at the formation of skilled technicians was not completely fulfilled, what was left is an efficient school for optometrists. The scientific side followed the personal taste of Ronchi, more and more oriented towards physiological optics in the last part of his life.

Remarkably, never the activity of INO crossed that of the Physical Institute in the period between the two world wars. A more productive relationship was maintained with the Observatory and the Italian astronomers. After the death of Ronchi in 1988, the INO underwent, under the direction of Tito Fortunato Arecchi, a considerable reorganization, with an extension of its scientific and applied landscape (dynamics of complex systems, lighting techniques; restoration and preservation of the cultural heritage).

Garbasso and Florence

Arcetri from the Beginning to the End of the 1st World War

As soon as Garbasso settled in Florence, in summer 1913, he backed a convention between the Administration (and banks) of the town and the Superintendent and Directorate of the Istituto di Studi Superiori, obtaining new positions for the Sezione di Scienze Fisiche e Naturali, and financial support for the renovation of the laboratories, in primis for the building of a new Physical Institute. The actual construction started quickly on the site Garbasso himself had chosen romantically on the hill of Arcetri, not by chance at walking distance from the site of the Astronomical Observatory, which was again in operating conditions after years of abandonment (see Sect. 6.3.1).

The following year 1914 marked the outbreak of the first World War. After one year of negotiation and fierce debate, Italy joined the Triple Entente and engaged in the war against the Austro-Hungarian Empire, in May 1915 (see Sect. 6.2.2). Garbasso, at the age of 54, joined immediately the front-line as a volunteer lieutenant in the Engineer Corps, setting up a system of phonotelemetry against the Austrian artillery units. But he remained in close contact with his institute and his plans for the development of an advanced scientific and technical Florentine centre. His interest in the technical side was stimulated by his war experience and by his acquaintance with Pasqualini. They had much to share, both physicists, innovators and involved in political life (at the time Pasqualini was also town councillor). Pasqualini visited Garbasso on the front-line sometimes in 1916. Then, on leave for the beginning of academic year 1916–1917, Garbasso sized the opportunity of the "opening address" to recall the convention of 1913 and to thank the administration (and the banks) for the generosity with which the Istituto di Studi Superiori had been endowed with new staff positions and with the almost completed new Physical Institute, with its arcade and cloister in "Tuscan" style, on the hill of Arcetri ([10], pp. 16–17).

But Garbasso had a wish which coincided with the wish of Abetti, namely the concentration of more Institutes in a common area. So, apart the abundant patriotic rhetoric of the speech, Garbasso presented in full his plan for the Physical Section of the Istituto di Studi Superiori, to be concentrated in Arcetri. In his mind, the hill was to become a kind of City of Science, as can be seen in the decoration of the hall of the Physics building. The ceiling shows, in Art Nouveau style, the Galilean discoveries: the Sunspots, Jupiter's four satellites, the phases of Venus, the ring of Saturn, the features of Moon surface. On the walls, two large frescos display allegories of Research and of Learning. Moreover, the bas-reliefs of the members of the Accademia del Cimento (1657–1667) are aligned around the central cloister and in the surrounding garden there was a bust (now lost) of Minerva, the goddess of knowledge. Besides the building for the Physical Institute, large enough for housing a number of researchers and technicians, the "pavilion" intended for Terrestrial Physics was already completed (see Sect. 6.3.2).

Furthermore, Garbasso urged also the creation of another Laboratory, better, a Research Institute, where (in his words) "people with scientific formation and aware of the needs of practical work would be prepared to help and advise shop foremen ... in view of the gigantic economic upheaval announced by the gigantic war". These were the premises of the Laboratorio di Ottica applicata and Meccanica di Precisione discussed in the previous section. Garbasso expressed also the hope to have in Arcetri the "Museo degli Strumenti Antichi" of Lorenese origin (partly dispersed by the Lorena themselves when they left Florence), to become a centre for the study of the History of Science. This part of the project was not realised, but one more thing shows the broadmindedness of Garbasso in envisaging a site devoted to Physical Sciences: he hoped that "the old, glorious Observatory of Donati and Amici would turn at least part of its activity to the studies of Astrophysics, as in the intention of his excellent colleague, professor Abetti". Antonio Abetti, who supported fully Garbasso's plan, indeed changed the name from Astronomical to Astrophysical Observatory, the second in Italy, mindful of the work of Father Secchi. The Observatory would have later an important part in the cultural environment of the Group of Florence (see Sect. 6.3.1).

The project of the Laboratorio di Ottica Pratica e Meccanica di Precisione (accounted for in previous section), was officially approved September 1918 as a body attached to the Physical Institute.[17] One of the problems was the director, who should have been in principle a technician with a good scientific background, not necessarily a university professor. The choice in the end was Battelli's pupil, Augusto Occhialini, co-author of the treatise on Radioactivity, the second "normalista", after Ròiti, entering the story, on the move from Pisa after the death of Battelli, and already father of Giuseppe, Peppino, not yet GPS or Beppo (see note ([15]) and Sect. 6.3.2). Garbasso was in touch with Occhialini while this one served at the UIR: Garbasso encouraged him to move from Pisa to Florence, where the position of "aiuto" was vacant (Lo Surdo had left for Rome, 1918), and was instrumental for his appointment as a member of the Italian War Mission in USA with the task of studying the techniques of optical glass and of setting up agreements of technical cooperation. Occhialini stayed in USA from June 1918 to February 1919 and had the opportunity of meeting several American scientists, among whom R. W. Wood, A. A. Michelson and R. A. Millikan. In the meanwhile he became "aiuto" and, October 1918, was appointed director of the Laboratorio. The official inauguration took place on November 24, 1918, twenty days after the collapse of the Austro-Hungarian Empire. The speech of Garbasso began with the words: "The war ended: we must rebuild the world". The task proved to be much more difficult and even painful than expected.[18]

---

[17] Many details with some errors and questionable opinions are in [22].

[18] Garbasso left the army as a major of the Engineers Corps and resumed eagerly his place at the Istituto di Fisica with particular care for his duties as a lecturer, but his interest shifted more and more towards public life and political commitment, with the aim of benefiting at the same time his adopted city and his institution in times of economic difficulties and of social unrest. It is not strange that, after years of direct engagement in warfare, patriot Garbasso joined the nationalist party of chauvinist Luigi Federzoni, ending into the fascist party seen as the defender of the values of the Risorgimento and of the sacrifices sustained by so many on the front line during the war. This

From the Institution of the University to Academic Year 1925–26

In spite of the circumstances, the scientific and didactic activity of the institute did not stop during the war and this happened by merit of Rita Brunetti, "normalista" and "laureata" in Physics with Battelli with a well recognised work in spectroscopy. After one more year of specialisation in Pisa she took up the position of assistant of Garbasso in Florence and started working on the Stark effect with Lo Surdo until he left for Rome. With Garbasso at war, Brunetti managed to keep going the Physical Institute, still in the old seat downtown, both in teaching and in research, working successfully in X-ray and visible spectroscopy.[19] Back from the States in spring 1919, Occhialini took up his appointments as "aiuto" and as director of the Laboratory during an exhausting time, when the Physical Institute and the attached Laboratory were replaced in the new buildings in Arcetri. He and Brunetti were helped in that job by Vasco Ronchi (see Sect. 6.3.2). In 1921 Occhialini went to his chair in Sassari. Brunetti became "aiuto" and Garbasso promptly filled the vacant position of assistant with a brilliant student of Puccianti, Franco Rasetti, "laureato" by the end of 1922 with a remarkable thesis in spectroscopy. In Arcetri Rasetti found, in his words, *"a very pleasant place ... with a pretty good equipment ... especially for spectroscopy ... and not much teaching ... because Garbasso gave the Physics course"*.[20]

---

attitude was common among ex-combatants, even among upright refined intellectuals like Garbasso. So he was elected mayor of Florence in 1920 and kept the position under the fascist government with the title of Podestà until 1928. At the same time, like Corbino, he filled important positions in the organisation and direction of scientific research, in particular in the CNR (see Sect. 6.2.2), supporting actively the financing of well equipped laboratories and promoting the cultural updating and qualification of students and young researchers with the institution of scholarships for stays in foreign advanced institutions.

[19] This was the beginning of a noteworthy career, which led Rita Brunetti (1890–1942) to become "aiuto" of Garbasso from 1921 to 1926, and then Chair Professor for two years in Ferrara, for eight years in Cagliari and from 1936 in Pavia. Her work covered spectroscopy from visible to X-rays, magnetic properties of matter, nuclear physics and its bio-medical applications, history of science, good popular works, two treatises at the didactical level. In an academic environment dominated by males Brunetti was the only Italian woman attaining the directorship of a Physical Institute. She died prematurely, probably because of a professional disease, but in the very last years she attempted to use photographic plates for the detection of cosmic rays.

[20] Rasetti gives an interesting account of his experience with Garbasso: *"he had been a good physicist, at the time he was only interested in politics"*, but *"he gave his course in elementary Physics and was quite intelligent at it. And later Fermi explained to him what we were doing and he understood..he followed what we were doing and he was a very pleasant person ... as for being fascist he was very moderate, in fact (Rasetti is sure that) had he lived longer, he would have become disgusted with Fascism. But in the first few years ... Fascism didn't seem very bad ... after 1924 ... people lost hope (that Fascism would become a reasonable dictatorship). Still, even in the States there was a lot of admiration for Mussolini."*, excerpt from [7].

Table 1 Teaching staff of the "corso di laurea" in Physics, academic year 1924–25

Courses	Teachers
Analisi Matematica (I e II)	F. Tricomi
Analisi Superiore	F. Tricomi
Geometria Analitica e Proiettiva	E. Ciani
Geometria Descrittiva	E. Ciani
Fisica Sperimentale (I e II)	A. Garbasso
Fisica Superiore	A. Garbasso
Esercizi di Fisica	A. Garbasso
Chimica Generale e Inorganica (I e II)	L. Rolla
Chimica Fisica	L. Rolla
Meccanica Razionale	E. Fermi
Fisica Matematica (Electromagnetism, Spectroscopy)[a]	E. Fermi
Astrofisica	G. Abetti
Disegno	R. Brizzi
Mineralogia (optional)	P. Aloisi
Chimica Organica (optional)	A. Angeli

([a]) The following year, the course was named Fisica Teorica and Fermi changed the program in topics of Fisica Statistica (Statistical Physics)

Actually Garbasso was succeeding in transforming the Istituto di Studi Superiori in a regular University, be it still of class B,[21] and to establish the regular "corso di laurea" in Physics (and Mathematics), with the pattern of teaching subjects provided by the national regulations originally set by Matteucci: it became possible to have students from the beginning of their curriculum. The first regular academic year began November 1924. The teaching staff of the "corso di laurea" in Physics was as per Table 1: notice the position of Enrico Fermi.

Indeed a turning point was the professorship "in charge" (Professore Incaricato) offered him by Garbasso for the teaching of Mathematical Physics and Theoretical Mechanics (Meccanica Razionale).

Apart the famous work of Fermi on Statistics (written in those years in Arcetri), he and Rasetti, old friends from the times of Pisa, initiated a very fruitful collaboration both on experimental (spectroscopy!) and theoretical subjects, the two being endowed with a vivid physical sense, the first adding his profound understanding of the new atomic Physics (and relativity), the latter his ability in devising and handling experiments. Both made friends with spectroscopist Rita Brunetti, exchanging ideas and experience. Later Fermi would quote Brunetti's results of those years.

A second turning point is 1926. Rita Brunetti won a competition for Experimental Physics and left Florence for Ferrara, destitute of a laboratory: she was hosted for

[21] In the opening address Garbasso underlined that the inauguration of the revived "Studio Generale" was greeted by the representatives of the same Communes already existing in the State of Florence in 1321, when the "Studio" came to existence for the first time.

Table 2 The parallel lives of Persico, Fermi and Rasetti as young men

Name	Born	Liceo	Laurea	1st Appointments
Persico	July 9, 1900 Rome	July 1917, Rome	Nov. 1921	'21–'24 Rome assistant
				(Corbino)
				'24–'26 Rome professor
				"in charge" (Corbino)
Fermi	August 10, 1901 Rome	July 1918, Pisa	July 1922[a]	'22–'24 Rome professor
				"in charge" (Corbino)
				'24–'26 Florence professor
				"in charge" (Garbasso)
Rasetti	July 10, 1901 Rome	July 1918, Pisa	Dec. 1922	'22–'26 Florence
				assistant (Garbasso)

([a]) "Normalista"

her experimental work by Quirino Maiorana in Bologna. At the same time Fermi and Enrico Persico won the first competition for Theoretical Physics,[22] a new entry in the set of physical teachings, strongly supported by both Corbino and Garbasso. Fermi was called by Corbino in Rome, Persico by Garbasso in Florence. It is worthwhile noting here the position of Pisa in the years following the end of the war, a point of excellence with the high-level teaching of Puccianti in Experimental Physics in the wake of Battelli and with the school of Mathematics conducted by Luigi Bianchi after Ulisse Dini. This favourable situation was rewarded by the presence of a number of very good students, of whom three were to play a key role in the development of the Italian school of Physics and in particular of the groups of Florence and Rome: Enrico Fermi, Franco Rasetti and later Gilberto Bernardini. The fourth personality in this context was Enrico Persico, "laureato" and assistant of Corbino in Rome, familiar with such mathematicians as Tullio Levi Civita and Guido Castelnuovo, and a theoretician with a sense for experiments.

Table 2 highlights the parallel lives of Persico, Fermi and Rasetti as young men.

The friendship between Persico and Fermi begins during the Liceo (secondary school) in Rome, that between Fermi and Rasetti during the University in Pisa. The

[22] The third winner was Aldo Pontremoli, called by the University of Milan, where he founded the Physical Institute, and disappeared in the Arctic in the disaster of the *Italia* dirigible in 1928. Pontremoli, born in 1896, was an assistant of Corbino around 1920 and signed a paper on the mass of radiation in an empty space with Fermi [23].

relationship among the three, practically self-taught in the fields of new Physics, is well described by the set of their scientific articles covering the years (1921–1926).

When in 1926 Fermi goes back to Rome (and Persico goes from Rome to Florence), Rasetti follows Fermi, as assistant and "aiuto" to Corbino, and in two years will become professor of Spectroscopy (with an important programme on Raman effect). Corbino will attract around the personality of Fermi more promising students: Emilio Segré (1905–1989), Ettore Majorana (1906–1938), Edoardo Amaldi (1908–1983), the group of Rome is formed.

The story of Florence is less simple, but also here Garbasso was able to attract outstanding young people and build a successful group. One must underline once again the action of the two men who were instrumental in those achievements. Both Corbino in Rome and Garbasso in Florence opened their institutes to the best young physicists emerging from Italian universities in those years, several of them from Pisa. This is a recognisable policy: both use their scientific stature and their position in public administration in order to build "schools of Physics" based on the work of young individuals of precocious capacity and qualification, with a keen interest and a fresh understanding of the "new" Physics, which placed them above the average culture of the contemporaneous academic establishment. In the fifteen years or so after the end of the war the two groups were unusually close, with an effective exchange of persons and of knowledge, setting up connections and friendships which would last in time.

The Group of Arcetri and the Dawn of Cosmic Ray Physics in Italy (and not only that)

A Good Teaching Staff and a Good Set of Students

4 December 1987 was the 80th birthday of Giuseppe Occhialini. On that occasion the Physics Department of the University of Florence organized a round table, with Paolo Blasi as moderator, with the presence of (in order of age) Bruno Rossi, Gilberto Bernardini, Giuseppe Occhialini and Daria Bocciarelli, the four surviving personalities of the "Group of Florence". Edoardo Amaldi took part in the round table and Manlio Mandò, a student in Florence from 1931 and a witness to the last part of the life of the group, opened the session illustrating the following Tables 3 and 4.[23]

All the contributions showed how deeply felt, after so long, was the recollection of that short stretch of years, short but so full of ambitions, hopes, strength, joy of being a part of a significant common effort towards "scientific truth" and overall friendship. Mandò and the external witness Amaldi defined that feeling "the spirit

[23] From the contribution of M. Mandò to the Round Table 1987, unpublished. The original Tables are integrated and slightly modified with added notes for the purpose of the present work.

Table 3 Teaching staff of the Physical Institute of Florence 1913–1937, with A. Garbasso director from 1913 to March 1933, L. Tieri from Fall 1933

Academic Years	"Aiuto"	Assistant	2nd Assistant	Other Teachers
1913–17	A. Lo Surdo[a]	Rita Brunetti	–	A. Abetti
1917–18	–	Rita Brunetti	–	A. Abetti
1918–20	A. Occhialini	Rita Brunetti	–	A. Abetti
1920–21	A. Occhialini[a]	Rita Brunetti	V. Ronchi	A. Abetti
1921–22	–	Rita Brunetti	–	G. Abetti
		V. Ronchi		
1922–24	Rita Brunetti	V. Ronchi	F. Rasetti	G. Abetti
1924–26	Rita Brunetti[a]	V. Ronchi	F. Rasetti	G. Abetti
				E. Fermi[b]
1926–27	–	V. Ronchi	–	G. Abetti,
				E. Persico[c]
1927–28	V. Ronchi	F. Olivieri	B. Rossi[d]	G. Abetti
				E. Persico
1928–30	V. Ronchi[a]	F. Olivieri	B. Rossi	G. Abetti
				E. Persico
				G. Bernardini[d]
1930–31	B. Rossi[e]	G. Bernardini	G. Occhialini	G. Abetti
1931–32	B. Rossi[e]	G. Bernardini[f]	G. Occhialini	G. Abetti
1932–33	G. Bernardini	G. Occhialini	L. Emo Capodilista	G. Racah[g]
			Daria Bocciarelli[h]	
1933–35	G. Bernardini	G. Occhialini	L. Emo Capodilista[i]	G. Racah
			Daria Bocciarelli	
1935–37	G. Bernardini[j]	G. Occhialini[k]	Daria Bocciarelli[l]	G. Racah[m]

([a]) The events concerning Lo Surdo, Brunetti, A. Occhialini, Ronchi, Rasetti, Fermi till Academic year 1925–26 have been accounted for in Sects. 6.4.1 and 6.4.2 (see Sect. 6.3.2 for Ronchi)
([b]) Professor "in charge" of Meccanica Razionale and Fisica Matematica (see Table 1)
([c]) Chair Professor of Fisica Teorica and "in charge" of Meccanica Razionale. From 1930–31 in Turin. From 1950 in Rome
([d]) See Table 5
([e]) Also professor "in charge" of Fisica Teorica in the place of Persico. From 1932–33 Chair Professor of Experimental Physics in Padua and director of the Institute of Physics. From 1938 in Copenhagen, guest of Niels Bohr, then in Manchester with P. M. S. Blackett and then in the United States
([f]) Professor "in charge" of Meccanica Razionale
([g]) Professor "in charge" of Fisica Teorica in place of Rossi
([h]) "Extra" assistant of Garbasso
([i]) From 1935 to 1946 in the United States with a scholarship at Berkeley. Back to Italy he leaves research
([j]) From 1937–38 Chair Professor of Experimental Physics in Camerino. From 1938 in Bologna, also director of the Institute. From 1947 in Rome
([k]) From 1937 in Brazil. From 1944 in Bristol and, 1948, in Brussels. From 1949 Chair Professor in Genoa and from 1951 in Milano
([l]) From 1937 at the Physical Laboratory of the Istituto Superiore di Sanità in Rome with G. C. Trabacchi
([m]) From 1937–38 Chair Professor of Fisica Teorica in Pisa. From the end of 1938 at the Weizman Institute in Israel

Table 4 "Laureati" in Physics from academic year 1928–29 to 1937–38, "Corso di Laurea" established in academic year 1924–25

Academic year	Names
1928–29	Londei Luisa, Marconi Rita, Panerai Tullia, Zini Rodolfo
1929–30	Colacevich Attilio[b], Occhialini Giuseppe[a], Romani Abigaille, Francesco Scandone[c]
	in Mathematics: Calamai Giulio[b]
	in Chemistry: Franchetti Simone[a]
1930–31	Genoviè Gino, Racah Giulio[a], Righini Guglielmo[b]
1931–32	Baroni Ermanno, Bocciarelli Daria[a], Caponi Pier Giovanni, Mari Giovanni Antonio
	in Mathematics: Foà Alberto
1932–33	Castellani Giuseppe, Cipriani Edvige, Crinò Beatrice[d],
	Emo Capodilista Lorenzo[a]
	in Mathematics: Sestini Giorgio
1933–34	De Benedetti Sergio[a], Francese Clara
1934–35	Mandò Manlio[a]
1935–36	Castelli Iris, Fracastoro Mario[b], Persano Aldo, Ricci Elena,
	Serafini Francesco
1936–37	De Seras Luigi
1937–38	Barsotti Nedda, Landini Oliviero[d], Orzatesi Giuseppe, Pagani Lina

([a]) Physical Institute (see Tables 3, 5 and text)
([b]) Astrophysical Observatory (see Sect. 6.3.1)
([c]) "Istituto Nazionale di Ottica" and then industry (see Sect. 6.3.2 and Table 5)
([d]) Industry

of Arcetri". What follows is an attempt to present the "administrative" scenario and the meaning of the word "school" as applied to the group.

Table 3 shows the evolution of the staff of the Physical Institute under the direction of Antonio Garbasso from 1913 to 1933, and of Laureto Tieri[24] to 1938, when the racist campaign sparked off by Mussolini led to the 1938 laws, which expelled Jews from one day to another from the Italian scientific community. The notes give details about the fate of Arcetri's actors when she or he left the group. Table 4 lists the "laureati" in Physics (plus some in Mathematics) after the coming into operation of the "corso di laurea" in 1924. The table shows also that the "corso di laurea" provided with fresh young personalities both the Physical Institute and the Astrophysical Observatory, this being one of the successful results of the policy of Garbasso and Abetti.

---

[24] L. Tieri (1879–1952) "laureato" in Rome 1903 and assistant of Blaserna and then of Corbino. Known for his experiments on the Hall effect in Bismuth, is co-author of the first paper (experimental!) of Persico [24]. From 1924 Chair Professor of Experimental Physics in Messina. From 1933 in Florence in the place of Garbasso. Ritired in 1949.

It is interesting to examine in Tables 3 and 4 the four academic years from 1926 to 1930, the years of Persico. For one year Ronchi is the only assistant of Garbasso, becoming later "aiuto", but he is engaged in his effort to revive the Laboratorio di Ottica e Meccanica (see Sect. 6.3.2) and is not in the least interested in the "new" Physics introduced by Persico. On the other hand the first students are already in their second year of the regular corso di laurea, among them Giuseppe Occhialini and Francesco Scandone, who will be joined year after year by Giulio Racah, Daria Bocciarelli, Beatrice Crinò, Lorenzo Emo Capodilista. Then the turning point of the arrival of Bruno Rossi, Fall 1927, and of Gilberto Bernardini, Fall 1928: the Group of Florence comes to existence.[25]

One after the other the best students get the "laurea" and find a position in the institute through the interest of Garbasso and of Persico. The theses of "laurea" are all on experimental subjects related to the researches initiated by Rossi and Bernardini (except Racah, see Table 5). But experimentalist Rossi and student Racah collect the first notes from the lectures of Persico, first published in Florence 1929.[26] As a result Rossi will be professor "in charge" of Fisica Teorica when Persico leaves for Turin, and Racah will inherit that position when Rossi wins the professorship and goes to Padua.

Table 5 summarises the initial steps of the "young Arcetrini": the names are those which appear in all the papers published from 1928 to 1937, when, after the death of Garbasso, the winning of professorships and the political situation led eventually to the dispersal of the group.

A facet of the behaviour of these young people is their quick integration in the group since students. This is not only because of the enthusiasm of the leaders, Rossi and Bernardini, who would share their work with the students. They took profit also of two important assets which are frequently referred to in their recollections: the weekly reading of the leading international journals promoted by Persico and sustained by Bernardini with his characteristic zeal, and the Astrophysical, Physical and Mathematical Seminar promoted by Giorgio Abetti. In this way all of them were made aware of the more recent developments in the ongoing physical research;

---

[25] Bruno Rossi (Venice 1905; Cambridge Mass. 1993). Among the rich set of biographic material one may choose the autobiography [25, 26]. From Venice to ill-equipped Padua and Bologna: the happy encounter with Rita Brunetti, *"the only person who taught him some Physics"* and supervisor of his "laurea". Brunetti recommends Rossi to Garbasso, who promptly accepts him as assistant.

Gilberto Bernardini (Fiesole 1906, La Romola (Florence) 1995). A good biography does not exist as yet, in particular for the first period of his scientific activity. See Mandò, [8] p. 613, the biographic sketch of the Accademia dei Lincei and Giorgio Salvini [27]. "laureato" *cum laude* in Pisa with Puccianti, 1928, "normalista", working at first in a small optical industry in Florence, unhappy researcher in the first year of the Istituto Nazionale di Ottica. Attracted by the lectures of Persico, gets a position as "extra" assistant to his course of Meccanica Razionale. In 1930 Ronchi quits the Physical Institute and Bernardini becomes assistant of Garbasso (Rossi becoming "aiuto").

[26] These notes will become the first draft of the well known treatise by Persico, *Fondamenti di Meccanica Atomica* (Zanichelli, Bologna) 1936.

Table 5 The initial steps of the "young Arcetrini"

Name	Born	Laurea	First appointment
Rossi	1905, Venezia	1927 Bologna	Fall 1927, 2nd assistant of Garbasso
Bernardini	1906, Fiesole	1928 Pisa	Fall 1928, "extra" assistant of Persico (see note ([25]))
Occhialini	1907, Fossombrone	1929 Firenze	From 1930 2nd assistant of Garbasso
Scandone[a]	1909, Firenze	1929 Firenze	INO, and then Officine Galileo
Racah[b]	1909, Firenze	1931 Firenze	Fall 1932, professor "in charge" of Fisica Teorica
Bocciarelli[c]	1909, Firenze	1931 Firenze	From 1932 "extra" assistant of Garbasso
Crinò[d]	1913, Firenze	1933 Firenze	Laurea in Chemistry and Officine Galileo
Emo Capodilista[e]	1909, Firenze	1933 Firenze	From 1933 2nd assistant of Garbasso
De Benedetti[f]	1912, Firenze	1933 Firenze	in Padua with Rossi
Mandò[g]	1912, Terni	1935 Firenze	in Palermo with Segré

The following list points out the papers which were the result or the premise of the thesis of "laurea" made under the supervision of members of the staff

([a]) Persico E. and Scandone F., "L'effetto Hall con elettrodi estesi", *Rend. Accad. Lincei*, **10** (1929). This paper was splitted into three parts: nota prima 238–249; nota seconda 361–368; nota terza (Scandone only author) 437–440. A very precocious student, after the thesis Scandone finds a position at the Istituto Nazionale di Ottica and then in Industry, becoming soon the director of the Officine Galileo

([b]) Rossi B., Racah G., "A proposito di un'osservazione di Stark sulla realtà del moto assoluto", *Il Nuovo Cimento*, **6** (1929) 317

([c]) Bocciarelli D., "A hard component of the beta-radiation of Potassium", *Nature*, **128** (1931) 347

([d]) Rossi B., Crinò B., "Le anomalie di assorbimento della radiazione penetrante", *Rend. Accad. Lincei*, **15** (1932) 741. A very precocius girl student, Beatrice Crinò shifted her interests to Applied Physics

([e]) Bernardini G., Emo Capolista L.,"Sulla radiazione gamma del Po+Be", *La Ricerca Scientifica*, **2** (1935) 17

([f]) Bernardini G., De Benedetti S., "Misure di assorbimento della radiazione penetrante secondo diverse inclinazioni zenitali", *La Ricerca Scientifica*, **2** (1933) 73

([g]) Bernardini G., Mandò M.,"Sulla disintegrazione del Berillio per azione dei raggi gamma", *La Ricerca Scientifica*, **2** (1935) 38

also they became acquainted with leading scientists who were happy to visit Arcetri through the international connections established by Abetti.[27]

A third important asset was the position of Garbasso in the Italian scientific environment, which gave him the possibility of providing scholarships for stays in leading foreign (mainly European) laboratories. It is likely that these circumstances largely compensated for the scarcity of financial means: those ambitious young men were striving after "fundamental" problems in the investigation of the physical reality ("... of the secrets of nature", as Bruno Rossi puts it humorously in his autobiography), but they were also enough well-informed, perceptive enough, and wise enough, to identify subjects which would not involve large expenses in costly instrumentation.[28]

Another facet of the behaviour of the "Arcetrini", to a certain point different from the behaviour of the group of Rome, is pointed out by Guido Tagliaferri who reports [28] a precise remark of Occhialini: *"The presence in Arcetri of Enrico Persico and the arrival of the newly "laureati" Bernardini from Pisa and Rossi from Bologna as assistants made possible the formation of a group of enthusiastic young physicists. The [scientific] interest of the Laboratory shifted from spectroscopy to nuclear physics and cosmic rays. So, 1927–1928, the School of Arcetri was born."* [29]. Tagliaferri writes: *"With the word "school" used by G. O. one should not understand a group of followers of a "maestro", but rather an informal community of scholars in the same discipline, who share the scope of its advancement, and to that scope they address the investigations of each one of them, using freely the results".* That this was the case is shown by considering the whole of the papers published by the members of the group from 1930 to 1937: most bear only one signature, but all represent the results of a shared knowledge. G. Occhialini provides an interesting addition to Tagliaferri's commentary. In his words: *"the absence of scientific guide by Garbasso was important to train the muscles of Rossi and Bernardini".*[29] Politician Garbasso was not only a passionate man of science, but also wise and generous enough as to let the intelligence and fantasy of his young researchers free, giving them his constant support in practical problems and encouraging them to publish quickly their results, which he was happy to present in the *Rendiconti dell'Accademia dei Lincei* and in the journal of the CNR, *La Ricerca Scientifica* (see note ([29])).

The Science and the Scientists

A proper account of the scientific results is beyond the scope of this paper. What follows is intended rather to shed some light on the attitude towards research and on the efficiency of the "modus operandi" of the "informal community" of Arcetri.

[27] G. Occhialini, private communication to A. B. and contribution to the Round Table 1987. Also Bruno Rossi, ibid. and [25, 26].

[28] G. Occhialini, private communication to A. B. and contribution to the Round Table 1987.

[29] G. Occhialini private communication to A. B., 1987.

Giuseppe Occhialini states that with the advent of Rossi and Bernardini the interests of the Laboratory shifted from spectroscopy to nuclear physics and cosmic rays (see [29] and note ([29])). The very fact of this reorientation is an indication of the quality (of the curiosity) and of the ambition of Bernardini and Rossi, in their attempt to attack research along "new" lines, new at least in the Italian environment, these lines being typically of an experimental kind, associated with the development of "new" instruments.[30]

The first attempts are daring but not successful: an experimental verification of the corpuscle-wave nature of electrons [30] and a spectroscopic determination of the chemical composition of the cometary tail, a subject which reveals an incipient astrophysical interest.[31] And then in 1929, the cosmic rays. The story has been told mainly in the recollections of Bruno Rossi and Giuseppe Occhialini,[32] but some details perhaps are still missing. It has not been possible up-to-now to find the original thesis for the laurea of Occhialini, not even the title (a research is in progress at the Archives of the University of Florence). But a testimony of Livio Scarsi, from the very words of Occhialini sometime in the late eighties [32], is that Augusto Occhialini suggested the subject of cosmic rays to his son Beppino. Augusto was familiar with several American as well as German physicists, in particular Millikan, busy in propounding his theory of gamma-rays emitted in primitive nucleosynthesis, and Kohlhörster, who was working with Bothe in experiments for cosmic ray detection with Geiger-Müller (GM) counters (and with a coincidence method). The point is that the thesis work of Beppino contained a "tesina" (extra contribution) presenting the results of Bothe and Kohlhörster, just appeared in *Zeitschrift für Physik* [33]. Rossi was not moved by Millikan's theory, but the paper of Bothe and Kohlhörster awoke Rossi's understanding of new, different, features of cosmic rays and of the possibility of performing new, critical, measurements on them. Rossi obtained quickly a scholarship from the CNR and spent the summer months of 1930 in Berlin at Bothe's Laboratory. Back to Arcetri with the good recipes, he "put himself immediately at work" with his mates in Arcetri, first of all on the production of GM counters,[33] Rossi understood that these were the right detectors apt to open a new field of research, a new chapter of Physics. Furthermore, they were not very costly, which fulfilled one of the requirements of an enthusiastic but poorly financed Laboratory. But Rossi did more

---

[30] This is not surprising in the case of Rossi, a grateful pupil of Rita Brunetti, who maintained that the history of instruments coincides with the history of Physics (see note ([19])). But also Bernardini was born in the experimentalist environment of the Pisa of Battelli and Puccianti.

[31] Minor contributions of Rossi in 1929 refer to the Raman effect, a spectroscopic subject well in the reach of the Laboratory in Arcetri: in that very year Rasetti would publish his important results on Raman effect taking up the field.

[32] A general information with extended references is found in Leonardo Gariboldi [31], who suggests that Giuseppe Occhialini was influenced in his intellectual formation by Battelli through the influence of his father. The suggestion appears to be correct, if only because of the strong feeling for the motherland, which is characteristic of the "marchigiani". G. O., native of Fossombrone like his father, was educated in Florence and was one of the first students in the "corso di laurea" just started up by Garbasso.

[33] A lively description is in [25, 26].

than that: with astounding efficiency he invented and realised within the year his coincidence method based on the use of thermionic valves, the "circuito alla Rossi" [34].[34] From then on the results follow one another, giving Rossi the possibility of defending publicly against Millikan the notion of the corpuscular nature of cosmic rays in the "International Conference on Nuclear Physics" held in Rome, October 1931.

In a few years Rossi, Bernardini and the younger co-workers, in primis Daria Bocciarelli, produce about thirty short notes and papers on the absorption of cosmic rays (the Rossi curve and the multiple production [35]), their behaviour in the Earth magnetic field (zenithal effect,[35] the first attempts to E-W effect) and related technical problems. The last paper of Rossi before leaving Arcetri for Padua is with Fermi: "Azione del campo magnetico terrestre sulla radiazione penetrante" [37]. This was also the first work of Fermi on cosmic rays: it also witnesses the connections of the Group of Florence with the Group of Rome.

The measurements on cosmic rays implied the use of ionisation chambers not only for the detection of the "primary" radiation but also for the measurement of environmental low-level ionising radiation. The GM counters offered a new efficient way for this kind of measurements: indeed they were already an important tool in the study of Radioactivity (see note ([34])). The quick learning of the technique of GM counters is at the basis of the shifting of the interest of the laboratory from spectroscopy to cosmic rays and nuclear physics. So while Rossi proceeded with tireless energy in cosmic-rays investigation, he encouraged just "laureato" Beppino Occhialini to study weakly radioactive substances making use of counters.[36] The result was the first paper of Occhialini, on the activity of rubidium with a magnetic spectrometer designed and built by him. The detector was a small counter with very thin (less than ten microns) Al wall [40]. The same apparatus was used by Daria Bocciarelli for her thesis and for her first paper on the radioactivity of potassium (see Table 5). In successive three papers Bocciarelli extends the measurements making use of a method of coincidences. The success of these measurements gives an idea of the skill in producing "refined" counters, in designing instruments and in conducting measurements.

---

[34] *"The first counting of the penetrating rays was in 1916 by Hess and Lawson, but Bothe and Kohlhörster used for the first time the wire corpuscle-counter, already in use in researches on radioactivity and extraordinarily useful. This device sends in a circuit a short electric signal whenever it is traversed by a fast charged particle. The signals can be amplified and the amplified current can reach a counting device. The method of the corpuscle-counter has been adopted in the researches carried out in the Physical Institute of our University and allows, by suitably connecting two or more devices, investigations on absorption, direction, nature of the cosmic rays, which may be very difficult or impossible to perform with electroscopes."*. This is how Persico announced the state of the art in Arcetri in the first days of November 1930, in his last opening address to the academic year before leaving for Turin. He quoted also by name "doctor Rossi of our Physical Institute" who was able to show "the formation of secondary electronic rays through a lead shield traversed by the primary radiation".

[35] The first work on this effect was by G. Bernardini [36].

[36] See [38]. Bernardini has a contribution on the technique of magnetic spectrometers for slow electrons [39].

In 1931 Bernardini was prevented by military duties to go to Cambridge, so Occhialini took his place and with a three-months scholarship of the CNR joined P. M. S. Blackett at the Cavendish Laboratory. His mission was to learn the technique of the Wilson cloud chamber mastered by Blackett. Occhialini added the technique of counters and of fast coincidences learnt from Rossi in Arcetri. The three weeks became three years and the results are contained in the four papers signed between 1932 and 1934 by Blackett, Occhialini[37] and later also by Chadwick [43]. Those in Arcetri joined in the enthusiasm for the success when reading Occhialini's letters describing his work (see note ([23])).

In the meanwhile the neutron had been discovered opening a new window in the study of the atomic nucleus. The testimony of Rasetti is interesting [7]. He spent one and half year at Lise Meitner's Laboratory in Dahlem between 1931–32. Back to Rome he found that during his absence there were *"only vague talks"* about leading the Laboratory towards nuclear physics (the experimental work in progress was essentially in spectroscopy). He found however that Fermi was ready to shift from spectroscopy to a more exciting field: the actual work started when Rasetti built the first apparatus on the basis of his experience in Dahlem.

The Group of Florence had already abandoned spectroscopy since 1930. Perhaps also because of the departure of Rossi in the Fall of 1932, the investigation in nuclear physics was accelerated, the subject chosen being the production of neutrons from berillium. After some preparatory work (cfr. M. Mandò [8]), Bernardini spent a few months in 1934 at Lise Meitner's Laboratory with a scholarship of the Academy of Lincei (also Emo Capodilista was there in the same year). The result was a study of the reaction ($Be + He \rightarrow C + n$) with several papers, mostly in collaboration with Daria Bocciarelli (also with Emo and Mandò, see Table 5). A result of Bernardini and Bocciarelli was also the study of proportional counters.[38]

But the balance of the Physical Institute was changing. A few months after the departure of Rossi for his professorship in Padua, badly ill Garbasso died, in March 1933. Rossi (already busy with the rebuilding of the institute and the preparation of the E-W experiment in Asmara) expressed later his deep gratitude to a man who had done so much for him and, to the last moments of his life, gave support to Bruno's project recommending its financing [25, 26]. After the death of Garbasso, Abetti became the provisional director of the institute with full satisfaction of the junior members of the staff, but the Faculty, suspicious of their independent attitude, after some hesitation called Laureto Tieri instead of waiting for Emilio Segré, the probable winner of the next competition (as suggested by Fermi). The direction of Tieri was not necessarily antagonistic towards the group [8].

But also the political situation was rapidly worsening. Differently from Rome, the members of the Arcetri Group were quite aware and felt politically involved, there were even harsh debates among them, that only the strong ascendancy of Garbasso had been capable of quenching. In the words of Occhialini: *"Garbasso was that*

---

[37] The first paper [41] was followed by [42]. From then on the signature of Occhialini as author becomes GPS.

[38] Contribution to the Round Table 1987 and Ref. [44].

not-existent animal, the intelligent, honest, good fascist ... with a smiling tolerance for the divergent opinions of the junior persons".[39] In a climate made uneasy by the death of Garbasso, Occhialini, back from Cambridge and the Cavendish, made a big but unsuccessful effort to obtain an adequate financing for the construction of a Wilson Chamber.[40] He would react more sharply than the majority of his friends to the cultural ambience which was leading Italy into wars and racism. So he decided to leave Italy for some time and joined Gleb Wataghin in Brasil, to lay the foundation of a modern school of Physics in São Paulo. He was back in Italy eight years later after his detours in Bristol and Brussels.

In fact 1937 was the year of the dispersal of the group. Bernardini became Chair Professor and, after one year in Camerino, settled in Bologna, where he was director of that Physical Institute till 1947. With his characteristic energy he succeeded in continuing his work in particular on cosmic rays. One of the last papers while still in Florence is with Simone Franchetti, a chemist (see Table 4) who was appointed assistant in 1937 by Tieri, and was forced to abandon by the racist persecution. He was back at the end of the war.

Racah had built a very successful career in Theoretical Physics also through an intense relation with the theoreticians of the Group of Rome. He became Chair Professor in Pisa just in time to be forced to leave his position in 1938 and to emigrate to Israel at the Weizman Institute. Once filo-fascist, Racah perhaps remembered the hot discussions with Occhialini.

Daria Bocciarelli found a position at the Istituto Superiore di Sanità with Trabacchi and contributed to the success of that Physics Laboratory during long years, first in nuclear physics (the million Volt accelerator for neutron production), and then in Electronic Spectroscopy.

Lorenzo Emo Capodilista went, in 1935, to the United States for a stay at the Stanford Laboratory in Berkeley. He came back after a couple of years as agent of a firm for scientific instrumentation, abandoning active research.

Mandò after his laurea joined Emilio Segré in Palermo. When Segré was forced to leave, in 1938, Mandò joined Bernardini in Bologna. Then there was the bracket of the war, including a period in a prisoners of war camp. Finally, he returned to Florence and contributed, with Simone Franchetti (successor of Tieri in 1949), to the development of the post-war institute. Arcetri was coming to life again, and this together with the rest of Florentine Physics, no longer concentrated on the hills through the activity of the CNR Microwaves Centre, a remarkably successful achievement of Nello Carrara (a "normalista" and student in Pisa with Fermi and Rasetti) and the outstanding scientific and didactical work of Giuliano Toraldo di Francia.

[39] G. Occhialini, private communication to A. B., 1987.

[40] G. Occhialini, unpublished document.

The "Spirit" of Arcetri

The time is now to leave the stage to the actors, picking up a few significant quotations from the notes (unpublished as yet) of the Round Table, December 4, 1987 (see Sect. 6.5.1). The quotations are from Bruno Rossi, Edoardo Amaldi and Giuseppe Occhialini.

Bruno Rossi: *"When I think back to my past life I feel as if the years in Arcetri were a dream, a magic experience, which left a permanent mark on all my life: I think that Daria, Gilberto and Beppo, here with us, can understand, can interpret what I am trying to say, and also those would understand that are not here any more, Racah, Emo, Scandone, Righini, Beatrice Crinò ... We were a small group of young people, just "laureati", in part still students. We were very different from one another as for familiar tradition, attitude, tastes, character, but we were united by strong friendship ties and by a common commitment to science. Other factors contributed to the special atmosphere created by these human relations, to what Mandò defined "the spirit of Arcetri". First the almost paternal attitude of the director, professor Antonio Garbasso, who would do his best to facilitate our work, using his political authority. Second, the proximity of the Observatory and specially of Giorgio Abetti who ... somehow ... took us physicists under his protection. Last but not least, the fascination of Florentine hills which would reassure our spirit and would allow our mind to sweep over the contingent practical problems of our work. I arrived in Arcetri ... full of enthusiasm for the new life which was to begin and with the ambitious intention of undertaking some kind of research which would contribute in a substantial way to the development of science. I found ... Gilberto who had the same intention..."*

Edoardo Amaldi: *"Corbino in Rome and Garbasso in Florence played a truly important role. Corbino was a self-made man ... of great intelligence and a clear vision of the scientific and organizational problems of the country. Garbasso was a man of remarkable culture ... very different from Corbino ... [But he had the same] very positive attitude towards the new Physics which was being born in Europe [in those years] ...and had the same will and ability to help the young physicists who would enter in those fields and produce scientifically ... Contacts and exchange of ideas between the groups of Rome and Florence were kept through relatively frequent visits of the Florentines in via Panisperna and of the Romans in Arcetri ... The Florentines would invite us to present our results in the Seminar ... established through the initiative ... of Giorgio Abetti ... I was particularly impressed [by him], an exceptional person endowed with an uncommon charm, who would ask appropriate and interesting questions on any subject with unsurpassable grace and politeness ... I was coming from the Institute of via Panisperna, which was beyond doubt a very well functioning and attractive place. In Arcetri the atmosphere was very different: the interest for music and beauty arts would appear frequently during the work ... or in intervals such as that for having tea, which was prepared by Daria for everybody. An almost imperceptible romantic climate would waft in Arcetri, while in via Panisperna extra-scientific interests were almost exclusively mountain trips and nature ... and*

Fig. 1 Beppo describing "male" (Bad) and "bene" (Good) of Arcetri (December 4, 1987) and the *mystic triangle* made of Abetti, Garbasso and Persico. The writing on the black board is as follows. Bad: accessibility, didactic material, laboratories, workshop, storeroom, general services; Good: library, proximity of the Observatory, accomodation for assistants, motorcars (Persico, Racah, Caponi, Emo: these persons would collect colleagues and students from downtown) Antico Crespino (a "trattoria" at walking distance from the institute), ecstasy of sunsets

international contemporary literature, a field in which Rasetti surpassed everybody ... Also the research subjects were rather different, but all these diversities between Rome and Florence were a reason of attraction between the members of the two groups" (Fig. 1).

G. Occhialini: *"Garbasso, Abetti and Persico ... these persons had in common very important qualities: they [belonged to the category] of professors and scientists who were loved and respected, with no fear, no feeling of awe in front of them ... [and furthermore] a common style, a common attitude towards what would be called Europe ... those aristocratic sages probably had an influence on the members of the laboratory in a notable lack of aggressivity ... [Abetti's] Seminar would bring the name of Florence where it was unknown ... people came from everywhere, such as Hans Bethe, same age as Rossi, already involved in what was to become the Physics of fields ... the Seminar was a high-level club ... but it was not only for senior or junior researchers, but also for students who were striving to become researchers ... So, together with the regular reading of journals promoted by Persico, junior people were put in the condition to have access with up-to-date scientific information to such exclusive Institutes as Rutherford's Cavendish."*

The contribution of Occhialini was made specially amusing and touching by his "Table list of Bad (Male) and Good (Bene) in Arcetri" that he draw on the blackboard in his characteristic humorous way: one misses Beppo's sharp to-the-point commentary.

General References

Dizionario Biografico degli Italiani (Istituto dell'Enciclopedia Italiana, Roma) 1960.
Dizionario Biografico degli Scienziati e Tecnologi (Zanichelli, Bologna) 1999.
Lessico Universale Italiano, (Istituto dell'Enciclopedia Italiana, Roma) 1975.
From *Storia d'Italia*, annali 3, *Scienza e Tecnica* (Einaudi) 1980: contributions of Pietro Redondi "Cultura e Scienza dall'Illuminismo al Positivismo"; Roberto Maiocchi "Il ruolo delle Scienze nello sviluppo industriale italiano"; Giacomo Cavallo e Antonio Messina "L'indagine fisica nel '900 e politica della ricerca".

Acknowledgements The intention of the authors was to give an idea, however sketchy, of the cultural premises from which the group of Florence emerged with its peculiar characteristics. A parallel reference to the group of Rome appeared necessary to make clear similarities and differences, also because of the rich human and intellectual exchanges between the two groups in the few years of a surprisingly productive revival of Italian Physical Sciences, with results lasting in time.

Our thanks are due to many people and organizations. First of all to DARIA BOCCIARELLI (and SERGIO STEVE) for their moving effort in recuperating eight years of old memories and references. The same to MARIA SERENA SCANDONE, acting on behalf of her father. A special acknowledgement for the professional, and generous, work of ANNA CORINNA CITERNESI, editor of the Italian Bibliography of History of Science, who carried out the bibliographic researches and made the consultation of original documents possible.

LUISA BONOLIS provided most of the references of the early scientific production of Bernardini and Bocciarelli and contributed to the understanding of historical and interpretative problems about the origin of the groups of Rome and Florence.

MILLETTA SBRILLI, director of the Centro Archivistico della Scuola Normale Superiore, kindly provided useful information on the scientific staff of the School from the Archivio Storico, minutes of the board of directors, and a list of students from the end of 19th century to early 1930s.

We are indebted to ROBERTO VERGARA CAFFARELLI for a nice biography of Luigi Puccianti and to VALERIA DEL GAMBA, who kindly provided us with information about the life of Rasetti.

The Archivio Storico dell'Università degli Studi Firenze allowed full access to the set of Annuari from 1870 to 1937.

References

1. DRAGONI G., "Augusto Righi", *Museoscienza* Nuova Serie, **6** (1994) 24.
2. MAIOCCHI R., "Le Istituzioni di Ricerca in Italia tra le Due Guerre", in "L'Ottica in Italia fra Otto e Novecento", *Atti della Fondazione Giorgio Ronchi*, **4–5** (2000) 691.
3. TOMASI T. and SISTOLI PAOLI N., "La Scuola Normale di Pisa dal 1813 al 1945" (ETS, Pisa) 1998.

4. GAMBA E., "Il trattato sulla radioattività, 1909, di Angelo Battelli" in *Angelo Battelli (1862–1916). L'uomo, lo scienziato, il politico: atti del Convegno di Studi*, edited by GORGOLINI L. (Società di Studi Storici per il Montefeltro, San Leo) 2005.
5. OCCHIALINI A., "Commemorazione di Angelo Battelli", *Nuovo Cimento*, **13** (1917) 1.
6. SEGRÉ E., *Enrico Fermi Fisico* (Zanichelli, Bologna) 1984.
7. RASETTI F., *"Interview by Judith R. Goodstein"*, CalTech Archives, revised version (Birkhauser Verlag) 2001.
8. LEOPARDI C., "L'Ateneo Fiorentino: dallo Studium Generale (1321) all'Istituto di Studi Superiori (1859)" and MANDÒ M., "Notizie sugli Studi di Fisica 1859-1949", both of them in *Storia dell'Ateneo Fiorentino* (Parretti Grafiche, Firenze) 1986.
9. MINIATI M., "L'Istituto e Museo di Storia della Scienza" in *A brief history of meteorology in Florence*, edited by MARACCHI G. (CESIA, Firenze) 1991.
10. GARBASSO A., "La tradizione del pensiero toscano" Opening Address, *Annuario 1916–1917* del R. Istituto di Studi Superiori Pratici e di Perfezionamento, Firenze (1917).
11. CANZI M. P., "L'evoluzione della spettroscopia in Italia. L'opera di Puccianti e di Garbasso", *Atti del XIV e XV Congresso Nazionale di Storia della Fisica* (1994) 277.
12. LEONE M., PAOLETTI A. and ROBOTTI N., "The Birth of Quantum Mechanics in Italy: the Stark-Lo Surdo Effect" in *History of Physics and Astronomy in Italy in the* 19th *and* 20th *Centuries: Sources, Themes, and International Context, COFIN 2001*, edited by GARIBOLDI L. and TUCCI P., s.n.t., 127–136.
13. BRUNETTI R., "Antonio Garbasso. La vita, il pensiero e l'opera scientifica", *Nuovo Cimento*, **10** (1933) 129.
14. CAMPRINI S. and PORCHEDDU G. B., "La storia degli strumenti di Fisica coincide con la storia della Fisica stessa. Rita Brunetti tra Fisica Sperimentale e Fisica Teorica", *Atti del XVIII Congresso Nazionale di Storia della Fisica e dell'Astronomia*, edited by TUCCI P. (Milano) 1999.
15. AA. VV., *Arcetri. Astronomia a Firenze e il colle di Arcetri* (Cassa di Risparmio di Firenze, Firenze) 1983.
16. CHINNICI I., "Pietro Tacchini (1838-1905): un protagonista dell'astronomia italiana post-unitaria", *Astronomia*, **4** (2005) viii.
17. GASPERINI A., MAZZONI M. and RIGHINI A., "L'evoluzione della Torre Solare di Arcetri nel carteggio Hale-Abetti", *Giornale di Astronomia*, **3** (2004) 23.
18. MAZZONI M., "1926: la rinascita dell'Ottica italiana e gli atti del Rotary Club Firenze", *Giornale di Astronomia*, **3** (2006) 33.
19. TOMMASSINI L., "L'Organizzazione della Scienza tra guerra e dopoguerra - Convegno: L'Ottica in Italia tra Otto e Novecento", *Atti della Fondazione Giorgio Ronchi*, **4–5** (2000) 653.
20. EMANUELLI G., *Scienziati fisico-matematici marchigiani 1846–1951* (STEU, Urbino) 1964.
21. LONGOBARDI G., BORCHI E., CETICA M., FRANCINI F. and LABATE A., *L'ottica e la Toscana* (Nardini Editore, Firenze) 2005. The book contains a wide survey on this topic.
22. RONCHI V., "Perché, quando e come nacque l'Istituto Nazionale di Ottica di Arcetri", *Atti della Fondazione Giorgio Ronchi*, **XLI** (1977).
23. FERMI E. and PONTREMOLI A., "Sulla massa della radiazione in uno spazio vuoto", *Rend. Accad. Lincei*, **32** (1923) 162.
24. PERSICO E. and TIERI L., "The Hall effect in Bismuth solidified in a magnetic field", *Rend. Accad. Lincei*, **30** (1921) 464.
25. ROSSI B., *Momenti nella vita di uno scienziato* (Zanichelli, Bologna) 1987.
26. ROSSI B., "Arcetri 1928-1932", in *Early History of Cosmic Ray Studies. Personal Reminiscences with Old Photographs*, edited by SEKIDO Y. and ELLIOT H. (D. Reidel, Dordrecht) 1985.
27. SALVINI G., "Gilberto Bernardini", *Rend. Suppl. Accad. Lincei*, **9/8** (1997) 75.
28. TAGLIAFERRI G., "Giuseppe Occhialini", *Istituto Lombardo di Scienze e Lettere, Rendiconti* **128** (1994) 231.
29. OCCHIALINI G., "Occhialini, Giuseppe", in *Scienziati e Tecnologi Contemporanei*, Vol. II (Mondadori, Milano) 1974, p. 322.

30. Bernardini G. and Rossi B., "Sull'azione fotografica degli elettroni lenti", *Rend. Accad. Lincei*, **10** (1929) 182.
31. Gariboldi L., "G. Occhialini, his training years (1907-1934)", in *History of Physics and Astronomy in Italy in the* 19th *and* 20th *Centuries: Sources, Themes, and International Context, COFIN 2001*, edited by Gariboldi L. and Tucci P., s.n.t., 137–149.
32. Scarsi L., contribution to *"A meeting in honour of G. O."* Arcetri, December 4, 2004, unpublished.
33. Bothe W. and Kohlhorster W., "Das Wesen der Hohenstrahlung", *Z. Phys.* **56** (1929) 751.
34. Rossi B., "Method of registering Multiple Simultaneous Impulses of several Geiger counters", *Nature*, **125** (1930) 636.
35. Rossi B., "Assorbimento e diffusione della radiazione penetrante nel piombo e nel ferro", *Rend. Accad. Lincei*, **13** (1931) 600.
36. Bernardini G., "The variation of penetrating radiation with zenith distance", *Nature*, **129** (1932) 578.
37. Fermi E. and Rossi B., "Azione del campo magnetico terrestre sulla radiazione penetrante", *Rend. Accad. Lincei*, **17** (1933) 346.
38. Degregorio A. and Sebastiani F., "Le sostanze debolmente radioattive e la radioattività indotta dai neutroni", *Giornale di Fisica*, **46** (2005) 97.
39. Bernardini G., "Sulla tecnica degli spettrografi magnetici per elettroni lenti", *Rend. Accad. Lincei*, **13** (1931) 361.
40. Occhialini G., "Uno spettrografo magnetico per raggi beta emessi da sostanze debolmente radioattive", *Rend. Accad. Lincei*, **14** (1931) 103.
41. Blackett P. M. S. and Occhialini G., "Photography of penetrating corpuscolar radiation", *Nature*, **130** (1932) 363.
42. Blackett P. M. S. and Occhialini G. P. S., "Some photographs of the tracks of penetrating radiation", *Proc. R. Soc. London Ser. A*, **139** (1933) 699.
43. Chadwick J., Blackett P. M. S., Occhialini G., "New evidence for the positive electron", *Nature*, **131** (1933) 473; Chadwick J., Blackett P. M. S., Occhialini G., "Some experiments on the production of positive electron", *Proc. R. Soc. London Ser. A*, **144** (1934) 235.
44. Bernardini G. and Bocciarelli D., "Una registrazione a valvole per coincidenze tra contatori a moltiplicazione", *Rend. Accad. Lincei*, **23** (1936) 265.

Giuseppe Occhialini and the History of Cosmic-Ray Physics in the 1930s: From Florence to Cambridge

Martha Cecilia Bustamante

Introduction

Giuseppe Occhialini's stay in Cambridge at the Cavendish laboratory spanned three years, from 1931 to 1934, although he originally had left Italy for England with the idea of staying only three months. This "Cambridge period" turned out to be most important in his scientific life and established him as a confirmed researcher. The work he performed on cloud chambers, cosmic rays and the positron, with the Cavendish physicist Patrick Maynard Stuart Blackett made him one of the leading figures in the international scientific scene of the thirties. It contributed an essential part of our present-day physical knowledge.

What did Occhialini find at his arrival in Cavendish? How did he organize his life? In what scientific and instrumental context did he work? What could he do with the scientific background he had and within the new Cambridge framework? How did he interact with Blackett? How did he come to contribute major discoveries? Without pretending to completely answer those questions, I will try to recall significant aspects and moments of Occhialini's visit to Cavendish [1]. I will end by referring to Occhialini's return to Italy and to his trip to Brazil.

To the Cavendish

It was in July 1931 that Giuseppe Occhialini eagerly joined the Cavendish laboratory. As a contemporary witness noted, in those days this laboratory was, "with its 'Nursery' and 'Garage'… a dingy, dirty and dismal sort of place built in the nineteenth century" [2]. It also was a place where "everything was improvised… the

M. C. Bustamante (✉)
Equipe Rehseis, CNRS UMR 7596-Université Paris7 Denis-Diderot, Paris, France
e-mail: mcbusta@Paris7.Jussieu.fr

L. Gariboldi et al. (eds.), *The Scientific Legacy of Beppo Occhialini*,
https://doi.org/10.1007/978-3-031-37034-2_7

budget was minimal and expenditure was very tightly controlled. In general, at the Cavendish all the physicists had to make all their apparatus for themselves" [3]. Occhialini easily adapted to these conditions. He had a special gift for constructing devices and had a good training in instrumental work.

Occhialini went to the Cavendish thanks to a CNR (Italy's National Research Council) fellowship that had been intended for Gilberto Bernardini, who instead had to serve in the Army. The idea of Occhialini's visit resulted from a meeting that had taken place in Berlin between Bruno Rossi and Blackett. These two physicists had developed a friendship in this occasion. As Rossi was interested in Blackett's knowledge of cloud chambers, they agreed to send one of Rossi's pupils to Cambridge to learn this technique. Occhialini enthusiastically seized this opportunity, since, following the Roman physicist Emilio Segré, he had been deeply interested in imaging devices and dreamt of building a big cloud chamber [4].

Occhialini was a vivid and perspicacious man, who had engaged scientific studies with some reticence but quickly convinced himself it was the right choice. He came from the University of Florence, where in 1929 he was a fresh undergraduate with a degree in Physics based on a dissertation work in spectroscopy [1]. This work was supervised by his father Raffaele Augusto Occhialini, a well-known physicist and university professor, whose strong character played an important role in young Giuseppe's intellectual and moral development. Occhialini then worked on Geiger-Müller counters and cosmic rays under the supervision of Bruno Rossi, a young and rising star who had joined the University of Florence. He was also initiated into the mysteries of the new quantum physics by Enrico Persico, another brilliant figure, schoolmate of Fermi and professor of theoretical physics at the same University of Florence.

After his graduation, Occhialini became a voluntary research assistant at the Institute of Physics located on the Arcetri Hill, a place Galileo Galilei had once called his home. He became a temporary research assistant the following year, and subsequently obtained a permanent appointment. That was a rather good situation for a young Florentine physicist in the late twenties. At that time the Institute of physics of the University was just emerging. The conditions were poor. The financial situation was precarious, despite the efforts of the director Antonio Garbasso. As Bruno Rossi, then Garbasso's assistant, recalled years later, the Institute was always late in paying its electric bills. The only reason why electricity was not cut off was that the Institute director was the Major of the city [5].

At the Cavendish, Occhialini found himself in a very new and different world. This laboratory was much bigger than the Institute of Physics of the University of Florence. Founded in 1874, it had soon become a place of excellence within the context of a flourishing British scientific development. By the end of the century, the Cavendish laboratory had become one of the premier world centres in experimental physics. When Occhialini arrived, the laboratory was under the direction of Ernest Rutherford, a very prominent scientific personality with a prestige comparable to that of his predecessors James Clerk Maxwell, Lord Rayleigh, and Joseph John Thomson. Rutherford dominated the Cavendish both physically and intellectually [6]. One can imagine that Occhialini, with his vivid nature and enthusiasm, was quite taken by

Rutherford's strong personality. We will see, however, that it was Blackett who deeply and permanently impressed him.

In a short autobiography of 1974, Occhialini recalled the Cavendish at the time of his arrival. Cavendish scientists pivoted around Rutherford and his assistant James Chadwick. Blackett already was amongst the most distinguished ones. Occhialini also noted that many researchers came there from all over the world. The laboratory had become a place of pilgrimage for outstanding young physicists interested in radioactivity and atomic physics [7].

Occhialini enjoyed a friendly community through the academic clubs that met at the Cavendish and elsewhere in Cambridge. These clubs consisted in selected people that met for the purpose of political, intellectual, or scientific discussion. Probably thanks to Blackett, Occhialini was admitted to the Kapitza Club and participated to its session of 1932–1933. This club was a seminar group founded by the Russian physicist Peter Kapitza in 1922. "The original idea (of the seminar) was to provide a forum for unfettered discussion of current developments in physics, freed from ambitions that tented to characterize laboratory discussions, and for theoretical debate, for which there was no provision in the formal university system." Membership was limited and was decided by election. Occhialini thus joined a list of prestigious names. The club also had distinguished foreign guests such as James Frank in 1924 and Niels Bohr in 1925. Originally the members were almost all, apart from Kapitza himself, primarily theoreticians, Paul Dirac and D. R. Hartree among them. As the years passed, membership increased and the Kapitza Club began to include eminent experimentalists. Blackett himself was a member. Contrary to other academic Cambridge clubs, such as the $\nabla^2 V$ Club, which was mainly for mathematical physics, the Kapitza club "was one in which the barrier between theory and experiment was all but obliterated" [8].

Occhialini was lucky to come to the Cavendish laboratory at the time he did, for Kapitza's stimulating presence at the club had little time left. As is well known, the Russian physicist went to visit his homeland soon after the summer of 1934 and was unable to return. When Occhialini arrived, the club had held 377 meetings, all of them with Kapitza in charge [8]. The club's activities consisted of informal discussions and readings of papers. Meetings took place in College rooms, often in Dirac's room in St John's College, and once a week, probably on Tuesdays and after dinner. One can imagine how stimulating these meetings were for Occhialini.

Electric Counters and Cloud Chambers

In the Cavendish Laboratory, Giuseppe Occhialini joined P. M. S. Blackett. It was maybe the first time the two physicists met. Occhialini was 24 years old and Blackett ten years older. Blackett was a member of Rutherford's teaching staff, having been appointed as a university lecturer. Occhialini and Blackett each had their own instrumental and experimental specialties. Occhialini was an expert on Geiger-Müller counters, Blackett was a world authority in cloud chambers.

Patrick Blackett was an experimentalist, a scientific "Jack of all trades", as he characterized himself. Instruments and devices were for him the means through which he could illustrate his gifts. So, for a physicist like him, Occhialini's arrival to Cavendish presented promising prospects. In Italy Giuseppe Occhialini had been trained in a very modern technique. Geiger-Müller counters had only been invented by H. Geiger and W. Müller in 1928. In Florence, Rossi had the idea of placing two counters in coincidence in order to detect cosmic ray events. Occhialini's early research and first paper, the only one he had published by the time he went to Cambridge concerned this subject. There Occhialini presented a magnetic spectrograph for β-rays emitted by weak-radioactive sources [such as Rubidium and Potassium], a device used in one of the first applications that the Florence group made of the counter technique [9].

Geiger-Müller counters consist of a metal cylinder with a thin metal wire stretched along the axis and held by insulators. The cylinder is filled with a gas at a pressure of a fraction of an atmosphere and a potential difference is applied between the tube and the wire. An energetic charged particle entering the cylinder can ionize it and induce an avalanche of electrons that reached the wire until and full discharge. As Bruno Rossi and his pupils in Italy quickly understood, Geiger-Müller counters are very sensitive, being able to react even to a single ionization anywhere in the tube. They were suitable for the detection of beta and gamma rays and were inexpensive. This is why the instrument was adopted by Bruno Rossi. It was just what a small laboratory like his needed [10].

In contrast, Geiger-Müller counters were not really in usage at Cavendish. However, as early as 1908 Rutherford and Geiger had collaborated in Manchester to devise an electric particle detector, which proved effective but not reliable [11]. Another detector of the same kind was used at the Cavendish towards 1930.

Blackett's own speciality, the cloud chamber, was a typical Cavendish technique. C. T. R. Wilson, its inventor, was a Cavendish professor when he invented this device towards the end of the nineteenth century. Blackett had attended Wilson's lectures on light just after the war [12]. The device resulted from the confluence of two distinct fields, ionic physics and meteorology. Wilson, who started his research in 1895, investigated the condensation of water drops in moist air and the reproduction of atmospheric phenomena: rain, hail, fog. Studying condensation phenomena he "intensified his efforts to see the drop formation itself" and thus invented a technique that rendered visible the path of ionizing agents [13]. The operation was based upon the property of ions to serve as centres for the formation of droplets from a supersaturated vapour. When a charged particle passes through a vessel containing a gas saturated with vapour, ions are formed all along the trajectory. By suddenly increasing the volume (sudden expansion), the vapour begins to condense around whatever ions are present in the chamber. A cloud of very minute droplets then appears showing the trajectory of the charged particle. This array of droplets can be observed visually and recorded photographically for quantitative work.

In 1923, Rutherford wanted Blackett to study nuclear reactions with the cloud chamber, hoping it would reveal more details of the nuclear collision process than the earlier technique of scintillations. This was a significant turn on Rutherford's part. Indeed he had been using the technique of scintillations since the beginning of the century and much of his Manchester work depended on it. This technique involved scintillations caused by the impact of the fast particles on a zinc sulphide screen and observed through a low-power microscope. Being visual, it predisposed Rutherford to use the equally visual cloud-chamber technique.

Blackett took over work of a Japanese student (Takeo Shimizu) who had already been looking for nuclear disintegrations with a cloud chamber. Blackett was a former naval officer with a strong education in the sciences and engineering arts. In the span of a few months in 1924 he managed to set-up a chamber which took half a million of photographs whereas Shimizu only had a few thousands. Blackett improved the suddenness and the repeatability of the expansion by means of a simple string device. He also arranged that the photograph was taken just as the expansion was completed. Among his numerous photographs he found a few representing alpha disintegration processes.

In the following years, Blackett devoted himself to the development of this automatic cloud chamber and to its application to the field of nuclear physics. By the end of the 1920s, he had become a leading figure among Cavendish physicists, and an international authority in cloud-chamber methods and nuclear physics. At the same moment, after almost ten years of collaboration, Blackett's collaboration with Rutherford came to an end [14]. So the way was free for him to switch over to cosmic radiation with the help of his Italian guest Occhialini.

Cosmic Rays

Perhaps Occhialini, who had been trained in cosmic-rays studies, suggested to Blackett to work on this subject. Or perhaps Blackett got the idea from Bruno Rossi during their Berlin meeting. Cosmic-ray physics was a relatively new branch of physics. It had not yet penetrated the Cavendish laboratory but had already received considerable attention in other centres.

Robert Millikan himself had given the name of cosmic radiation to the phenomenon discovered at the beginning of the century by the Austrian physicist Victor Hess. Millikan was a world authority on that matter. He had founded a laboratory in the California Institute of Technology in order to study the ultra-penetrating radiation as it was also named at the time. The idea was to improve observations on balloons with electroscopes and ionization chambers. Millikan's program, launched in the mid-1920s, required important resources. The observations were made out of doors, at sea level, at high altitude, on the top of the California mountains, and in the upper atmosphere. It was necessary to transport very fragile and heavy devices. In order to account for the wealth of results obtained in his powerful laboratory, Millikan proposed what he called the "birth cry theory", a set of more or less scientific conjectures about the origin and the nature of cosmic radiation. In particular, he believed that the cosmic rays could only be hard gamma rays. If the cosmic rays were charged particles, Millikan reasoned, they would not have sufficient energy to penetrate, as they did, the air and water equivalent of a few centimetres of lead. This hypothesis roughly agreed with empirical results.

Occhialini had another conception of cosmic rays to transmit to Blackett. Rossi and his students were aware of the experiment performed in Germany by Bothe and Kolhörster in 1929 that seemed to refute the gamma-ray hypothesis and instead suggested that the cosmic rays were mainly composed by charged particles. Part of their program was to confirm the latter hypothesis by improving the counter technique.

In their experiment, Bothe and Kolhörster had placed two Geiger-Müller counters one above the other a small distance apart and recorded the simultaneous pulses (coincidences) due to the passage of individual particles through both counters. They recorded the coincidences by connecting the wires of the two counters to two separate electroscopes. They obtained only doubles coincidences and the method was very cumbersome. In order to improve this technique, Rossi devised a neat valve circuit that easily recorded coincidences of any order [10]. It was mainly from this work that Occhialini acquired the expertise that he brought to the Cavendish.

Collaboration with Blackett

Occhialini and Blackett together launched a new Cavendish project focused on cosmic-ray studies by means of cloud chambers. Physicists elsewhere had already demonstrated that cloud chambers were well adapted to the observation of the ultra-penetrating radiation. In 1927, Russian physicist Dimitri Skobeltzyn, while investigating the β-rays from radioactive sources by means of a cloud chamber and a deflecting magnetic field, noticed a few tracks of unusually high-energy, "penetrating rays", which occurred in small groups [15]. He thus opened the door to programmatic observations of cosmic rays by using cloud chambers in magnetic fields. Pierre Auger in Paris, and Millikan and his student Carl Anderson in Caltech, launched such programs shortly afterwards.

To follow the way that Skobeltzyn's observations had suggested was not an easy task for Blackett and Occhialini. In the area of cosmic radiation the received cloud chamber method had serious drawbacks. As it had a very short sensitive time and as the expansions were made at random, cosmic ray tracks were found only on a very small fraction of the photographs. Blackett and Occhialini's new set-up made use of Geiger-Müller counters for triggering the chamber, so that "particles of high energy took their own photographs" (see Figs. 1 and 2). Occhialini brought to the project his expertise with counters and the coincidence method. The two physicists placed one Geiger-Müller counter above the chamber and another below, and arranged their coincidence circuit so that only particles passing through both counters, and therefore lying in the plane of the chamber, triggered the expansion. The average wait after the mechanism was set up was about 2 min. In a laboratory where electric counters were merely beginning to be used, it was a very innovative idea. It proves

Fig. 1 Blackett and Occhialini's counter cloud chamber

immensely successful one. The efficiency of the instrument in observing cosmic rays was improved by a factor of about 1000. We will see later which results they obtained.

No doubt, Blackett's background knowledge in cloud chambers together with Occhialini's knowledge in tube counters and coincidence method proved decisive to the outcome of the project. For instance, the circuit associated with the counters still presented difficulties and to make them work was not easy, furthermore, in the counter-controlled technique, the ray crosses the chamber before the expansion and distortion occurs before the drops are formed. The best way of dealing with this source of distortion is to make sure the ray has little time to diffuse before the expansion is complete. It requires the chamber to expand promptly after the passage of the cosmic particle. Blackett made theoretical considerations about ion diffusion to know the time required to obtain tracks of no more than a given width.

Historical studies on the positron's discovery give us a detailed account of the work Occhialini and Blackett made at Cavendish with the novel chamber [16]. Fortunately, Occhialini could see his Italian grant extended (he also received an occasional financial support from Rutherford when he showed him the photographs of cosmic-ray showers). Blackett and Occhialini published four papers together, the two last of which were co-signed by James Chadwick. The first one is a letter to "Nature" dated 21 August, 1932 and entitled "Photography of Penetrating Corpuscular Radiation" [17]. This letter was the second Occhialini's scientific publications, appearing almost one year after the arrival of the physicist at the Cavendish.

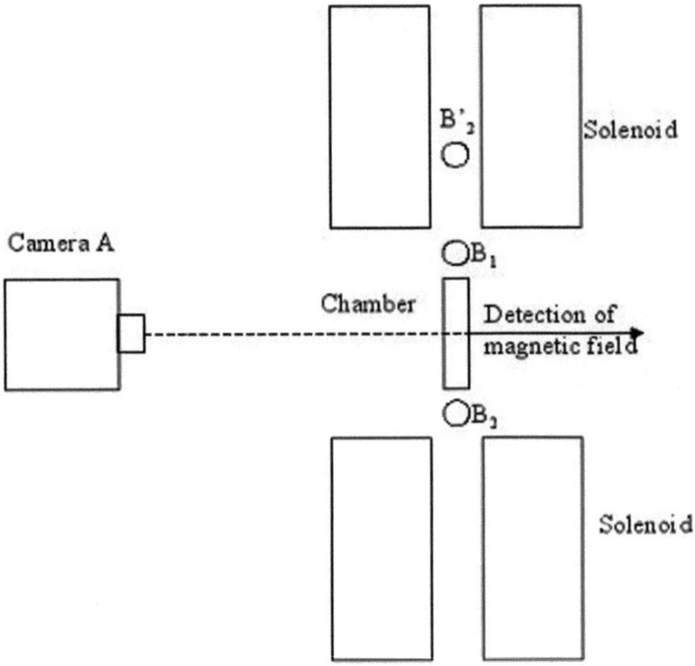

Fig. 2 Blackett and Occhialini's experimental set-up. B1 and B2 indicate the positions of the counters. (See P. M. S. Blackett and G. P. S. Occhialini, "Some photographs of the tracks of Penetrating radiation", *Proceedings of the Royal Society, Ser. A*, **139** (1933) 699–727)

The authors describe the new cloud chamber triggered by counters and present a set of results. They do not talk about the several instrumental and experimental tests they must have done so the instrument would work as they wanted. They emphasize the speed of their technique: with the chamber placed in a magnetic field of 2000 gauss the average wait for the mechanism, as noted above, was about 2 min, at that time they obtained one hundred stereoscopic pairs of photographs, among them 59 showed the track of a single high-speed particle passing through both counters; 17 showed either multiple tracks of varying degrees of complexity, or else a single track passing through one and not through both counters; 24 showed no tracks. None of the straight tracks showed a measurable curvature thus indicating that their energy must have been greater than 600 MeV (if electrons) or 200 MeV (if protons). In 17 of the photographs showing multiple tracks, only about 10% were bent by the field, indicating energies of the order 2–20 MeV (if electrons) [17].

Blackett and Occhialini's second paper was communicated by Rutherford to the Royal Society five months later. It is an extension of the first work, with some very new results. It is a long article, 27 pages including fifteen photographs. This is the historical paper, "Some photographs of the tracks of Penetrating radiation" [18] (see Fig. 3).

699

Some Photographs of the Tracks of Penetrating Radiation.

By P. M. S. Blackett and G. P. S. Occhialini, The Cavendish Laboratory, Cambridge University.

(Communicated by Lord Rutherford, O.M., F.R.S.—Received February 7, 1933.)

[Plates 21-24.]

1. *The Experimental Method.*

We have recently developed a method by which the high speed particles associated with penetrating radiation can be made to take their own cloud photographs.* By this means it is possible to obtain these photographs very much more speedily than by the usual method of making expansions at random. For when this latter method is used it is only on a small fraction of the photographs that a track will be found. The average number of photographs required to obtain one track will depend on the size and orientation of the chamber and on the effective time of expansion. The latter is not likely to be more than 1/20 second. From measurements with counters it is known that about 1·5 fast particles fall, from all directions, on 1 sq. cm. per second. Roughly consistent with these figures are the results found with cloud chambers. Skobelzyn† has obtained as many as one track every ten expansions, but in the work of Anderson‡ the number of tracks which were long enough to be suitable for energy measurements was only about 1 in 50 photographs. By our method, tracks are found on 80 per cent. of the photographs. We intend to give a full account of the technique of this method of photography in a separate paper, confining ourselves here to a rough outline only.§

A cloud chamber of diameter 13 cm. and depth 3 cm. is arranged with its plane vertical and two Geiger-Müller counters, each 10 cm. by 2 cm. are placed one above and one below the chamber so that any ray which passes straight through both counters will also pass through the illuminated part of the chamber, fig 2. An alternative arrangement is described on p. 719. The

* 'Nature,' vol. 130, p. 363 (1932).

† 'C. R. Acad. Sci. Paris,' vol. 195, p. 315 (1932).

‡ 'Phys. Rev.,' vol. 41, p. 405 (1932).

§ Mott-Smith and Locher 'Phys. Rev.,' vol. 38, p. 1399 (1931); vol. 39, p. 883 (1932) have found a correlation between the occurrence of these tracks and the discharge of a counter, and Johnson, Fleischer and Street, 'Phys. Rev.,' vol. 40, p. 1048 (1932) have used coincidences to operate the illuminating flash of a continuously working closed chamber.

VOL. CXXXIX.—A. 3 A

Fig. 3 Front page of Blackett and Occhialini's historical paper, *Proceedings of the Royal Society, Ser. A*, **139** (1933) 699–727

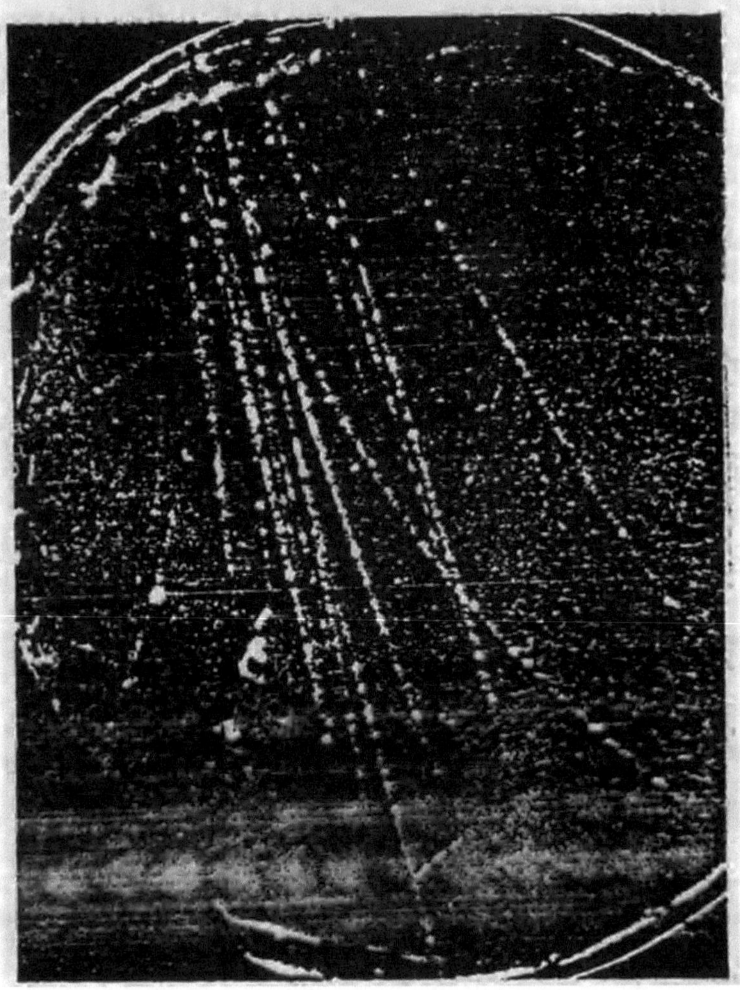

Fig. 4 Stereoscopic counter cloud chamber photograph of a shower. Reprinted from P. M. S. Blackett and G. P. S. Occhialini, "Some photographs of the tracks of Penetrating radiation", *Proceedings of the Royal Society, Ser. A*, **139** (1933) 699–727, Plate 22. Figures 2, 3 and 4 are reproduced with kind permission of the Royal Society

As is well known, Blackett and Occhialini obtained photographs showing multiple tracks caused by showers of particles diverging from a region over the chamber (see Fig. 4). Among them were tracks which the two physicists interpreted as being due to positively charged particles with a mass comparable with that of an electron rather than with that of a proton. By systematically analysing the curvature, range and ionization, they eliminated many other likely hypotheses and obtained evidence supporting the existence of positive electrons. They pointed out arguments in favour of the view that in the showers the main beam of downward-moving particles consisted mainly of positive and negative electrons in about equal numbers. At the same time, Blackett and Occhialini pointed out the relation between the particle they had observed and the anti-electron predicted by Dirac's relativistic theory of the electron.

As is also well known, Carl Anderson, Millikan's assistant at Caltech, preceded Blackett and Occhialini by a few months by presenting in a short article in the August 1932 issue of "Nature" experimental evidence for a positive electron [19]. In a non-triggered cloud chamber he had observed positive tracks corresponding to a mass much less that that of a proton. Contrary to Blackett and Occhialini, Anderson did not refer to Dirac's theory. Although this theory was not easily accepted by Rutherford, Chadwick and other Cavendish physicists, Blackett and Occhialini, being familiar with Dirac's ideas and being in frequent contact with Dirac in the Kapitza club, did not hesitate to take it into account. As for the mechanism by which the showers might be produced, they suggested that the electrons were created during collisions and that non-ionizing links existed in the showers [18].

A few months later, Blackett and Occhialini oriented their search towards the production of positive electrons by means of radioactive sources. Chadwick joined them and brought his expertise on neutron sources [20]. The three physicists looked for positive electrons produced in the interaction of neutrons with matter. Their results appeared in the April 1933 issue of "Nature". A lead target (2mm thick) fixed inside the chamber was irradiated with a polonium source and a piece of beryllium, contained in a small container, placed close to the wall of the cloud chamber. The lead was thus exposed to the gamma rays and neutrons emitted from the beryllium. Measurements of the ionization and curvature of the tracks supported the view that positive electrons were being created in the lead target. No evidence was obtained about the manner of their production, whether caused by the neutrons or by the gamma-rays hitting the target [21]. In a matter of days, this question was settled by Curie and Joliot in Paris: their absorption experiments proved that the gamma-rays, not the neutrons, were responsible for the production of the positive electrons [20].

A second paper of Blackett, Occhialini and Chadwick, published in 1934, dealt with the positive electrons produced by the interaction of the nearly homogeneous gamma radiation from thorium C with matter. This was a natural extension of the earlier work using polonium as a source with beryllium. With the thorium C, they obtained 4000 tracks of positrons were studied [22]. This was the last Cavendish work to which Occhialini contributed.

Indubitably, Giuseppe Occhialini's visit to Cambridge and his collaboration with P. M. S. Blackett were very successful. As he noted at the Memorial meeting for the English physicist held at the Royal Society in October 1974 and one of the very few non-scientific papers he wrote, it was "a wonderful time in Cambridge." Occhialini thus befriended Blackett, a man he greatly admired. In his autobiography, he judges that besides his father, the person who most influenced him was Blackett. Occhialini spoke of Blackett's clarity and integrity, his independence of mind and passionate sense of justice. Recalling his time at the Cavendish, he noted: "What not everyone had the chance to see was the passionate intensity with which Blackett worked. I can still see him, that Saturday morning when we first ran the chamber, bursting out of the dark room with four dripping photographic plates held high and shouting for all the Cavendish to hear *one of each, Beppe, one of each!*" [23].

The feelings between Blackett and Occhialini were reciprocal. Blackett recognized Occhialini's scientific qualities as those on an experimentalist of the first order, with physical intuition and great skills and capacity at imagining devises. No doubt he also felt Occhialini's brilliance, enthusiasm and charisma. In 1948 the physics Nobel Prize was awarded to Blackett, "for the development of the Wilson cloud chamber and his discoveries therewith in the fields of nuclear physics and cosmic rays". Occhialini went unrewarded. The attitude Blackett had at this occasion sheds light on his feelings towards Occhialini. He sent a telegram to Occhialini telling him "Caro Beppo we are very happy but it would never have happened without you" [1]. To Occhialini's father he wrote: "I would have been happier if Beppe had been honoured at the same time. For it was certainly his arrival in Cambridge which stimulated my embarking on the field of cosmic rays, which I have never left. And our work together in 1932-1933 was a real collaboration of the happiest kind" [1]. Actually, Occhialini's contribution to the positron discovery did not go entirely unrecognized by peers. He was nominated for the Nobel Prize. In 1948 the prize was awarded to Blackett, but initial nominations of Blackett occurred years before, in 1935, as they did for Anderson. On the occasion of a nomination of 1936, French physicist Louis de Broglie suggested one-half of the prize to Anderson and one-half to Blackett and Occhialini[1] [24].

To Brazil

Occhialini came back to Italy in 1934. He returned to the Institute of physics of the University of Florence where he had a permanent position. He did some work on radioactivity [25]. He began also to teach as a professor in the faculty of architecture (1936) of the university. For his positron work with Blackett, he was awarded the Premio Sella dell'Accademia dei Lincei and the Premio Vallauri dell'Accademia delle Scienze di Torino [1].

However, at Arcetri the institute was no longer as before. Garbasso had passed away prematurely. Rossi held a professorship at Padua and Persico at Torino. Globally, things had changed in Italy. Owing to the rise of fascism, the political climate had become unbearable to Occhialini. He was a liberal and had become markedly anti-fascist, although he had adhered to the fascist party in the 1920s. In 1932, at the moment of the Spanish and Abyssinian events he had to stay in the party in order to save his job, but it was a tormenting decision [1].

In 1937 Occhialini accepted a professorship in Brazil offered by the Italian government and perhaps mediated by his father. He joined the University of São Paulo which had being recently created in 1934. The physics department was under the direction of Gleb Wataghin, an Italian-naturalized physicist, born in Ukraine. With

[1] During the Cambridge period, Occhialini and Chadwick visited together Parisian laboratories.

Wataghin, whom he had already met during a visit to the Cavendish, Occhialini worked to develop Brazilian scientific activity. The first as an experimentalist and the second as a theoretician, they launched cosmic-ray physics in this country. This was an adequate research topic for a country like Brazil, as it had been for Italy, because it only required small laboratories.

Although the conditions were not to be as favourable as in Europe, Occhialini had a real opportunity to extend his scientific career. He held several chairs at the University of São Paulo, on general and experimental physics. Giulio Cesare Lattes soon became his most prominent student. Other gifted students of his were Mario Schönberg, Marcelo Damy de Souza Santos, Paulus Aulus Pompéia, some of them coming from the Ecole Polytechnique which had been annexed to the new university. Occhialini joined his forces to those of Wataghin in forming these young pupils and creating São Paulo's school of physics. They succeeded in creating an atmosphere of intellectual pleasure and excellence. They saw to it that the university had a good scientific library and followed Europeans trends in its scientific activities [26]. Although the task was not easy, Occhialini and his students built a controlled cloud chamber.[2] With Wataghin he shared an interest for large cosmic showers. With his students, Occhialini focused on the utilisation of counters of particles for absorption measurements.[3]

Occhialini, Wataghin, and their students contributed to the success of a symposium on cosmic rays held in Rio de Janeiro on August 1941. This symposium was held under the auspices of the Brazilian Academy of sciences, on the occasion of the visit of an American scientific delegation headed by Arthur H. Compton. Since the late 1920s Compton was a world authority on cosmic rays as was the case for Millikan, although they were working in an opposite direction because Compton, like Rossi, favoured the corpuscular hypothesis for cosmic rays rather than the γ-ray hypothesis. The delegation was travelling in South America to measure cosmic radiation as part of the program of research of the Ryerson Physical Laboratory of the University of Chicago where Compton held a chair.

The symposium was a large and successful one. The proceedings published by the Brazilian Academy reflect contributions by members of the American delegation and by specialists from several Brazilian academic institutions: the University of São Paulo (with Occhialini, Wataghin and their students), the National Institute of Technology, and the Naval School among others. Occhialini himself presented several papers: "The influence of a solar eclipse on the cosmic ray intensity", "On the ultra-soft component of the cosmic radiation", and "On the properties of cosmic radiation in the temperate and equatorial zones" relating absorption experiments that he and his students performed with counters in the laboratory or outdoors. In collaboration with de Souza Santos, he also reported on "Two useful gadgets for controlled Wilson chambers", that is, an asymmetrical multivibrator circuit and a reliable valve system

[2] Interview of G. P. S. Occhialini by Martha Cecilia Bustamante and Antonio Augusto Passos Videira, Paris 1992.

[3] Besides working on cosmic rays, Occhialini went on to be interested in studies on Rubidium β-Radioactivity. It is an extension of the work he had made in Italy from 1934 until he left.

for controlled cloud chambers that improved over the conventional systems earlier described by Blackett [27].

Soon after this symposium, the international political situation degraded. In August 1942 Brazil entered the war, the group held by the two Italians at the University of São Paulo dissolved, and Wataghin left the direction of the Physics Institute. Out of consideration for Occhialini he could not take position against fascist Italy, for Occhialini's father was living there and risked reprisals. But Occhialini became an enemy alien and had to give up his position. He took a pleasant refuge in the Itatiaya Mountains near São Paulo. There he lived in a meteorological hut, as an expert mountain climber and as a guide. Reportedly he wrote "an excellent [unpublished] guide book on that range" [4]. After the Italian armistice in September 1943, C. Chagas, Nobel-Prize winner in Medicine in 1921, invited him to the Biophysics Laboratory in Rio de Janeiro. During this stay in Chaga's laboratory, Occhialini worked on techniques of photography and plate processing. At the end of 1944 he left Brazil and returned to England.

During his stay in Brazil, Occhialini's largely contributed to the new scientific impulsion desired by the Brazilian government in the early 1930s. Yet the Brazilian episode, like the Cambridge one, was just another chapter in his scientific life. Others events, others circumstances and discoveries that significantly contributed to nuclear and cosmic-ray physics were lying before him. Occhialini left Brazil for England because Blackett had invited him to join the British team working with Americans on the atomic bomb. Although he made the trip, he declined participation in that military research, owing to Italian nationality. Instead he went to Bristol to collaborate with Cecil Powell, again thanks to Blackett's help. As in his collaboration with Blackett, Occhialini once again left his mark. He was the decisive partner in the research work that led to the discovery of the pion in the cosmic radiation in 1947 and played a very important part in the development of the emulsion technique and its effective utilization by Powell's group. With Occhialini, nuclear emulsion technique came into great prominence.

References

1. The Occhialini Archive, deposited at the University of Milan in the Physics department has no documents on the Cambridge period. In a letter to Carl D. Anderson, Occhialini explained: "I thought that I had some documentation of the Wilson chamber work in the early days with Blackett, but during my peregrinations in England, Belgium and Brazil, this material has been lost." (Letter to C. D. Anderson, Occhialini Archive). However, the whole of Occhialini's correspondence had not yet been examined. We may hope that in this supplementary source, more information about Occhialini's days in Cambridge will be unearthed.
2. Hendry J. (Editor), *Cambridge Physics in the thirties* (Adam Hilger, Bristol) 1984, p. 66.
3. Blackett P. M. S. "The Old Days of the Cavendish", *Rivista del Nuovo Cimento*, **1**, Special Issue (1969) xxxii–xxxix.
4. Bignami G. "Giuseppe Paolo Stanislao Occhialini. 5 December 1907 - 30 December 1993", *Biographical Memoirs of Fellows of the Royal Society of London*, **48** (2002) 218–222.
5. Rossi B. *Moments in the Life of a Scientist* (Cambridge University Press, Cambridge) 1990.

6. HENDRY J., *op. cit.*, p. 68.
7. See the entry "Occhialini Giuseppe", in *Scienziati e Tecnologi Contemporanei*, vol. **2** (Mondadori, Milano) 1974, pp. 322–324.
8. HENDRY J., *op. cit.*, pp. 109–110. On Kapitza Seminar see also: Kragh H. S. *Dirac: A Scientific Biography* (Cambridge University Press, Cambridge) 1990.
9. OCCHIALINI G. "Uno spettrografo magnetico per raggi β emessi da sostanze debolmente radioattive", *Rendiconti della Reale Accademia dei Lincei*, **14** (1931) 103–107.
10. BUSTAMANTE M. C. "Bruno Rossi au début des années trente : une étape décisive dans la physique des rayons cosmiques", *Archives internationales d'histoire des sciences*, **44** (1994) 92–115.
11. HENDRY J., *op. cit.*, p. 111.
12. LOVELL B. "Patrick Maynard Stuart Blackett, Baron Blackett, of Chelsea", *Biographical Memoirs of Fellows of the Royal Society of London*, **21** (1975) 1–115.
13. GALISON P. and ASSMUS A. "Artificial Clouds, real Particles", in *The Uses of Experiment. Studies in the Natural Sciences*, edited by Gooding D., Pinch T. and Schaffer S. (Cambridge University Press, Cambridge) 1989, pp. 225–274.
14. BUSTAMANTE M. C. "Blackett's Experimental Researches on the Energy of Cosmic Rays", *Archives internationales d'histoire des sciences*, **47**, No. 138 (1997) 108–141.
15. SKOBELTZYN D. "The Early Stage of Cosmic Ray Particle Research", in *Early History of Cosmic Ray Studies*, edited by Sekido Y. and Elliot H., (D. Reidel Publish. Co., Dordrecht) 1985, pp. 47–52.
16. DE MARIA M. and RUSSO A. "The Discovery of the Positron", *Rivista di Storia della Scienza*, **2** (1985) 237–286.
17. BLACKETT P. M. S. and OCCHIALINI G. P. S. "Photography of Penetrating Corpuscular Radiation", *Nature*, **130** (1932) 363 (in Letters to the Editor).
18. BLACKETT P. M. S. and OCCHIALINI G. P. S. "Some Photographs of the Tracks of Penetrating Radiation", *Proceedings of the Royal Society of London, Ser. A*, **139** (1933) 699–727.
19. ANDERSON C. D. "The Apparent Existence of Easily Deflectable Positives", *Science*, **76** (1932) 238–239.
20. ROQUÉ X. "The Manufacture of the Positron", *Studies in History and Philosophy of Modern Physics*, **28** (1997) 73–129.
21. CHADWICK J., BLACKETT P. M. S. and OCCHIALINI G. P. S. "New Evidence for the Positive Electron", *Nature*, **131** (1933) 473.
22. CHADWICK J., BLACKETT P. M. S. and OCCHIALINI G. P. S. "Some Experiments on the Production of positive Electrons", *Proceedings of the Royal Society, Ser. A*, **144** (1934) 235–249.
23. OCCHIALINI G. P. S. contribution to the "Memorial Meeting for Lord Blackett, O.M., C.H., F.R.S. at the Royal Society on 31 October 1974", *Notes and Records of the Royal Society of London*, **29** (1975) 144–146.
24. NYE M. J. *Blackett. Physics, War, and Politics in the Twentieth Century* (Harvard University Press, Cambridge) 2004.
25. OCCHIALINI G. P. S. "La radiazione gamma del Polonio-Berillio", *Rendiconti della Reale Accademia dei Lincei*, **25** (1937) 188–194.
26. PASSOS VIDEIRA A. A. and BUSTAMANTE M. C. "Gleb Wataghin en la Universidad de São Paulo: un momento culminante de la ciencia brasileña", *Quipu*, **10**, no. 3 (1993) 263–284.
27. See DAMY DE SOUZA SANTOS M. and OCCHIALINI G. P. S. *Symposium sôbre Raios Cosmicos, Rio de Janeiro, agôsto 4-8* 1941 (Imprensa Nacional) 1948.

Occhialini's Trajectory in Latin America

Ana M. Ribeiro de Andrade

Introduction

Giuseppe Occhialini treated mathematics with limited deference but was a master in conducting experiments, which revealed essential issues resulting from the observation of nature. His role in the training of the first generation of Brazilian physicists at the recently created Universidade de São Paulo (USP) was an important one. The study of cosmic rays, which at the time was considered an area of very high-energy nuclear physics, fascinated experimental and theoretical physicists at the university. Since cosmic rays are present anywhere in nature and since it is possible to detect them at any altitude, research in this area could be conducted in small laboratories at any location, and with limited financial resources. It was therefore well suited to Brazil's economic-financial reality and research could be conducted at a similar level to that of other countries.

It is significant that foreign professors have been present at USP since its creation in 1934. Competence was an important consideration in the hiring of the first foreign professors in Europe, but ideology and social relations interfered in the process. The Italian colony in the city of São Paulo was quite influential, so in order to protect the social and human sciences from fascist influence the organizers at the Faculdade de Filosofia, Ciências e Letras (USP) reserved the mathematics and physics courses for Italian professors. In this manner, pressured by his family to stay away from political movements in Italy, it was not hard for Occhialini to exchange the Università di Firenze for USP. Not only was his father friends with Gleb Wataghin, head of the Physics Department, but the process was made easy because of his scientific credentials, obtained after his collaboration with Patrick Blackett at the Cavendish

A. M. Ribeiro de Andrade (✉)
Museu de Astronomia e Ciências Afins, MAST/MCT, Rio de Janeiro, Brazil
e-mail: anaribeiro@mast.br

L. Gariboldi et al. (eds.), *The Scientific Legacy of Beppo Occhialini*,
https://doi.org/10.1007/978-3-031-37034-2_8

Laboratory in Cambridge in the perfection of the Wilson Chamber using the coincidence counter technique.[1]

With his introduction of a new utilisation method of the Wilson Chamber at USP, he not only strengthened the group led by Wataghin but also became a leader in the experimental area and redirected the scientific career of at least two Brazilian students, Ugo Camerini and Cesar Lattes, with whom he later worked at the University of Bristol and the Centro Brasileiro de Pesquisas Físicas. Because of his charisma, coupled with his interest for art and literature, he made many friends in São Paulo and Rio de Janeiro, and also contributed to setting up the Laboratorio de Fisica Cósmica de Chacaltaya in Bolivia.

Physics at USP

In contrast with the precarious conditions in the Physics Department, by the end of the 1930's the two approaches to physics, experimental and theoretical, had already been consolidated at USP and research results were comparable to those obtained in European and U.S. laboratories. During Occhialini's five years at the institution (1937-1942) its dynamic scientific and intellectual environment was an uncommon feature in science in a peripherical country [8]. Aside from Wataghin and Occhialini, the theoretical physicist Mario Schenberg was emerging in the field after having worked with G. Gamov and others in the U.S. He states that Occhialini was very influential in his training [2].

In this manner the activities in theoretical physics were led by Wataghin and Schenberg—who offered courses in celestial mechanics, quantum electrodynamics, etc.—while the activities in experimental physics reached a much higher level under Occhialini's management. Unlike professors who taught in a traditional manner, they based their courses on seminars about recent themes and subjects published in international journals [9]. There was a close relationship between the two groups, both of which worked very hard. While on a ship headed for Italy in 1938, where Schenberg was going to participate in an internship with E. Fermi and later with W. Pauli, and where Occhialini was going on holiday, they wrote a paper together on the variation of the intensity of cosmic ray showers with latitude (Table 1, article 1). The following year Schenberg engaged in a theoretical study of the phenomenon of penetrant showers or the multiple and simultaneous production of mesons which Wataghin and his assistants, along with the experimental physicists Marcello Damy de Souza Santos and Paulus Pompéia, had shown experimentally. The group's work [10] resulted in an international cooperation between USP and the University of Chicago, the purpose

[1] With the exception of Occhialini, the biographies and the scientific trajectories followed by the pioneers of physics are registered in interviews. They are an important source of facts, for the accounts allow one to become acquainted with the professional environment and to identify the networks of scientific sociability. About Occhialini's activities in Brazil see: the interviews with G. Wataghin, M. Schenberg, M. Damy de Souza Santos, J. Leite Lopes and other Brazilian physicists [1–4]. About USP see also [5–7].

Table 1 Occhialini's scientific production in Brazil

1. Occhialini G., Schönberg M., "Sobre uma componente ultra molle de radiação cósmica", *Annaes da Academia Brasileira de Ciências*, **11**, No. 4 (1939) 351–355; **12**, No. 3 (1940) 197–202
2. Monteux Y., Occhialini G., "Sur un nouveau type de compteurs plans", *Annaes da Academia Brasileira de Ciências*, **12**, No. 2 (1940) 125–129
3. Occhialini G., "Contributo allo studio dell'effetto di latitudine per gli sciami", *Annaes da Academia Brasileira de Ciências*, **12**, No. 1 (1940) 39–44
4. Occhialini G., "Sur la radioactivité beta du rubidium", *Annaes da Academia Brasileira de Ciências*, **12**, No. 2 (1940) 155–158
5. Occhialini G., Pompéia P. A., Saboya J. A. R., "Nota sobre a estabilização de tensão em corrente alternada", *Annaes da Academia Brasileira de Ciências*, **12**, No. 4 (1940) 349–352
6. Occhialini G., Santos M. D. S., "On a method of recording random events", *Annaes da Academia Brasileira de Ciências*, **13**, No. 1 (1941) 57–62
7. Occhialini G., Santos M. D. S., "Two useful gadgets for controlled Wilson Chambers", in *Symposium sobre Raios Cósmicos, agôsto 4–8, 1941, Academia Brasileira de Ciências, Rio de Janeiro, 1943* (Imprensa Nacional) 1948

of which was to measure and detect cosmic rays in the stratosphere. In 1941 an expedition was organised, in which the American physicist Arthur Compton took part, to measure cosmic radiation in the interior of the state of São Paulo using hydrogen balloons carrying Geiger counters [3, 4].

That very year the first international physics event to take place in Brazil was held at the Academia Brasileira de Ciências, the *Symposium sobre Raios Cósmicos* (Rio de Janeiro, 1941). Participants included Occhialini, Compton, Wataghin, José Leite Lopes, biophysicist Carlos Chagas Filho, mathematicians, students of the USP Physics Department and of the Universidade do Brasil (RJ) and military school professors. The Academia Brasileira de Ciências was the locus for scientific discussions; and the *Anais da Academia Brasileira de Ciências*, where Occhialini published seven papers, most of them in co-authorship with USP colleagues, was the oldest and the most important national journal [2], renowned in the international scientific community.

Even though Occhialini and Wataghin lived far from science production centers, they never ceased to keep in contact with Heisenberg, Niels Bohr, Heitler, Blackett, Dirac, Rutherford, Fermi, Max Born, Wolfgang Pauli, Ernest Lawrence, among other elite members of European and U.S. physics. This uninterrupted communication gave impetus to their students' advancement and made it easier for them to have access to foreign laboratories to complete their training. Wataghin and Occhialini's leadership also contributed to the maintenance of the tradition, popular at the time, in which experimental physicists made and improved their research apparatuses themselves. For this reason, because of the experience he gained at the Cavendish Laboratory and also because of his magnetic personality, Occhialini attracted theoretical physics students to his group. The experimental approach came into focus in a definitive manner with his development of a method in which two coincidence Geiger counters automatically recorded the passage of a particle by triggering a photographic camera that registered the vapour track left by the particle as it crossed the cloud chamber. Lattes says that he abandoned theoretical physics, Wataghin's and Schenberg's field, due to Occhialini, who taught Camerini, Andréa Wataghin and himself to build and operate a small Wilson Chamber.[2]

The intensification of World War II inverted the situation. In Europe and the U.S., physics laboratories turned to the development of military technology, while at USP research practically came to a full stop: as a consequence of his Italo-Ukrainian origins, Wataghin was asked to resign from his post of director of the department; the majority of professors chose to secretly build artefacts for the Brazilian navy and army; and shortly thereafter Schenberg became engaged in party politics and was elected federal congressman for the Communist Party. Wataghin, Lattes and a few recent graduates were therefore the only ones who continued to dedicate themselves exclusively to cosmic ray research [2, 3].

When Brazil declared war on the axis powers in August 1942, Occhialini feared being deported to Italy and preferred to resign from USP to live incognito, possibly in the newly created Parque Nacional de Itatiaia. It is not known why he chose this region, what he did when he lived there and for how long he had been hiding there.[3] Aside from its exuberant beauty and numerous mountains, Itatiaia city is strategically located near the main road linking São Paulo and Rio de Janeiro, and a few German and Swiss families lived in this region. It is presumed he remained in this region until Italy's unconditional surrender in September 1943. He eventually left Brazil to co-operate with English war efforts, but before leaving he collaborated with biophysicist Carlos Chagas Filho at the Universidade do Brasil and gave a course on X-rays at USP, which enabled Lattes to use photographic film in physics [11].

Occhialini arrived in England on "(...) an invitation from the British Government to join the British Atomic Energy team, but with the success of the bomb trial in Arizona, policy changed and foreigners were excluded. We took on Occhialini in the first place partly to relieve the DSIR [Department of Scientific and Industrial Research] from an embarrassing situation, but the following spring we agreed to finance him from department funds [Physics Department of the University of Bristol]".[4] In June 1945 he started to work with Cecil Powell at the small H. H. Wills Physical Laboratory. Powell, a pacifist, and also considered a leftist, did not take part in the war efforts and preferred to lead a modest research project.

---

[2] Lattes' interview, see [3, 11].

[3] It is hard to believe that Occhialini worked as mountain-hiking guide. The region was uninhabited and visited only by botanists.

[4] Reference [12], quotation on p. 30.

Reunion with the Brazilian Physicist

Occhialini met Powell when he was working on neutron-proton scattering at approximately 10 MeV. Occhialini's unique ability to perfect research techniques helped him to convince C. Waller, a chemist at Ilford Ltd., to prepare emulsions plates with a silver density around six times higher than that normally used in nuclear physics. Many attempts were necessary before the plates acquired the desired density without increasing the background residues, the undesirable grains, which masked the tracks of protons and other particles under study.

On the other side of the Atlantic, Lattes was impressed to receive Occhialini's positive prints of photomicrographs of protons and α-particles obtained with the new emulsion produced experimentally by Ilford. As he realised that this high-density silver plate was more powerful and more adequate for the study of cosmic rays than his small cloud chamber, he immediately wrote to Occhialini inquiring about the possibility of going to Bristol himself. Occhialini predicted that the new concentrated emulsion could lead to scientific discoveries, so he suggested Powell to invite Lattes [13].

It was not hard to "join youthful forces" at the H. H. Wills Laboratory. "Since 1946 the contribution by DSIR to the work of Powell has been substantial (...)",[5] aside from the financial support provided by the Iron Stell Federation, Anglo-Iranian Oil, the Electrical Research Association, the Diamond Corporation, Eastman Kodak and the Royal Society. However, it would be difficult to find adhesions among junior England physicists who had been recruited for war and had been encouraged by the government to return to scientific activities. During those years the English paid little regard to experimental physics and to universities in the interior, since undergraduate courses had been redirected during WWII towards the training of electronics technicians. Bristol's main attraction was the theoretical physicist N. F. Mott [11, 13].

In January 1946, after Occhialini had seen a γ-ray photograph by Lattes of the slow-meson triggered Wilson Chamber, Powell made a formal invitation for him to join the Bristol group. Lattes, who typically acted quickly, immediately arranged to travel to distant and cold England, accepting the fact that he would live on a research associate scholarship, and knowing quite well that in the winter of 1946 a meal could consist of a bowl of soup and a slice of bread [14–16]. As is well known, laboratory-based development of new techniques attracts qualified researchers because of the possibility of obtaining promising scientific results, as well as a rapid return in intellectual investment in the form of *recognition, credit* and *scientific credibility*.[6] Lattes found Powell and Occhialini working on n-p scattering using common photographic emulsions, such as Powell himself had done long before. Having just left a group conducted by Wataghin's theoretical rigour and shaped by a quantum-oriented view of radiation and matter and by Occhialini's own experimental creativity in the subject

[5] Reference [12], quotation on pp. 31–32.

[6] The meaning here is that used in the sociology of science, respectively, in Merton's, Bourdieu's and Latour & Woolgar's work.

of cosmic rays, Lattes stated that seeing a proton would require a lot of imagination. According to him, Powell was very conservative in science, but he allowed his subordinates much freedom. Compared to Lattes' work in USP, where he had been criticised for excessive dedication to theoretical physics, his work was now almost entirely experimental [14, 15, 17]. This motivation led Lattes to suggest that Ugo Camerini join the Bristol group. He was in fact investing in the future, so he could later work with cosmic rays, his favourite subject. He worked intensively in calibrating Ilford's new plates in order to obtain range-energy relations [18]. He kept in touch with C. Waller at Ilford, convinced that the fading image problem might be solved by modifying the composition of the emulsions by adding borax, following the guidance of Johnny Williamson from the Chemistry Department. Later, Camerini collaborated in this investigation as well.

In Cambridge, studying reactions produced by the 1 MeV deuteron beam of the Cockroft-Walton accelerator, Lattes was interested in light targets—D, Li, Be, B, F—in order to understand obscure issues relating to these reactions. He insisted in determining the range-energy relation for α-particles, protons and deuterons in the new nuclear emulsion. In the same experiment, he placed borax-loaded plates, which Ilford had prepared at his request [18]. As he predicted to Leite Lopes, the new emulsions called C_2 would soon be adopted in nuclear physics and in the study of cosmic rays: "(...) a real race is on the way"[7] Lattes, Fowler and Cuer—the latter two Powell's students—managed not only to establish the range-energy relation protons through the analysis of tracks on those plates but also to make an important correction: the relation between the range in the air and in the emulsion is not constant.[8] Still with Cuer, he investigated the disintegrations produced by neutrons in the emulsion by loading it with D, Li, Be, etc. salts. In the theoretical domain, he was studying the action of nuclear particles in photographic emulsions, along with Peter Burton, a student of Mott's.

The Construction of the π-Meson[9]

Whereas WWII led to the meeting of Powell, Occhialini and Lattes—all of whom engaged in scientific research rather than taking part in the war effort in their countries—the great battle to mobilise heterogeneous elements and to find mesons in cosmic rays started when Occhialini and Lattes met again. Together they transformed life on the fourth floor of the so-called *cigarette tour*, where the H. H. Wills Laboratory was located. Happy and carefree, they restarted cosmic rays research and became experienced in the use of photographic emulsions, as had Powell. As a result of this, said Occhialini, English physicists and technicians broke out of their torpor, having been frustrated and feeling guilty because they had not participated in war

[7] See Ref. [15].

[8] See Table 2, articles 3 and 4; and [11].

[9] This part relies mainly on [19], pp. 23–53 and [20], pp. 313–321.

Table 2 The discovery of the π-meson

1. Powell C., Occhialini G., "Multiple disintegration processes produced by cosmic rays", *Nature*, **159** (1947) 93–94
2. Powell C., Occhialini G., "Nuclear disintegrations produced by slow charged particles of small mass", *Nature*, *159* (1947) 186–190
3. Lattes C., Fowler P., Cuer P., "A study of a nuclear transmutations of light elements by the photographic method", *Proceedings of the Physical Society*, **59** (1947) 883–900
4. Lattes C., Fowler P., Cuer P., "Range-energy relation for protons and α-particles in the new Ilford 'Nuclear Research' Emulsions", *Nature*, **159** (1947) 301–302
5. Lattes C., Occhialini G., "Determination of the energy and momentum of fast neutrons in cosmic rays", *Nature*, **159** (1947) 331–332
6. Lattes C., Muirhead H., Occhialini G., Powell C., "Processes involving charged mesons", *Nature*, **159** (1947) 694–697
7. Lattes C., Occhialini G., Powell C., "Observations on the tracks of slow mesons in photographic emulsions", *Nature*, **160** (oct. 1947) 453–456, 486–492
8. Lattes C., Occhialini G., Powell C., "A determination of the ratio of the masses of π- and μ-mesons by the method of grain-counting", *Proceedings of the Physical Society*, **61** (1948) 173–183

activities against nazi-fascism. They reacted when faced with the pair's intellectual excitement and were able to present civilised scientific contributions to the world.[10]

Obsessed with the idea of eliminating fading in the emulsions, Lattes and Occhialini decided to expose some plates at Pic du Midi. Occhialini was going on holiday in the French Pyrenées, where Powell and he had already exposed the old plates, so the Observatoire du Pic du Midi (2850 m) turned out to be doubly convenient. Two researchers at the observatory, Hugon and Max Cosyns, helped conduct the experiment for approximately one month.[11] All the plates were calibrated, that is the range-energy relation had already been determined, but some had been borax-loaded by Lattes. The hypothesis was that the anti-fading effect of the borax would allow the detection capability to last longer. Without the borax, on the other hand, the plates would suffer too much fading and would lose their detection capabilities in about one week [17, 22]. On the same evening of January 1947, when he returned to Bristol, Occhialini processed the emulsion and sent *Nature* a note praising the advantages of nuclear emulsions in the study of cosmic radiation (Table 2, article 1). The photomicrographs confirmed the hypothesis that at high elevations borax's antifading effect permitted the registration of a number of events and minute details, such as accompanying the trajectory of lithium-8—which Occhialini called *tarelo*—emitting a β-particle. The lithium-8 then decayed to beryllium-8, which in turn emitted two α-particles, which he called *martelo*. The detection of the neutron energy was part of the result. The photomicrography mosaic, which had been prepared by Powell, mobilized the entire laboratory into studying the normal low-energy events of cos-

[10] See [13], p. 7.

[11] This support is registered, e.g., in Table 2, article 5. See also [21], pp. 368–370.

mic rays. The work required many hours of exhausting and patient activity at the microscope. The microscopists were all women, some of which were physics graduates. A few days later, Marieta Kurz found a strange event: a line which was more crooked than that of a proton, less dense, with many changes in direction, and about one-eighth the mass of a proton. A similar event relating to a "double" meson (π-μ decay) was observed in the following days [17].

Mass measurements were of two sorts. Occhialini and Lattes started refining the technique of grain counting and energy balance on secondary tracks in the observed reactions. Hugh Muirhead and another three students dealt with the problem by using multiple Coulomb scattering [14, 17, 18]. He believed that "the experience of our Brazilian friends in cosmic rays at this time was invaluable and on many occasions I heard the name Wataghin mentioned".[12] Altogether, the methods were sufficient to guarantee that Yukawa e Anderson's mesons were present (Table 2, articles 5 and 6).

"The excitement of the of the discovery of π-mesons was intense and the occasion was such that Occhialini, of undoubted rationalist outlook, could only express his feelings by going into the R. C. Cathedral to light a candle!",[13] remembered Tyndall, the H. H. Hills director. Particle physics, so far subordinated to nuclear physics, was about to be formally recognised as an independent field.

There was one problem, however. How could the Bristol group guarantee their interpretation of the phenomenon if the number of analysed events did not allow the statistical reduction of errors in each of the methods used for mass measurement?

This difficulty reveals the uncertainty of the first results of scientific investigation, and the necessity of obtaining more events, for the discovery only acquires scientific value when singular observations become regular. An isolated event is insignificant, and the experiment must be carried out obstinately, with courage and vigour, before the evidence can be consolidated. Perhaps only an ex-student of Occhialini's would have thought of exposing the borax-loaded plated at peaks higher than those in the Pyrenées, so as to detect the mesons more credibly.

Lattes went to the Geography Department in search of an accessible and safe location to repeat the experiment. A meteorology station in the Andes, 20 km from La Paz and 5600 m high, was located. The geographers probably had help from the Spanish meteorologists Mario Porto (director of the Irish Meteorological Service) and Duperier (Blackett's collaborator in London), who knew Ismael Escobar, a political refugee from the Spanish Civil War responsible for Bolivia's meteorological network.[14] Lattes had to act quickly, according to Occhialini and Powell, so that the Bristol group would not be left behind by D. H. Perkins' group at the Imperial College of Science and Technology, a group which had already exposed nuclear emulsion plates at about nine thousand meters from Royal Air Force planes. But in the articles published in *Nature* (Table 2, articles 1 and 2) they had already revealed their addition of borax to the emulsions, so they could lose their leadership in the race, for in such cases the first to arrive becomes the winner.

---

[12] See [18], quotation on p. 15.

[13] Reference [12], quotation on p. 32.

[14] See [23], especially p. 102.

Powell, spokesman for the group of twenty young scientists and technicians, had no trouble convincing the laboratory's director, Professor Tyndall, to obtain funding for the journey from the British government and industries. After WWII, the discovery of new subatomic particles contributed towards a greater understanding of nuclear forces, possibly leading to an increased political and military advantage for the promoting countries and for shareholders of companies directly involved in scientific applications. Lattes left on April 7, 1947, for Rio de Janeiro, from where he crossed Brazil to reach Bolivia's capital.

From La Paz, in the company of Spanish meteorologist Ismael Escobar, he departed for the meteorological stations. Despite the discomfort and adversities encountered at 5600 m of altitude—precarious access, low temperatures and scarce oxygen—Lattes placed small piles of sensitive photographic plates at the second meteorological station. It was in fact a tiny and crude installation comprising four pieces of wood, lost amidst Mt. Chacaltaya's magnificent scenery of eternal snow. There the nuclear-emulsion plates received a number of particles thousands of times greater than in the Pic du Midi experiment.

One month later, at Escobar's house, he processed one pile of plates and found a complete "double" meson. Back in Bristol, about 30 mesons were found on the other plates. Even negative mesons were found, which at the end of their trajectory resemble the drawing of a star. After obtaining the list of the mesons' mass from counting the grains in the lines, Lattes' result confirmed the first experiment.[15]

The outcomes of the two experiments did not originate from natural evidence, primary observation or from an isolated action. A handful of multiple and interconnected factors propitiated the success of the two experiments: besides previous experience with cosmic rays accumulated by Occhialini and Lattes, and Powell's experience in nuclear physics using photographic plates, the discovery was due to the technical improvement of Ilford's so-called C_2 nuclear emulsion plates. They represented the result of combining Occhialini's skills in experimental physics, Lattes' clarity of thought and strong theoretical background[16] and Powell's alliance with representatives of other institutions. "It was a reality of intense, arduous and continuous work of deep excitement and incredibly fulfilled dreams", said Occhialini.[17]

The process of construction of the meson is well-documented in six articles published in *Nature* and in two articles in *The Proceedings of the Physical Society*, the authors of which appear in alphabetical order, as was usual at the time (Table 2). Occhialini and Lattes did not wait for the acclamation of results in Bristol, which concentrated around Powell in 1950, after his own name.[18] Lattes surprised many when he decided to leave Bristol. In an apparent whim, he went to the Berkeley Radiation Laboratory with the intention of trying to detect mesons produced at the 184-inch Lawrence's syncrocyclotron. Fifteen days after his arrival in the U.S., mesons pro-

[15] See [17] p. 3. Lattes' measurement calculations are kept at the Wills Memorial Library.

[16] See [24], p. 11.

[17] Reference [13], quotation on p. 7.

[18] See Powell letters to N. Bohr and Rosenfeld Papers between 1948–1951 to apprehend Powell's movement.

duced artificially at this machine were detected by Lattes and E. Gardner. The Bristol group took almost one year to observe thirty "double" mesons, while in Berkeley they detected this same amount in one day, for they knew where the source of the mesons was, where their destination was and at what angle they reached the emulsion plates.

Lattes and Gardner's work caused great impact, for Lawrence transformed the scientific event into a press fanfare [25, 26]. The π-meson detection is registered in two articles: one published in March 1948, in *Science*, and the following year another in the *Physical Review*, the same year that Lattes returned to Brazil to invest his prestige in the creation of the Centro Brasileiro de Pesquisas Físicas (CBPF), the Conselho Nacional de Pesquisas (CNPq) and the Laboratorio de Fisica Cósmica in Chacaltaya.

Also in 1948, Occhialini went to the Université Libre de Bruxelles to organize a research group to improve the method for developing thick emulsions uniformly (Ilford G_5). That same year he invited the Brazilian theoretical physicist Schenberg to work at the center for nuclear research at that university. Schenberg stayed there for five years, working on cosmic rays and statistical mechanics with Prigogine, Cosyns and evidently with his friend and colleague Occhialini [2].

The Return to Brazil

Since the mid-'40s, professors of physics and mathematics at the Faculdade Nacional de Filosofia (FNFI) of the Universidade do Brasil, influenced by Wataghin and Occhialini and motivated by Schenberg's and Damy de Souza Santos' performance, wished to bring teaching and research together, as was done at USP and universities abroad.[19] Nonetheless, before this could be done the proper conditions for scientific work had to be created: laboratories, a library, full-time work hours and technician support. The project gained momentum in 1945 with the creation of the Instituto de Biofísica by Carlos Chagas Filho, who supported the claims made by his colleagues at the FNFI. They invited Lattes and convinced the director to create a nuclear-physics chair and to include the subject in the curriculum. But without funds for the installation of the laboratories needed for Lattes' work, they concluded that the battle in favour of research at the Universidade do Brasil had been lost.

The year 1948, when this defeat occurred, was an exceptional year. Democratic ideas were spreading in the large urban centers, the Sociedade Brasileira para o Progresso da Ciência was created and Lattes was being given wide coverage by the national press. This group of professors at the FNFI then made their last risky attempt, namely to engage in science outside the university with private capital.

The *scientific capital* accumulated by Lattes was applied towards the creation of the Centro Brasileiro de Pesquisas Físicas in 1949 and of the Conselho Nacional de

[19] The level of the majority of foreign professors was very poor and Italian professors at FNFI were accused of excusing the fascist regime instead of engaging in science, see [19, 27], pp. 55–66.

Pesquisas in 1951.[20] Both institutions relied on political support from congressmen, ex- and future presidents of Brazil, the military, and government technicians interested in the production of nuclear energy in Brazil. As in the financial issues present in the game of political interests, the scientific patrons supported future negotiations, including those which took part in other arenas and joined forces to finance the CBPF, becoming its members. Alliances were inevitable. The physicists and mathematicians who founded the CBPF belonged to a generation whose social, political and philosophical options had been conditioned by WWII and therefore wanted to work on science in Brazil rather than on their careers at American institutions [14].

Support for the CBPF was not unanimous, however. In particular, some FNFI professors and competitors, such as USP nuclear physicists, particle physicists and cosmic ray physicists, regarded CBPF as a rival. Before, there had been a large difference between the level of courses offered in São Paulo and in Rio de Janeiro, so all the CBPF pioneers, with the exception of Hervásio de Carvalho, had attended USP.

The usual competition between colleagues and institutions present in science becomes more serious or more apparent when new disciplines, groups or areas are institutionalised, or when new research equipment is needed. Disputes typically occur between researchers belonging to established scientific fields and recent rivals, especially when the latter already start off with solid political support. This was precisely the case with CBPF, which emerged in cooperation with the FNFI Physics Department. The corresponding department at USP had already lost Occhialini, Lattes, Camerini and Schenberg, and still had to face the departure of Wataghin, who returned to Europe in 1949. It was therefore struggling to recover its previous productivity level. Scientific research is not any more rational and logic than other human activities, so it is easy to understand the reaction and competition at the beginning of the '50s, specially in the experimental area, when the first three researchers worked at CBPF, where the theoretical physicists Leite Lopes, G. Beck, J. Tiomno e R. Feynmann also worked.

The Brazilian delegate at Unesco, Paulo Berredo Carneiro (scientist and a founding member of CBPF), was enthusiastic about the repercussion of Lattes' work and was responsible in April 1951 for the success of the negotiations of the first Technical Assistance Treaty between Unesco and the Brazilian government. The treaty favoured CBPF by granting scholarships to visiting fellows, allowing Brazilian researchers to continue their training abroad and providing aid in the acquisition of research material and journals. This type of collaboration had become common at the time. Twenty such treaties had been made official with the Unesco Member States, eight of which benefited Latin America.[21]

Foreign investment and financing in underdeveloped countries after the war was scarce, so Unesco support allowed the rebirth of international cooperation in the realm of the exact and natural sciences. The so-called "Physics and Chemistry Science"

[20] About the creation and the activities of the CBPF in the 1950s, and politics of the CNPq, see [19], pp. 55–142 and [28].

[21] See official letters in [29–31] rapport of Unesco's director [32], p. 190; fellowships [33].

field was the greatest beneficiary in the 1950–52 period, having received 27.78% (1950), 28.03% (1951) and 24.37% (1952) of Unesco's total budget.[22]

The first visiting researchers selected by Unesco arrived in 1951 to work at CBPF: Occhialini, who brought with him a nuclear-emulsion processor and twenty photomultipliers; the Dutch Gerard Hepp, a specialist in fine electronics at Philips Research Laboratory; the German Helmut Schwartz, a specialist in vaccuum technology; and the Brazilian Ugo Camerini, who was considered a special case. Gert Molière, a physicist at the University of Tübingen, arrived in 1952. Aside from the Unesco scholarship, all of them received daily stipends from CNPq.[23]

Paulo Carneiro visited CBPF in October 1951 and, fascinated with Camerini and Hepp's activities, he wrote to the general director making other requests: "La importance grandissante des travaux de CBPF dans le cadre de l'economie du Brèsil me fait esperer (...) que vous donnerez votre apui".[24] CBPF's dynamism was reflected in the *Simpósio sobre Novas Técnicas de Física* (RJ e SP, July 1952), which had support from the CNPq and the Centro de Cooperação Científica da América Latina, a Unesco organization based in Montevideo. Of the 74 works presented in the Rio de Janeiro sessions by 32 Brazilian researchers or foreign researchers with links to national institutions, two themes predominated: particle accelerators and cosmic rays. Theoretical and experimental issues on cosmic ray physics were dealt with by Occhialini, Lattes, G. Molière and Sandoval Vallarte from Mexico.[25] Occhialini impressed the participants in the event and was honoured with the "Medalha da Ordem do Cruzeiro do Sul", the most important distinction bestowed on foreigners in Brazil.

Nuclear physics, including the cosmic ray physics, field theory and quantum mechanics, was the first sub-field of physics in Brazil to receive systematic support from CNPq and the national budget. This happened because in the 1950s nationalist-developmentalist economic policies influenced the State's actions and experimental physics was within reach of technological applications. In return for the ambitious project of building cyclotrons, introduced by the generals who headed CNPq and wanted to produce nuclear energy, Lattes demanded that a laboratory be built in Mt. Chacaltaya.

The Laboratory on Mount Chacaltaya[26]

Since Brazil has no high mountains of its own, and Chacaltaya was considered a privileged location ever since *Nature* revealed its existence, it was the logical choice for cosmic-rays research. Lattes' scientific capital led to the guaranteed cooperation

---

[22] See [34], p. 43 and [35], p. 46.

[23] See [19, 36], p. 100.

[24] Quotation in [29].

[25] See [19], pp. 135–137; [37], pp. 211–212 and [35], p. 62; [38, 39].

[26] This part relies mainly on [40–42].

between CBPF and the Universidad Mayor de San Andrés (UMSA) in the construction of a "(...) centro de investigaciones, ensenãnza y observaciones meteorológicas".[27]

Ismael Escobar, the director of the Servicio Meteorológico, was appointed director of the Laboratorio de Fisica Cósmica in January 1952. That very year American and Brazilian physicists started travelling to Bolivia to expose plates on Mount Chacaltaya and Lake Titicaca. Lattes and Roberto Salmeron, who were, respectively, the director of research and head of the CBPF Divisão de Raios Cósmicos, went there to evaluate the conditions for research. Faced with many obstacles, Salmeron preferred to obtain his PhD in Manchester on a Unesco scholarship [43, 44], while Lattes dedicated himself to formalizing a UMSA/CBPF cooperation program.

CBPF was a powerful ally, given the converging interests in science and external politics. For the production of science in Brazil, Chacaltaya was ideal, and for Brazilian diplomacy science could be used as a form of state propaganda in which politics is often above scientific and technical decisions, even though the conjunction of interests in science and international relations was not in fact part of a larger political project by the state. The Foreign Ministry took advantage of a program of scientific cooperation to expedite the execution of international treaties, while the Brazilian physicists looked at the state as a guarantor of investment in scientific capital. At first the interaction between the partners in this scheme went well, but once objectives started to differ, the lack of equilibrium in this power relation became apparent and the scientists had to yield to the diplomats.

Events occurred quickly. Scientific instruments made in the CBPF workshops were taken to Bolivia and the UMSA and CBPF representatives signed a scientific cooperation agreement between the two countries. In brief, the agreement dictated that the CBPF could use the Chacaltaya installations for ten years and that it could erect buildings, which would become UMSA property right away; in return, it would offer physics and mathematics courses at UMSA and would give two annual scholarships in Brazil to Bolivian students specializing in cosmic rays [45–47].

The period when Lattes was leading the enterprise was one of intense activity. Much was accomplished: scientific politics, investments in infrastructure, engineering and science, while research groups from MIT, USP and the University of Chicago participated as well. The projects and investigations were compatible with the technical and financial possibilities and had precise objectives: determining the half-life of the pion; measuring the density and the energy spectrum of air showers, among which were the V-meson and other little-known particles; and determining the second maximum of the Rossi curve.[28]

CNPq granted regular scholarships and aid for CBPF's activities, since the latter was not able to survive on aid from private enterprises. In fact, the help was even greater, for CNPq took care of all administrative expenses, technical and scientific personnel expenses, daily stipends, financial aid for researchers and their families, and granted scholarships to Bolivian students in Brazil. While the Laboratorio de

[27] Reference [41], quotation on p. 14.

[28] See [37], pp. 226–227.

Chacaltaya's infrastructure was being erected, there were five research groups: USP, MIT, Chicago (already mentioned), CBPF and Escobar himself. Interests converged, and Lattes interacted with all the groups, while Escobar dedicated himself to negotiations with the Bolivian authorities, substituting the allies of the government deposed in the 1952 *Revolução Boliviana*.

Aside from dozens of Geiger-Müller detectors and highly sensitive electronic circuits taken to Bolivia, CBPF put up three buildings in Chacaltaya to house technical and scientific personnel during the week, laboratories and workshops. Adaptation by foreign researchers was difficult, because of the cold weather, the poor quality of the meals and the scarcity of oxygen, at half the amount present at sea-level. Aside from all this, the economy was fragile and the politics was unstable, not to mention other material and cultural obstacles. Bolivia had shortages in many things, such as electrical lighting, libraries, bread, meat, and the cinema, but the trip was considered very picturesque.

Under Camerini's and Lattes's leadership, Alfred Hendel, Theodore Bowen and four students were part of the CBPF group between 1953 and 1955.[29] Lattes attempted to conciliate research with his positions as scientific director of CBPF and of the Cyclotrons Project, as counsellor for CNPq, and as professor at the Universidade do Brasil. Meanwhile the research group dedicated itself to setting up and testing the operation of the large Wilson Chamber built at the University of Chicago for Marcel Schein. Brazilian engineering students had the opportunity to learn a great deal of electronics with Brown and Hendel, but local conditions were inadequate for solving the complex daily problems related to the cloud chamber. Hepp went to Bolivia as an assistant, but also could not do anything. The observation of rare events, that is, high-energy events related to mesons or to the so-called Rochester and Butler V particles, was not feasible because of errors in the instrument's technical project. This experiment was conducted by other groups using similar equipment operating at sea-level, but only at high altitudes was it possible to observe the initial collisions without the occurrence of many secondary particles [48–50].

Occhialini participated actively in the development of the CBPF cosmic ray research program and was in Bolivia; he left in Chacaltaya the nuclear-emulsion processor he had built in Italy. The fact that after his departure nobody was able to operate it brings into evidence the distance between the periphery and a science production center, namely Italy. Technical problems created obstacles for the production of science by the CBPF group in Chacaltaya,[30] and cooperation between foreign and Latin-American institutions in the 1950s was not sufficient for all phases of the process of creation of scientific knowledge to be successfully implemented.

[29] A. Hendel belonged to UMSA but was hired by CBPF to administer the project; T. Bowen was a member of the cosmic ray group led by Marcel Schein of the University of Chicago, who collaborated with Lattes. Bowen impressed Brazilian students (e.g., Fernando and Susana Souza Barros) because of his abilities as an experimental physicist and also because of his complete integration with the aimarás culture [48, 49].

[30] UNESCO kept track of the agreement by means of detailed reports. See [34], pp. 154–155; [32, 51].

Occhialini returned to Europe in 1952, while the senior researchers of the CBPF group exchanged Bolivia for the U.S., without abandoning cosmic ray and particle physics. Some adopted the new technique of exposing nuclear emulsions, consisting of sending equipment bearing plates to high altitudes in stratospheric balloons. Lattes performed this experiment at the University of Chicago in 1955; Camerini settled at the University of Wisconsin in Madison; Bowen left the University of Chicago and went to the University of Michigan; and Hendel went to the University of Arizona.

Starting in 1958, CBPF decided to cancel the agreement with the Universidad Mayor de San Andrés. Conflicts resulted and CBPF has to step back when faced with pressure from the Foreign Ministry. Assuming a leadership position in Latin America was part of Brazil's foreign policy, as was the idea of revising the country's relationship with the U.S. Nonetheless, diplomacy was important for Brazil's interests and for those of the great European powers, and the country still considered pan-americanism more important than latin-americanism. The negotiation between CBPF and the Foreign Ministry reflected the existing tension between internal and relative autonomy in science and the pressure which is external to the production of scientific knowledge. That is, since science cannot be dissociated from other power systems and movements in society, the scientists yielded in order to guarantee financing for other research. In practice, the links with the Laboratorio de Fisica Cósmica de Chacaltaya were maintained until Escobar's final move to the U.S. As can be seen, countries with no scientific tradition face tremendous challenges to produce knowledge, even when internationally renowned scientists take part in the efforts.

Final Considerations

Far away from Brazil, Occhialini met his Brazilian colleagues many times at international events. Until recently he was remembered often by his colleagues for his intuition and creativity in the production of science and for his role in the institutionalisation of experimental physics at USP, as well as for his picturesque stories, his sense of humour and his talents in winter sports at Chacaltaya. Most people admired him. He not only made great friends, particularly Lattes, Schenberg, Camerini, Leite Lopes, J. Danon and Damy Souza Santos, but he also helped the career of many people who worked with him, directly or indirectly, at USP and CBPF. Aside from issues relating to his scientific credibility, which allowed him to be hired to work at the Universidade de São Paulo, at the H. H. Wills Laboratory, the Université Libre de Bruxelles and the CBPF, on a Unesco scholarship, three issues are prominent in Occhialini's trajectory in Brazil: the contribution made by foreign professors in the training of Brazilian scientists; the social networks of science; and the transference of techniques and scientific instruments from central countries to countries in the scientific and political periphery.

Occhialini generously lent his academic prestige to the scientific institutions through which he passed, transferring knowledge and therefore contributing to the creation of a new generation of Brazilian, English, Belgian, Italian, etc. physicists. As

Fig. 1 The physicists Joaquim Costa Ribeiro, Cesar Lattes, Giuseppe Occhialini and Lattes's wife, Martha Lattes, for occasion of the "Simpósio sobre Novas Técnicas de Física" (Rio de Janeiro, 1952). *Credit* Archive MAST/CNPq

was shown, he co-operated closely with Brazilians in four institutions in the 1940s and 1950s, at the same time elevating their level significantly in the international science scenario. In this way he allowed for the introduction and utilisation of new methods, techniques and research equipment, and also managed to attract scientists of other nationalities and produce new knowledge. Like nobody, he knew how to *move* and *mobilise* the world in the laboratory, he knew how to obtain data and, in spite of the controversies, how to represent nature.

Finally, when examined from the perspective of social history, Occhialini's contribution to the development of science and the training and establishment of researchers in Latin America allows us to consider the value of science and its social functions in the contemporary world, whether we take into account the increase in material riches or the makings of science in different societies (Fig 1).

References

1. CPDOC - Centro de Pesquisa e Documentação de Histria Contemporânea do Brasil, Programa de Histria Oral (Archives CPDOC/ FGV).
2. Schenberg M., Entrevista a Carla Costa e Tjerk Franken (Recife) 1978. (Archives CPDOC/ FGV).

3. Santos M. D. S., Entrevista a Ricardo Guedes F. Pinto e Tjerk Franken (São Paulo) 1977. (Archives CPDOC/ FGV).
4. Wataghin G., Entrevista a Cylon Eudóxio Silva (Campinas) 1975. (Archives CPDOC/ FGV).
5. Petitjean P., "Autour de la mission française pour la création de l'Université de São Paulo" in *Science and empires: historical studies about scientific development and European expansion*, edited by Petitjean P., Jami C. and Moulin A. M. (Kluwer Academic Publishers, Turnhout) 1992.
6. Petitjean P., "Entre a ciência e a diplomacia: a organização da influência das tradições científicas européias" in *As ciências nas relações Brasil-França*, edited by Petitjean P., Dantes M. A., Hamburguer A. and Paty M. (EDUSP, São Paulo) 1996.
7. Schwartzman S., *Formação da comunidade científica no Brasil* (FINEP/ Cia Ed. Nacional, Rio de Janeiro) 1979.
8. Leite Lopes J., Interviews to Ana Maria Ribeiro de Andrade (Rio de Janeiro) 1993 and 1996.
9. Bassalo J., "Cesar Lattes: um dos descobridores do méson pi" in *Cesar Lattes: A descoberta do méson π e outras histórias*, edited by Caruso F., Marques A. and Tropper A. (CBPF, Rio de Janeiro) 1999.
10. Pompéia P. A., Santos M. D. S. and Wataghin G., *Physical Review* **57** (1940).
11. Lattes C. Interviews to Ana Maria Ribeiro de Andrade (Rio de Janeiro) 1996 and 2003.
12. Tyndall A. M., *A history of the Department of Physics in Bristol, 1876–1948* (Bristol) 1956. Bristol Library.
13. Occhialini G. B., "Cesar Lattes: the Bristol years" in *Topics on cosmic rays. 60 th anniversary of C. M. G. Lattes*, edited by Bellandi Filho J., Chinellato C. and Pemmaraju A., vol. **1** (Unicamp, Campinas) 1984, pp. 6–8.
14. Nussenzveig M., Vieira C. and Barros F. S., "Cesar Lattes. Modéstia, ciência e sabedoria", *Ciência Hoje*, **19** (1995) 10–22.
15. Lattes C., Letter from C. Lattes to Leite Lopes, April 21, 1946 (Archive Leite Lopes).
16. Lattes C., Entrevista a Maria de Lourdes Fávero e Ana Elisa Gerbasi da Silva (Rio de Janeiro) 1990. (Archive Proedes/UFRJ).
17. Lattes C., "My work in meson physics with nuclear emulsion" in *Topics on cosmic rays. 60 th anniversary of C. M. G. Lattes*, edited by Bellandi Filho J., Chinellato C., Pemmaraju A., vol. **1** (Unicamp, Campinas) 1984.
18. Muirhead H., "Encouters with Giulio Lattes", in *Topics on cosmic rays. 60 th anniversary of C. M. G. Lattes*, edited by Bellandi Filho J., Chinellato C. and Pemmaraju A., vol. **1** (Unicamp, Campinas) 1984.
19. Andrade A. M. R., *Físicos, mésons e política: a dinâmica da ciência na sociedade* (Hucitec, São Paulo) 1999.
20. Andrade A. M. R., "The socio-historical construction of the π-meson" in *Proceedings of the XX International Congress of History of Science*, vol. **14** (Kluwer Academic Publishers, Turnhout) 2002, pp. 313–321.
21. Davoust E., *L' Observatoire du Pic du Midi* (CNRS Éditions, Paris) 2000.
22. Powell C., Letter to C. Lattes, 1947 (Archive Leite Lopes).
23. Aguirre C., "Cesar Lattes y el desarollo de la ciencia en Bolivia" in *Cesar Lattes 70 anos: a nova fisica brasileira*, edited by Marques A. (CBPF, Rio de Janeiro) 1994.
24. Schenberg M., "Cesar Lattes, grande físico e personalidade extraordinária" in *Topics on cosmic rays. 60 th anniversary of C. M. G. Lattes*, edited by Bellandi Filho J., Chinellato C. and Pemmaraju A., vol. **1** (Unicamp, Campinas) 1984.
25. Andrade A. M. R., "O Cruzeiro e a construção de um mito da ciência", *Perspicillum*, **8**, No. 1 (1994) 107–137.
26. Andrade A. M. R. and Muniz R. P. A., "The quest for the Brazilian synchrocyclotron", *Historical Studies in the Physical and Biological Sciences*, **36** (2006) 311–327.
27. Frota Pessoa E., Entrevista a Ana Elisa Gerbasi da Silva e Lizete Castro Pereira Nunes (Rio de Janeiro) 1990. (Archive Proedes/UFRJ).
28. Andrade A. M. R., "Ideais políticos. A criação do Conselho Nacional de Pesquisa", *Parcerias Estratégicas*, **11** (2001) 221–242.

29. Carneiro P., Official letter from Paulo Carneiro to Unesco's director, T. Bodet, 16 Nov. 1951. 53 (81) A 031 TA 115 AMS - CBPF (Archive Unesco).
30. Carneiro P., Official letter from Paulo Carneiro to Unesco's director, T. Bodet, 31 Jan. 1951. 53 (81) A 031 TA 115 AMS - CBPF (Archive Unesco).
31. Carneiro P., Official letter from Paulo Carneiro to the Brazilian Foreign Minister, April 20, 1951. Livro de Ofícios da Unesco 1949–54. Cod. 80/4/2 (Arquivo Histórico do Itamaraty).
32. UNESCO, *Rapport du directeur general sur l'activité de l'organisation pendant l'année 1953. Presenté aux états membres et à la conference générale lors de sa 8eme session* (Montevidéu) nov.-dec. 1954. (Archive Unesco).
33. FELLOWSHIPS 1946–1956. 376 (81) (Archive Unesco).
34. UNESCO, *Rapport du directeur general sur l'activité de l'organisation lors de sa 4eme session* (Paris) set.-oct. 1949. (Archive Unesco).
35. UNESCO, *Rapport du directeur general sur l'activité de l'organisation lors de sa 4eme session* (Paris) nov.-dec. 1952. (Archive Unesco).
36. Carneiro P., Official letter from Paulo Carneiro to Unesco's director, T. Bodet, on 24 May 1951. 53 (81) A 031 TA 115 AMS - CBPF (Archive Unesco).
37. Ribeiro J. C., "A física no Brasil" in *As ciências no Brasil*, edited by Azevedo F., 2. ed., vol. **1** (UFRJ, Rio de Janeiro) 1994.
38. LASCO — Symposium Modern Research Techniques in Physics (Rio de Janeiro) 1952 - org. by Lasco. 53 A 064 (81) "52". (Archive Unesco).
39. Ehresmann C., Rapport from Charles Ehresmann, 378.4 (81) TA / A 187 (Archive Unesco).
40. Andrade A. M. R., "Os raios cósmicos entre a ciência e as relações internacionais" in *Ciência, política e relações internacionais*, edited by Maio M. (FIOCRUZ, Rio de Janeiro) 2004, pp. 215–242.
41. Aguirre C., *Medio siglo de ciencia en Bolivia* (Fundación Universal Simón Patiño, La Paz) 1996.
42. Bowen T., "The world's highest mountain", *Physics Today*, **9**, No. 7 (1956) pp. 14–16.
43. Salmeron R., Interview to Ana Maria Ribeiro de Andrade (Paris) 1998.
44. FELLOWSHIPS 1946-1956 376 (81) "56" Salmeron / TA (Archive Unesco).
45. CBPF - Centro Brasileiro de Pesquisas Físicas, Atas da Diretoria 1951–1952 (Archive CBPF).
46. Livro de Cartas, Telegramas. Itamaraty/ La Paz 1949-1953. Cod. 23/518 (Arquivo Histórico do Itamaraty).
47. Livro de correspondência diversa recebida pela Embaixada do Brasil em La Paz. 1952–53. Cod. 453/ 4/ 17 (Arquivo Histórico do Itamaraty).
48. Barros F., Interview to Ana Maria Ribeiro de Andrade and Elaine Rezende de Oliveira (Rio de Janeiro, 2003).
49. Barros S. S., Interview to Ana Maria Ribeiro de Andrade and Elaine Rezende de Oliveira (Rio de Janeiro, 2003).
50. Marques A., Interview to Ana Maria Ribeiro de Andrade and Elaine Rezende de Oliveira (Rio de Janeiro, 2003).
51. DOSSIÃŠ 53 (81) A 031 TA 115 AMS (Archive Unesco).

Occhialini's Scientific Production Between the Two English Periods

Leonardo Gariboldi

Introduction

Occhialini's scientific production between his stay at the Cavendish Laboratory in Cambridge in the early '30s and his stay at the Wills Laboratory in Bristol in the mid '40s is maybe the less known. The recollection of his papers was at the base of a reconstruction of his scientific activity and showed his work in a standard kind of researches on radioactivity and cosmic rays by means mainly of G-M counters and Wilson chambers,[1] showing a status of continuity with his precedent activity in Florence and Cambridge. It was a particularly striking fact to note Occhialini's technical ability in designing new kinds of simple devices to detect cosmic radiation in a situation of scarce financial and technological support.

New Kinds of Instruments

One of the most important topics in particle physics had been the development of new and better detectors and counters. In particular, the sensibility of the counters had been developed along three objectives:

(1) to detect single particles;
(2) to detect simultaneously arriving particles (coincidences);
(3) to get a prompt output of the device.

[1]On two devices by Occhialini and Damy de Souza Santos to be used with a controlled cloud chamber, see [1]

L. Gariboldi (✉)
Istituto di Fisica, Milan, Italy
e-mail: leonardo.gariboldi@unimi.it

Dipartimento di Fisica "Aldo Pontremoli", Università degli Studi di Milano, Milan, Italy

L. Gariboldi et al. (eds.), *The Scientific Legacy of Beppo Occhialini*,
https://doi.org/10.1007/978-3-031-37034-2_9

On the basis of a system of two electrodes with a potential difference between them, physicists developed different kinds of counters: Geiger point counters, Geiger-Klemperer proportional counters, Burcham-Lewis loop counters, etc. Cosmic ray physicists improved, in particular, the construction of G-M wire counters.

In order to have at proper disposal a counter sensible to particles arriving from one direction from a weak-emitting source, and with a large useful surface, Occhialini and Yolande Monteux[2] suggested a new kind of counter [2], derived from the coaxial G-M counter, to be used with the anti-coincidences method. Since cylindrical counters had to be very small to discriminate among the direction of provenience of a particle, Occhialini and Monteux suggested then to substitute the cylinder with a series of parallel molybdenum wires, stretched on glass spikes, between two aluminium plates in a rectangular knockdown ebanite frame, in order to be able to detect particles within a large solid angle and with a good individuation of the direction of origin of the particles. Occhialini-Monteux counter was built in order to detect and study short-living weak-emitting β-sources with absorbers of different thickness.

One of the problems met with in the use of G-M counters lay in the necessity of using resistors of at least 1 GΩ, which clearly showed a lack of stability over long periods of time. Among the possible solutions advanced by some physicists, Cosyns' one consisted in the use of non-ohmic resistors [3], in particular saturated photo-cells. A practical difficulty in the employment of Cosyns' circuit was the high price of photo-cells having a sufficiently high resistance.

Occhialini developed a less efficient and simpler method [4], which was also cheaper than Cosyns' photo-cells. He replaced Cosyns' photo-cell with a thermoionic valve, used with its filament cold but illuminated from the outside so as to act as a photo-cell. Occhialini was able to see that his non-ohmic resistor was much better than an ohmic one, but it could not attain complete saturation.

Occhialini, Pompéia, and Saboya analysed a technical aspect concerning the stabilisation of tension in alternating current [5], a subject of great interest in the researches based on the use of G-M counters, being the instruments with the strongest need of voltage stability. Some problems concerning high-pressure counters could be solved, according to Occhialini, thanks to the properties of synchronous motors, or of particular induction motors, connected to an electrical network with an alternating current at constant frequency. Occhialini's method permitted to use an electric power larger than the one usable with other circuits. This method could thus be used anywhere there was a distribution network that, although it could have slow or fast variations of tension, had very small variations of frequency, as it was in the case of São Paulo. A synchronous motor acted by an alternator could avoid variations of tension, the tension of the alternator being a function of the network frequency only. This looked to be a quite good solution for counters used in the São Paulo laboratory.

Occhialini and Damy de Souza Santos suggested in 1941 a new method of recording the arrival of cosmic rays particles [6]. A typical problem physicists met in their

[2] Yolande Monteux was the first woman who graduated in physics in Brazil. She was a member of the group of research in cosmic rays physics led by Gleb Wataghin and was, thus, a colleague of Damy de Souza Santos and Schönberg.

experiments on cosmic rays and radioactivity was the necessity either to record or count particles arriving at random and at a very high rate. A colleague of Occhialini at the Cavendish laboratory, Wynn-Williams, built maybe one of the most satisfactory devices for fast counting: the thyratron "Scale of Two" counter [7], that is thyratron tubes connected in such a way to construct a binary digital counter which gave reliable information of pulses due to particles arriving at a rate up to 2000–3000 per minute.

Occhialini and Damy de Souza Santos' ingenious device was a cutter connected to a vibrator circuit and a gramophone. Their method consisted in recording, with the cutter, sharp and short electrical pulses produced by a multivibrator circuit on an acetate record operated at 78 r.p.m. No appreciable distortion concerned the shape of the engraved pulses in the tests, so that Occhialini and Damy de Souza Santos' method seemed to be very useful for counting purposes, and for having a reliable picture of both the size and the shape of the electric pulses. A possible application of their device, because of this high fidelity reproduction, could be its use as an efficient and trustworthy tool for recording physiological phenomena. The main application of Occhialini and Damy de Souza Santos' device had been four: the measurements of the variations of the intensity of cosmic radiation during the solar eclipses [8] of October 1st, 1940; the study of the statistical distribution of cosmic ray particles; the observations of echoes from the ionosphere; the determination of the disintegration constant of short-lived artificially-produced radio-elements.

The Study of Cosmic Radiation

In São Paulo, with the help of Schönberg, Occhialini worked on a project of researches on the ultra-soft component of cosmic radiation, that is, cosmic radiation whose energy is less than a few MeV. The typical studies on cosmic radiation generally concerned the high-energy component. The high-energy component was of the foremost interest since it was not possible to obtain artificially such energetic particles. Knowledge about the soft component was very less rich, and was limited to the observation of the existence of particles with energies typical of nuclear reactions. Studies on the soft component started in the late '30s and were found to be strictly bound to three important problems:

(1) the determination of the mass of the mesotron[3];
(2) the study of nuclear evaporation by cosmic rays collisions;
(3) comparative measurements made with counters and with ionisation chambers.

Clay and Jonker were the first physicists to study the soft component of cosmic radiation [9], while the Italian physicists Bernardini and Ferretti found a large component of ultra-soft cosmic radiation at sea level. Occhialini and Schönberg researches [10] had the aim to complete Bernardini and Ferretti's results, by means of a telescope

[3] The mesotron was thought to be Yukawa's meson; it is known nowadays as the muon.

of two G-M counters in coincidence, with Cosyns' non-ohmic resistance photo-cells, movable to all possible zenith angles. Occhialini and Schönberg confirmed that the ultra-soft radiation had actually a small energy and was about $12 \pm 6\%$ of the total ionising radiation, but were not able to detect the zenithal effect, within the experiment errors. With the same kind of device, Occhialini and Schönberg tried also to study the neutral component of cosmic radiation, but their results were quite different from those obtained by Clay and Jonker, and Bernardini and Ferretti [11].

Since solar phenomena have an influence on the terrestrial magnetic field and since the intensity of cosmic rays at sea level depends on this field, physicists engaged in geomagnetic research were looking for solar phenomena acting on the intensity of cosmic rays observed at sea level. Occhialini and Damy de Souza Santos' aim [12] was to show the solar effect on the intensity of cosmic rays during the above-mentioned total solar eclipses on October 1st, 1940, that was visible from São Paulo. They actually used two different devices. The first one was a set counters, controlled by a multi-vibrator circuit, which had to count single pulses with, as a counter, a pick-up to record on gramophone plates its vibrations. The second device was a cosmic rays telescope, used to record the coincidences. The telescope showed a raise of 12–15% of the intensity of cosmic rays during the eclipses. The effect started sharply about three hours before the beginning of the astronomical eclipses, when the Moon's shadow was still thousands of kilometres away from the Earth.

Studies on the latitude effect had been made on the whole spectrum of cosmic radiation with ionisation chambers, and on single particles with counters [13]. Knowledge about the latitude effect of showers gave useful information on the soft component and, by comparison with the total effect, some indications on the hard component too. Theoretical studies on the soft component at sea level began in the late '30s. Observational results on the latitude effect for the soft component were discordant [14]. Occhialini's aim was to give his own contribute with a new set of measures made during a journey he made with Schönberg from Salvador do Bahía to Trieste, on January 3rd–19th, 1938 [15]. The detector was a set of three counters at triangle with different thicknesses of lead above. Occhialini found an equatorial variation of about 10% for single particles, and a much lower variation for the showers. He could thus deduce that soft showers and radiation did not show a noticeable latitude effect.

The Studies on Gamma and Beta Rays

At the International Congress of Physics-Chemistry-Biology, held in Paris in October 1937, Occhialini submitted the results of his studies on the diffusion of γ-rays emitted by a source of thorium C" (^{208}Tl) [16]. The aim of his work was to determinate the precise nature of the radiation emitted under great angles by different elements irradiated by the 2.6 MeV thorium C" γ-rays. The targets were made of carbon, aluminium, silver, and lead, and secondary radiation was detected by a radiation telescope. Occhialini confirmed the already known works on the study of such a subject. By comparison of the absorption curves for silver and lead, Occhialini directly saw the growing of anomalous radiation as a function of the atomic number, and the plateau typical of a Compton effect in heavy elements. A lead diffuser gave origin to four different kinds of radiation: the radiation corresponding to the Compton effect; a radiation corresponding to positrons annihilation at rest; a high energy radiation with a breakdown at 1.2 MeV; a small component very high energy radiation corresponding to positrons stopping and annihilation.

An interesting note [17] by Occhialini had again the γ-rays as the main subject. He was interested in the very hard radiation emitted by beryllium when irradiated by the α-particles emitted by a polonium source. After its discovery, the main aim was to measure exactly its energy. Occhialini tried an evaluation of the energy of beryllium radiation by the study of the photographs he took in 1933, with Blackett and Chadwick, with the Cavendish controlled cloud chamber in a magnetic field [18]. He chose only the tracks left by electrons whose origin was in the chamber or in a very thin plate able not to brake electrons originated inside it, so that to be sure of their energy. Unfortunately no photograph at all satisfied these conditions, but it was, however, possible to obtain useful information. From the selected tracks it was possible to see that most of the them had $E < 4$ MeV, corresponding to an energy for the beryllium radiation of about 4.3 MeV.

The discovery of the spontaneous decay of the mesotron was seen by the physicists as a mean to verify experimentally the different theories about the emission of β-rays. Schönberg advised Occhialini of the existence of a disagreement on the results of the β-spectrum of rubidium, as obtained by coincidence counters, or by a magnetic spectrograph. The β-spectrum of rubidium was particularly noteworthy for two reasons: the β-decay of rubidium does not involve the emission of γ-rays, and there are also lines of emission over the continuous spectrum. Occhialini decided to continue his studies on rubidium and exposed some considerations on the relative experimental technique [19]. Occhialini had already demonstrated that the magnetic rigidity of β-rays of rubidium was not higher than 1350 gauss-cm, corresponding to a maximum energy of the β-rays of about 140 keV. His colleague of Arcetri, Daria Bocciarelli, found similar values of magnetic rigidity (1296 gauss-cm) and β-rays energy (132 keV) [20], but other physicists found different (lower) results. Occhialini was particularly stricken by these so different values and tried to throw light on the problem. A careful analysis of the various sources of error and of the parameters of the energy spectrum confirmed Occhialini's previous results.

On the basis of this short analysis of Occhialinis scientific production during this period, we have to note his valuable skills in the production of new kinds of instruments, which could be produced in a laboratory with scarce financial supports, and his interest in the researches on typical problems in cosmic radiations and radioactivity studied in the '30s. Although the results cannot be compared in importance to those of other periods of his career, Occhialini always tried to be, with his students and colleagues, on the crest of the wave, with his passion and his deep interest in the contemporary most advanced problems in the physical science.

References

1. Occhialini G. P. S. and Damy De Souza Santos M. "Two Useful Gadgets for Controlled Wilson Chambers", *"Symposium Sôbre Raios Cósmicos", Rio de Janeiro, agôsto 4-8, 1941* (Imprensa Nacional) 1948, pp. 165–168.
2. Monteux Y. and Occhialini G. P. S. "Sur un nouveau type de compteurs plans (I)" *Annaes da Academia Brasileira de Ciências*, **12**, No. 2 (1940) 125.
3. Cosyns M. G. E. and De Bruyn J. "Notes sur le fonctionnement des Compteurs Geiger-Müller", *Académie Royale de Belgique, Bulletins de la Classe des Sciences*, **20** (1934) 371.
4. Occhialini G. P. S. "A Simple Type of Non-Ohmic Resistance for Use with Geiger-Müller Counters", *Journal of Scientific Instrument*, **15** (1938) 97.
5. Occhialini G. P. S., Pompéia P. A. and Saboya J. A. R. "Nota sobre a estabilização de tensão em corrente alternada", *Annaes da Academia Brasileira de Ciências*, **12**, No. 4 (1940) 349; Occhialini G. P. S. "Metodo per la stabilizzazione di alte tensioni", *La Ricerca Scientifica*, **13**, giugno-luglio (1942) 319.
6. Occhialini G. P. S. and Damy De Souza Santos M. "On a Method of Recording Random Events", *Annaes da Academia Brasileira de Ciências*, **13**, No. 1 (1941) 57.
7. Wynn- Williams C. E. "A Thyratron 'Scale of Two' Automatic Counter", *Proceedings of the Royal Society of London Ser. A*, **136** (1932) 312; Wynn- Williams C. E. "The Scale-of-Two Counter" in *Cambridge Physics in the Thirties*, edited by J. Hendry (Adam Hilger, Bristol) 1984, pp. 141–149.
8. Compton A. H. "Effect of an Eclipse on Cosmic Rays", *The Physical Review*, **58** (1940) 841.
9. Clay J. and Jonker K. H. J. "The Penetration of Corpuscular Cosmic Rays in Matter", *Physica*, **5** (1938) 81.
10. Occhialini G. P. S. and Schönberg M. "Sobre uma componente ultra molle da radiação cosmica (I)", *Annaes da Academia Brasileira de Ciências*, **11**, No. 4 (1939) 351; Occhialini G. P. S. and Schönberg M. "Sobre uma componente ultra-molle da radiação cosmica (II)", *Annaes da Academia Brasileira de Ciências*, **12**, No. 3 (1940) 197; Occhialini G. P. S. "Contributo allo studio della componente ultramolle della radiazione cosmica", *La Ricerca Scientifica*, **12**, novembre (1941) 1193.
11. Bernardini G. and Ferretti B. "Sulla componente elettronica della radiazione penetrante", *La Ricerca Scientifica*, **10** (1939) 39; Bernardini G. and Ferretti B. "Sulla radiazione mollissima", *Il Nuovo Cimento*, **16** (1939) 173.
12. Occhialini G. P. S. and Damy De Souza Santos M. "Effetto dell'eclissi totale di sole del 1° ottobre sull'intensità della radiazione cosmica", *La Ricerca Scientifica*, **11**, ottobre (1940) 792.
13. Clay J. "Die Korpuskulare Natur der Ultrastrahlung und ihr erdmagnetischer Effekt", *Die Naturwissenschaften*, **21** (1933) 43; Clay J. "Results of the Dutch Cosmic-Ray Expedition 1933", *Physica*, **1** (1934) 363; Compton A. H. and Stephenson R. J. "Cosmic-Ray Ionization at High Altitudes", *The Physical Review*, **45** (1934) 441; Hoerlin H. "Die Breitenabhängigkeit der Ultrastrahlung", *Die Naturwissenschaften*, **21** (1933) 822.
14. Auger P. and Leprince- Ringuet L. "Variation du Rayonnement cosmique suivant la Latitude", *Nature*, **135** (1934) 138; Johnson T. H. "Evidence for a Positron-Negatron Component of the Primary Cosmic Radiation", *The Physical Review*, **47** (1935) 318; Pickering W. H. "The Geographical Variation of the Cosmic-Ray Showers", *The Physical Review*, **49** (1936) 945; Johnson T. H. and Read D. N. "Unidirectional Measurements of the Cosmic-Ray Latitude Effect", *The Physical Review*, **51**, (1937) 557; Neher H.V. and Pickering W.H. "The Latitude Effect for Cosmic-Ray Showers", *The Physical Review*, **53** (1938) 111.
15. Occhialini G. P. S. "Mesures de l'effet de latitude pour les gerbes", *Comptes Rendus Hebdomadaires des Séances de l'Académie des Sciences*, **208** (1939) 101; Occhialini G. P. S. "Contributo allo studio dell'effetto di latitudine per gli sciami", *Annaes da Academia Brasileira de Ciências*, **12** (1940) 39; Occhialini G. P. S. "Sull'effetto di latitudine degli sciami", *La Ricerca Scientifica*, **11**, aprile (1940) 231.

16. Occhialini G. P. S. "Diffusion des rayons gamma du thorium C″" *Réunion international de Physique-Chimie-Biologie, Paris, octobre 1937* (Herman et C., Paris) 1938.
17. Occhialini G. P. S. "La radiazione gamma del Polonio-Berillio", *Rendiconti della Reale Accademia dei Lincei*, **25** (1937) 188.
18. Chadwick J., Blackett P. M. S. and Occhialini G. P. S. "Some Experiments on the Production of Positive Electrons", *Proceedings of the Royal Society of London Ser. A*, **144** (1934) 235.
19. Occhialini G. P. S. "Sur la radioactivité beta du rubidium", *Annaes da Academia Brasileira de Ciências*, **12**, No. 1 (1940) 155.
20. Occhialini G. P. S. "Uno spettrografo magnetico per raggi β emessi da sostanze debolmente radioattive", *Rendiconti della Reale Accademia Nazionale dei Lincei*, **14** (1931) 103; Bocciarelli D. "A Hard Component of the β-Radiation of Potassium", *Nature*, **128** (1931) 374; Libby W. F. and Lee D. D. "Energies of the Soft Beta-Radiations of Rubidium and Other Bodies. Method for Their Determination", *The Physical Review*, **55** (1939) 245.

Occhialini's Contribution to the Discovery of the Pion. An Interview by L. Gariboldi

William O. Lock and Leonardo Gariboldi

The paper "Processes Involving Charged Mesons" [1, 2] signed by Cesare Mansueto Giulio Lattes,[1] Hugh Muirhead, Giuseppe Paolo Stanislao Occhialini, and Cecil Frank Powell[2] was published in the May 24th, 1947 issue of *Nature*. In the introduction to this paper we can read the following announcement:

> we have found evidence of mesons which, at the end of their range, produce secondary mesons.

The primary mesons, whose discovery was announced with these very words, were at first thought to be the long searched for pions, the particles responsible for the strong interaction predicted by Hideki Yukawa in 1935,[3] the secondary mesons being the muons discovered by Carl Anderson and Seth Henry Neddermeyer in 1937[4] and identified with a particle different from Yukawa's meson by Marcello Conversi, Ettore Pancini and Oreste Piccioni in Rome [18–26].

The research of Powell's group[5] at the Wills Laboratory in Bristol leading to the discovery of the pion had been the second fundamental contribution by Occhialini to the physical sciences, after his work at the Cavendish lab with P. M. S. Blackett in

[1]On Lattes, see [3–5].
[2]On Powell, see [6, 7].
[3]On Yukawa's meson, see [8–15].
[4]On the discovery of the muons, see [16, 17].
[5]On the Bristol school of cosmic-ray physics, see [27]. On the discovery of the pion, besides the other works cited in this chapter, also see [28, 29]. A fundamental book on this context is [30].

W. O. Lock
CERN, Geneva, Switzerland

L. Gariboldi (✉)
Istituto di Fisica Generale Applicata, Sezione di Storia della Fisica,
Università degli Studi di Milano, Milano, Italy
e-mail: leonardo.gariboldi@unimi.it

L. Gariboldi et al. (eds.), *The Scientific Legacy of Beppo Occhialini*,
https://doi.org/10.1007/978-3-031-37034-2_10

the early thirties.[6] The series of important results obtained by Powell's group using the nuclear emulsion technique, thus improving the physicists' knowledge of cosmic radiation, made Bristol a capital in these studies, "the big sun surrounded by little satellites", according to an expression used by Louis Leprince-Ringuet at the 1952 Rochester Conference and a capital in the network of universities and laboratories that joined together in European collaborations.[7]

The nuclear emulsions used at the Wills Laboratory had been experimentally produced by Ilford Ltd. and exposed on mountains at high altitude to detect the interactions due to cosmic rays. Powell had been interested in the use of nuclear emulsions from 1938 on, by a suggestion from Walter Heitler, and preferred them to the cloud chamber because of the superiority of their results, and began to increase their ability to detect particles other than low-energy protons.

In 1944, Blackett invited Occhialini, then in Rio de Janeiro, to work with the Americans on the atomic-bomb project. Blackett's suggestion was kept secret to avoid repercussions on Occhialini's relatives in Italy. Occhialini arrived in the United Kingdom on January 23rd 1945, but, because of his nationality, he was not permitted to participate in war research. He thus spent some time in London, working with Edward Appleton, and then a few days at the General Electric Company. He was then invited by A. M. Tyndall, the director of the Wills Laboratory, to join Powell's group.

In the meantime, Blackett, via the Cabinet Advisory Committee on Atomic Energy, set up an "emulsion panel", chaired by Joseph Rotblat, to support and encourage industrial firms to produce better nuclear emulsions, such as Ilford and Kodak for more sensitive emulsions and better precision microscopes such as Cooke, Troughton and Simms. Among the members of the panel were Powell (Bristol), Livesey (Cambridge), Perkins (Imperial College, London) and May (King's College, London). Occhialini was not a member of the panel but he had a considerable influence on its work [32].

Occhialini, Powell, and their collaborators developed new kinds of plates such as the Ilford "Nuclear Research Emulsions", able to record the tracks of particles of more than six times the minimum ionisation. They were joined in winter 1946 by Lattes from São Paulo. The first exposures on the Pic-du-Midi were made by Occhialini, who, with the help of his friend Max Cosyns during a speleological campaign in the Bas-Pyrenées in August 1946, aided the transport to the Pic-du-Midi of a stack of Ilford C2 plates. The discovery of the first two "double meson" events happened probably in November 1946, although the publication was delayed until May 1947.

[6] It is an interesting fact that both Blackett and Powell had been Rutherford's students at the Cavendish laboratory in Cambridge.

[7] "Without doubt Powell's main contribution to cosmic ray studies was the development of the nuclear emulsion technique into a precise tool for identifying the characteristics of ionizing particles traversing them. Pre-eminent was his discovery, with colleagues, of the π-meson, or pion, a result of fundamental importance for the development of our ideas about nuclear forces [...]. The desire to observe the primary cosmic rays themselves led to the use of balloons to carry the nuclear emulsions to great heights; in turn this gave rise to big international collaborations, collaborations which played a big part in the eventual creation of CERN." Quoted from [31], p. 32.

The first observation was made by Marietta Kurz, and within a few days Irene Roberts found an event with the two tracks completely inside the emulsion.[8]

The search for a confirmation of the discovery of the π-meson made on the Pic-du-Midi took Lattes to the Chacaltaya meteorological station in Bolivia at high altitude (5500 m).[9] The first plate was developed in La Paz and scanned in Rio de Janeiro and a complete track of the "double meson" was found. With the discovery of the pion, a further step was made in the study of cosmic radiation thanks also to Occhialini's ability in developing a new kind of physical instrumentation.

***

Dr. Lock, in 1947 the group of physicists working at the Wills Laboratory in Bristol announced the discovery of the pion. The Group was headed by Cecil Frank Powell and the pion was a particle which had been predicted by Hideki Yukawa in 1935. The discovery of the pion and the contemporary discovery of the V-particles [35] *were maybe the last major discoveries made in Europe before the 1973 discovery of the neutral current. Four physicists of Bristol signed the paper published in "Nature" in May 1947: Powell himself, Occhialini, Muirhead and Lattes. The pion was discovered among cosmic-ray particles, but Powell used nuclear emulsions also with particles produced artificially before 1947. Before the war he exposed emulsions to a proton beam and during the war Powell studied neutron spectroscopy. Why did they choose to study interactions of cosmic rays and not only artificially produced particles?*

Lock: Because the intensity is greater. The number of interesting events you find in a plate was much higher the higher you went. In the original exposures you only find an event of interest very rarely, whereas when you are at mountain altitudes then the flux of incoming cosmic rays is very much greater. It was a question of intensity.

They thought they were ready to use nuclear emulsions with particles that they were unable to control, of course, both in energy, number, and flux, didn't they?

Lock: I'm not completely sure about this. I think the point is that because of the need to have a higher intensity of incoming particles you had to manufacture and fly balloons, The people who pressed for balloons in Europe were Muirhead and Camerini, with Powell originally being hesitant. I was there, so I can remember this very well. It took some time to convert Powell to balloon flying, but then when he did then of course he was immediately very effective.

Why did Powell not agree with the use of balloons? In 1947, Bradt and Peters in the US used balloons with emulsions, and Donald Perkins flew them on an airplane.

Lock: Yes, that's right. A big company called General Mills was making balloons in America much sooner than in England, but you're right, Bradt and Peters were the key people. With the balloon flights an exposure of several hours was all that was needed. Again, I remember very well when we started very early on in the balloon flying game, being on the back of the motorbike of the technician of the group,

---

[8] On the discovery of the π-meson, see also [33].

[9] On cosmic-rays studies at Chacaltaya, see [34].

Max Roberts, following the balloon in the sky, and after three or four hours, the laboratory sent a signal to cut the parachute off and down came the balloon with the emulsions. Of course these eventually ended up in the G-Stack in Italy. Originally, I think, Powell was at least hesitant but he didn't stop it, he didn't stop Camerini. We had the advice of a naval research man, Tom Coor, who was at that time attached to the U.S. embassy in London as O.N.R.L. representative.

Were they looking for the pion or was it a case of discovery by accident?
Lock: I think it was a discovery by persistance. It was just a question of going on to greater sensitivity, greater thicknesses and discovering more events.

With the 1947 paper, did they feel that the search for the pion was at its end or not?
Lock: No, I don't think so.

The discovery of the π-μ-e decay in 1948 maybe was an end point?
Lock: Yes. That was the first piece of research work that Muirhead and I did together. The electron decay, the spectrum of the electrons from the decay of the muons. Then you have to remember that the tau mesons with the 3π decay mode were then found almost immediately. Once the electron-sensitive emulsions were available, this 3π decay was then observed. Then came the discovery of $K_{3\pi}$. This meant that the doors were open to discover more things day by day, starting with this 3π decay. This was one of the first interesting events and it happened to be mostly within the layers of emulsion. You didn't lose the particles, they stayed in the layer. What we did for the electron decay was to use the Telepanto, which was Occhialini's invention, which was a precision microscope, that could protect the image within the emulsion upon a big screen.[10] So grain by grain you would copy the grains of the emulsion and then we used a set square. We had a draughtsmen's drawing board and plotted the grains along the track of the electron or positron with a protractor and then we simply measured the segments of the electrons, and that was early in '49. There was a Como conference on cosmic-ray physics in '49, and I was the second speaker to give the first results of the decay spectrum of the electrons. I remember very well the only person who asked a question was Pontecorvo. Blackett was chairman of that opening session and he made me stop. I asked "Two more minutes, two more minutes!" And he gave me the two more minutes.

[10] "[...] visitors were more frequently shown a whole gamut of the events on a screen in the darkroom by a microprojection arrangement set up by Occhialini and christened by him "the Telepanto" from (he said) "Tele: I see, Panto: Everything!". In it the stage of the microscope was given a slow transverse motion by clockwork, and at the same time focusing in depth was put into regular slow oscillation, in order that the viewer could follow tracks dipping into the emulsion or out towards its surface." Quoted from [36].

Did you know the result of the experiment of Conversi, Pancini and Piccioni?

Lock: I think the answer is yes, we did. Because, also, if you look in the paper by Davies, Lock and Muirhead [37] you'll see that other references are mentioned. Later, others in the States measured the decay spectrum with a falling cloud chamber, not a bubble chamber, and others used counters and we used emulsions.

Did you know the theory on two mesons of the Japanese physicists Tanikawa, Sakata and Inoue?

Lock: Not immediately.

Maybe after the discovery of the pion?

Lock: I think so, because you have to remember that the Japanese theories only became known in the West with a considerable time gap due to the war [38, 39].

Can you describe the role played by Occhialini in the studies with nuclear emulsions, which was his particular contribution concerning the discovery of the pion?

Lock: I would say that there were two things. First of all, his enthusiasm and his long working hours and his commitment to the discovery of the pion was really very great. In particular, of course, this question of processing emulsions was not entirely new to him but in a major part it was his pressure on Ilford, pressure on Kodak, to come up with more ideas. His commitment to the physics, I would say, was very marked and it is a pity that it was to some extent marred by the eventual conflict between Powell and Occhialini. They were rather different people.

What can you tell us about the development of thick emulsions? Who had particular ideas concerning the way to get thicker emulsions?

Lock: I think in a way it was self-evident that if you want to be able to measure things in emulsion you've got to have enough track length. It was realized early on, probably by Powell himself, that the way to the future was thicker emulsions. Because of this belief, naturally people pressurised the manufacturers to make them and then it became standard. Occhialini's major contribution was the elaboration of the essential two-stage "temperature development method for the processing of thick emulsions" [40].

The study of cosmic rays with nuclear emulsions implied a collaboration with the industry, with the industrial producers of photographic emulsions. Can you describe the collaboration of the group with Ilford and Kodak?

Lock: The collaboration took the form via what was known as the "emulsion panel" mentioned before. This consisted of representatives from industries and all the major laboratories using emulsions and wishing to improve them. This panel met, I think, only once a year, maybe twice. Livesey was the representative from Cambridge. You could see from the attendance list that this panel was a connecting voice of the different elements to put forward recommendations to the Government and to say: "please, send a contract to Ilford, give a contract to Kodak, give a contract to Cooke, Troughton & Simms, with time limits and cost limits". It was this informal panel, which existed for about six years, which was absolutely essential to make sure that things were done systematically, and whenever a new step was taken then of course this emulsion panel would follow what was being done. It was a very fruitful interaction between the industry and the scientific world.

And which was the role played by Joseph Rotblat, in particular? He chaired the panel?

Lock: He was the chairmen of the panel. He and Powell were good friends, that's for sure. Rotblat started to write a book on nuclear physics, but he got so tied up with other things—I don't remember the details—that he never wrote it. When the Second World War broke out, Rotblat was in Poland and two days before the Germans invaded Poland, 1939 he came to England to work with Chadwick on the atomic bomb and he's one of the few, if not the only one, of atomic scientists who refused to work on the project and left. As I say, he left Poland, came to England, and two days later, after his arrival in England, Hitler invaded Poland. He never saw any of his family again, all were wiped out. He of course stayed in England, instead of going back to Poland, where he had a permanent job. He developed his career on the Liverpool machine, the Liverpool accelerator. It's a tragic story, really. Poor Rotblat, but he was a very nice man, he was special.

How was the collaboration between Occhialini and Ilford and Kodak? For instance with Cecil Waller, with Berriman?

Lock: I don't know but he was not a member of the "emulsion panel", at least not as far as I could see. On the other hand, he must somehow have interacted with Powell, Waller and Berriman. But there's nothing official about it, not that I know of.

Were there any problems concerning industrial secrets?

Lock: Not that I know of, no.

As for the microscopes, what about the collaboration with the producers of microscopes, with Cooke, Troughton & Simms? How was this collaboration?

Lock: Again, I have no direct knowledge. It's a good study, this interaction as I said before between pure science and materialistic industries.

There were young physicists in Powell's group, some of them coming from abroad, for instance Lattes in '46, Camerini in '47 and so on. Who generally decided to call whom, and why? For instance, we know that Lattes was suggested by Occhialini.

Lock: Camerini was suggested by Lattes. I think what happened was that, quite early on, Powell's laboratory became a place people wanted to go to, so it wasn't so much a question of going to look for somebody, it was a question of making a choice between different people wanting to come. For example, when I graduated from Bristol and I had a good degree, I was given a choice of working with Powell and his group or working in the low-temperature group. So, because I knew already something about Powell's group and I'd been there as a summer student for a month or so, and I was already attracted by the idea of a multinational grouping of people, I chose Powell's group. I was following Muirhead who had come the year before. People just wanted to come, so there was no need for advertising for whom we wanted, we had plenty of them.

In 1947 how was Powell's group formed? How many people were there and with which role? How many physicists, for instance?

Lock: I think you only have to look at the people in the classical picture of the group: Powell, Occhialini, Muirhead, Lattes.

Let us say about ten physicists, professors and students?

Lock: I would say it was less to start with, I would say six.

And what about the scanners, the microscopists: how many people were there at the time?

Lock: Well, they steadily increased. One of them, my wife, is here. How many scanners were there? About fourteen at a guess. That was in 1950.

And in '47, maybe?

Lock: There were a few. There was Mrs. Powell, Marietta Kurz, Irene Roberts, Mrs. Andrews. If you look at a group photo taken in 1949, of the scanners, you see, there are twelve to fifteen scanners. It was quite a big group.

What training did the scanners have?

Lock: Nothing formal. They learnt on the job.

But were they trained to recognize particular tracks or not?

Lock: They knew what they were looking for and they had also to be aware of unusual things. To ask, not just think "oh, that's nothing". They always had to go to a physicist and ask. All the things there, it was the scanners who found them.

The scanners' contribution was recognized in the articles published at the time, and Powell wanted this fact. Can you tell why? After a few years, in the fifties, we did not have the name of the scanners on the photographs when they were published.

Lock: Oh, I see. I always thought it was a rule that they had to put all their names. Well, I think it was Powell's general policy to acknowledge the work done by putting the name of the person who found it, because of his attitude to life, that "people-were-equal" sort of thing.

Were most of the scanners women because of economical problems?

Lock: I don't know. Maybe it was considered a woman's job. On the other hand, Freddy Hertz, when he came to Geneva, had men scanners, he brought them from Rome. Don't forget that this was fifty-five years ago, when women were the workers but I think Powell was the driving force to name the scanners who found the events. There was a good atmosphere amongst the scanners. There were fourteen on the fourth floor at one stage when I was there. With fourteen of them you would think there'd be tension, but it was quite the opposite. Because we all made allowances for each other. We had some good parties together, scanners and physicists.

How was the life in the laboratory?

Lock: To some extent, of course, the cosmic-ray people were cut off from the rest because they were doing something rather different to low-temperature physics, or electronics or something like that. Also you happened to be on the top floor of the building, which physically put you at a distance. In order to help people mix between the different specialities there was the event of afternoon tea and that was another way in which people would meet each other and there was a good spirit in the laboratory because Tyndall had got together some very good people.

What about the time of work? Occhialini once wrote that you were often unshaved, maybe unwashed.

Lock: That's right. Yes, I think I was unshaved. I would say the life in the laboratory was very pleasant. There was a very friendly atmosphere. The fact that so many physicists married the scanners shows there was a good atmosphere. I think practically all of us are still together. I can't think of one marriage that has broken up.

How many rooms were there in the lab on the fourth floor?

Lock: Oh, I have no idea. Of course the physics building has changed now. The fourth floor was really in the roof, because the floor of the fourth floor was glass. If you're on the third floor working on the corridor, above your head the ceiling was glass. Powell's group was in this space, attic space, underneath the real roof, and actually that gave the length of free space to make the balloons, because otherwise how do you make a balloon? You need space. The rooms were more like plywood cubicles. For the research staff there was a dark room in the middle of the floor, then maybe eight rooms devoted to Powell.

And what kind of instrumentation had you got in the laboratory: microscopes, devices to process the emulsions? Can you describe it?

Lock: I think the important thing was to be well provided with microscopes, and there Cooke, Troughton & Simms filled the job very well. There were quite ten microscopes for the scanners, because there were four in the room where everyone used to come for tea, there were two where Peter Fowler was, two where Perkins was, two where Dainton was—and a projection microscope for Dainton, a Telepanto—there were eight microscopes on the glass floor, and then another four.

And did you process the emulsions by yourselves?

Lock: Yes, but not always. We had a dark room which was in the middle of the floor, part of the floor upstairs. At some stage, when we were at the very beginning of what we were doing, people used to sleep in the dark room on a simple bed in order to get up in the night and change the fixing solution. I remember doing that. So it was usually the physicists who were involved.

How did you spend your time outside of the laboratory? Was there a life together?

Lock: There were the evening parties, there was the balloon flying which is not a pastime. We went on holidays together to the Dolomites. Powell and Mrs. Powell came with us, Goku Menon, Dainton, Mulvey. We all came to the Dolomites in 1950 and had a mountain guide for two weeks. This shows you Powell wasn't stand-offish, he was very much friendly with everybody.

Where did Occhialini live in Bristol? How was his life?

Lock: I couldn't tell you how his life was. But I think he had an apartment somewhere near the university. He was a great smoker. There were very few cigarettes in those days, it was just after the war, and he and Camerini used to pick up fag-ends, and then make cigarettes with the bits left over that the other people were throwing away. Very unhygienic. Occhialini was quite the expert at that.

What about the financial support? Who paid for this research?

Lock: The Department of Science and Industrial Research, which is obviously a government agency. It was for some time the Ministry of Supply. It was a bit complicated. I don't remember who the DSIR (Department of Science and Industrial Research) reported to. There must have been a minister above, there must have been some minister for science. Powell managed to get a block grant year by year on a well-planned basis, given the importance, perhaps, of Blackett in the wheels of government. For example, people like myself had a government grant. I had a fixed amount, and when I came to the end of three years I stayed on another year with Powell as a researcher, but again, paid out of his government budget. So there was no problem then with money, it was not easy perhaps, but the work was regarded as something important, as Powell was a Nobel Prize winner.[11] That's another example the atmosphere...When Powell got the Hughes Medal[12] of the Royal Society, the year before he got the Nobel Prize, what he did was but take all of the physics group out for dinner to celebrate his award. When it came to the Nobel Prize, we had a party for all the physics department.

---

[11] Powell was awarded the 1950 Nobel Prize in Physics "for his development of the photographic method of studying nuclear processes and his discoveries regarding mesons made with this method.".

[12] Powell was awarded the 1949 Hughes Medal "for his distinguished work on the photography of particle tracks, and in connection with the discovery of mesons and their transformation.".

Occhialini spent three years at the Cavendish in 1931–34 with Patrick Blackett, and Powell himself worked at the Cavendish with Wilson. Which are the similarities with the laboratory practice at the Cavendish and in Bristol? Can we speak of a Rutherford school coming from the Cavendish to Bristol? Was there a peculiar way to make physics, maybe?

Lock: Well, it's a difficult question, but I think the Cavendish was an old, established, rather conservative place. For example, you couldn't stay working in the evening after six o'clock under J. J. Thomson, and certainly when Powell was there as a young researcher, that was one of the problems. They could only do their work during the day time, they could not stay in the evening. That makes things a bit too formal and it was a bit stupid, really, if you're in the middle of an experiment and all of a sudden they shut the doors. So I think that was the attitude of Powell, that he was very open. Some people got a DSIR grant, just as I have said, Goldschmidt-Clermont for example, but others would come with their own funds, they would have no parachute. It was a question of being accepted by Powell—or not, as the case went. So there were quite a few cases of Indians who were not that well trained in physics, who would want to come and had to be refused in some way, although Powell tried very hard, I think, to have this international mix. So, for example, we had an Indian from Chandigarh, in Northern India. There was Menon, of course. Menon was, later on in his career, Minister of Science and Technology in India.

Which was the relation of your group in Bristol with Blackett's group in Manchester? Was there any collaboration?

Lock: Well, don't forget that I was only in Manchester a short time, seven months. Again, there was the problem of the Nobel Prize, that Blackett got the Nobel Prize[13] and Occhialini didn't. Then, when he came to Powell, Powell got the Nobel Prize and Occhialini didn't. This caused bad feeling.

They were studying particles with cloud chambers.

Lock: Cloud chambers, that's right. But the emulsion people in Manchester... there were only two of us, John Major and myself. Blackett had set up a cloud chamber group with Barrister and Butler. At some stage he became convinced that the people should do emulsion work as well, and so there was in Manchester a very small emulsion group of about four young post-graduate students and three scanners. My wife was one of them. We had got married, but left in March '53. We went in September '52 and we left seven months later to go to Birmingham.

On your opinion, why did Occhialini leave Bristol?

Lock: I don't know for sure, but I think it was his unhappiness with being to some extent neglected, in a sense that he didn't share the Nobel Prize.

[13] Blackett was awarded the 1948 Nobel Prize in Physics "for his development of the Wilson cloud chamber method, and his discoveries therewith in the fields of nuclear physics and cosmic radiation."

But he left in '48, two years before Powell was awarded the Nobel Prize.

Lock: Yes, he left in '48, before that. The very fact that the book "Powell and Occhialini" [41] was not credited to him in the alphabetical way—Occhialini and Powell, as it should have been, had quite a substantial effect on him, that he wasn't being recognized. In fact, I spoke to Freddy Hertz recently, who said that after being in Bristol and in Brussels Occhialini was really just unhappy that he didn't get a share of the Nobel Prize. So, I think he was perhaps too eccentric. Brilliant, but too eccentric. On the other hand, to work with two Nobel Prize people, Blackett and Powell and not be a laureate too...

Do you know why he chose to leave in particular to Brussels? He was invited by Max Cosyns to build a new group in Brussels, but maybe he could have chosen other destinations, maybe again Brazil or Italy?

Lock: Good questions, but I don't know the answer.

Do you want to say a last thing about the Nobel Prize?

Lock: It is something which still is widely recognized as being only given to people who deserve it. Of course, people's ideas as to what is good physics and what is bad, vary, but I think the Nobel Prize is worth keeping, it's been a recognition of merit and in the case of Powell, the Nobel committee realised that they were neglecting a field which was important and growing. I remember we were so excited when we heard that the Nobel Prize had been awarded, we thought we were the best in the world. We probably were, for a bit.

As a conclusion, would you like to tell us a personal reminiscence or comment concerning Occhialini in Bristol?

Lock: Well, just thinking it out, I think Occhialini was a good thing, because you need people to have wild ideas, not in agreement with everyday beliefs, and he got things done. I think Occhialini was just showing this enthusiasm for something new and he was ready to work all hours, day and night, in order to take the next step. What was the pion really, was it a low-mass meson or was it a high-mass meson? There were in fact two groups at one time using different techniques to find out what the mass of the pion was. You need a few people like Occhialini to push things and trigger things. The steady step after another approach is equally needed. So in a sense they were a good combination when they weren't arguing. There were in fact two Bristol groups working at the same time using different techniques to find the mass of the pion. Both techniques found an answer but the Muirhead group were the closest to the present-day value.

* * *

Acknowledgements O. L. wishes to thank his wife for her help in typing and revising this document.

References

1. Lattes C. M. G., Muirhead H., Occhialini G. P. S. and Powell C. F. "Processes Involving Charged Mesons", *Nature*, **159** (1947) 694–697.
2. Lattes C. M. G., Occhialini G. P. S. and Powell C. F. "Observations on the Tracks of Slow Mesons in Photographic Emulsions", *Nature*, **160** (1947) 453–456, 486–492.
3. Nussenzveig M., Leite Vieira C. and De Souza Barros F. "Cesar Lattes. Modéstia, ciência e sabedoria", *Ciência hoje*, **19** (1995) 10–22.
4. Caruso E., Marques A. and Troper A. (Editors), *Cesar Lattes, a descoberta do meson π e outras histórias* (Centro Brasileiro de Pesquisas Físicas, Rio de Janeiro) 1999.
5. Bellandi J. and Pemmaraju A. (Editors), *Topics on Cosmic Rays. 60th anniversary of C. M. G. Lattes* (Unicamp, Campinas) 1984.
6. Frank F. C. and Perkins D. H. "Powell, Cecil Frank", *Biographical Memoirs of the Fellows of the Royal Society*, **17** (1971) 541–563.
7. Powell C. F. "Fragments of Autobiography", in *Selected Papers of Cecil Frank Powell*, edited by Burhop E. H. S., Lock W. O. and Menon M. G. K. (North Holland, Amsterdam) 1972, p. 17.
8. Brown L. M. "Hideki Yukawa and the Meson Theory", *Physics Today*, **39**, No. 12 (1986) 55–62.
9. Brown L. M. and Rechenberg H. "Quantum Field Theories, Nuclear Forces, and the Cosmic Rays (1934–1938)", *American Journal of Physics*, **59** (1991) 595–605.
10. Hayakawa S. "The Development of Meson Physics in Japan" in *The Birth of Particle Physics*, edited by Brown L. M. and Hoddeson L. (Cambridge University Press, Cambridge) 1983, pp. 82–107.
11. Mukherji V. "A History of the Meson Theory of Nuclear Forces from 1935 to 1952", *Archive for History of Exact Sciences*, **13** (1974) 27–102.
12. Yukawa H. "On the Interaction of Elementary Particles. I" *Proceedings of the Physical-Mathematical Society of Japan*, **27** (1935) 48–57.
13. Rechenberg H. and Brown L. M. "Yukawa's Heavy Quantum and the Mesotron (1935-1937)", *Centaurus*, **33** (1990) 214–252.
14. Yukawa H. "Meson Theory in Its Developments. Nobel Lecture, December 12, 1949", *Nobel Lectures, Physics 1942–1962* (Elsevier, New York) 1964, pp. 128–134.
15. Yukawa H. "Cosmic Rays and the Beginning of the Meson Theory" in *Early History of Cosmic Ray Studies. Personal Reminiscences with Old Photographs*, edited by Sekido Y. and Elliot H. (D. Reidel Publishing Company, Dordrecht) 1985, pp. 133–135.
16. Neddermeyer S. H. and Anderson C. D. "Note on the Nature of Cosmic-Ray Particles", *The Physical Review*, **51** (1937) 884–886.
17. Neddermeyer S. H. and Anderson C. D. "Cosmic-Ray Particles of Intermediate Mass", *The Physical Review*, **54** (1938) 88–89.
18. Bernardini G. "The Intriguing History of the μ-Meson" in *The Birth of Particle Physics*, edited by Brown L. M. and Hoddeson L. (Cambridge University Press, Cambridge) 1983, pp. 155–172.
19. Conversi M. "L'intricata storia del muone", *Il Nuovo Saggiatore*, **1** (1985) 33–40.
20. Conversi M., Pancini E. and Piccioni O. "On the Decay Process of Positive and Negative Mesons", *The Physical Review*, **68** (1945) 232.
21. Conversi M., Pancini E. and Piccioni O. "On the Disintegration of Negative Mesons", *The Physical Review*, **71** (1947) 209–210.
22. Conversi M. and Piccioni O. "Misura diretta della vita media dei mesoni frenati", *Il Nuovo Cimento*, **9** (1944) 40.
23. Conversi M. and Piccioni O. "Sulla disintegrazione dei mesoni lenti", *Il Nuovo Cimento*, **9** (1944) 71.
24. Conversi M. and Piccioni O. "On the Mean Life of Slow Mesons", *The Physical Review*, **70** (1946) 859–873.
25. Conversi M. and Piccioni O. "On the Disintegration of Slow Mesons", *The Physical Review*, **70** (1946) 874–881.

26. Galison P. "The Discovery of the Muon and the Failed Revolution against Quantum Electrodynamics", *Centaurus*, **26** (1983) 262–316.
27. Lock W. O. "Origins and early days of the Bristol school of cosmic-ray physics", *European Journal of Physics*, **11** (1990) 193–202.
28. Galison P. *Image and Logic. A Material Culture of Microphysics* (The University of Chicago Press, Chicago), 1997.
29. Ribeiro De Andrade A. M. "The socio-historical construction of π-meson" MAST, Notas técnico-científicas, 001/97.
30. Foster B. and Fowler P. H. (Editors), *40 Years of Particle Physics*, (Adam Hilger, Bristol) 1988.
31. Wolfendale A. W. "History of British Contributions to the Astrophysical Aspects of Cosmic Rays and Gamma Rays since the Second World War", *Quarterly Journal of the Royal Astronomical Society*, **29** (1988) 27–37.
32. Rochester G. D. "The Development and Use of Nuclear Emulsions in England in the Years 1945-50", *Colloque International sur l'Histoire de la Physique des Particules / International Colloquium on the History of Particle Physics, Paris, France, 21-23 juillet 1982, Journal de Physique, Colloque* **43** (1982) C8–89–90.
33. Perkins D. H. "The Birth of Pion Physics" in *The Restructuring of Physical Sciences in Europe and the United States 1945-1960*, edited by De Maria M., Grilli M. and Sebastiani F. (World Scientific, Singapore) 1989, pp. 585–604.
34. Ribeiro De Andrade A. M. "Os raios cósmicos entre a ciência e as relações internacionais", in *Ciência, política e relações internacionais*, edited by Maio M. (Fiocruz, Rio de Janeiro) 2004, p. 215–242.
35. Rochester G. D. and Butler C. C. "Evidence for the Existence of New Unstable Elementary Particles", *Nature*, **160** (1947) 855–857.
36. Tyndall A. M. *A History of the Department of Physics in Bristol 1876-1948 with Personal Reminiscences*, in *Histories of Physics in Bristol*, edited by Chambers R. G. and Hart M., 2005. http://www.phy.bris.ac.uk/history.html.
37. Davies J. H., Lock W. O. and Muirhead H. "The decay of μ-mesons", *The Philosophical Magazine*, **40** (1949) 1250–1260.
38. Sakata S. and Inoue T. "The correlations between mesons and Yukawa particles", *Progress of Theoretical Physics*, **1** (1946) 143–150.
39. The two-meson decay was pointed out also by Marshak R. E. and Bethe H. A. "On the Two-Meson Hypothesis", *The Physical Review*, **72** (1947) 506–509.
40. Dilworth C. C., Occhialini G. P. S. and Payne R. M. "Processing Thick Emulsions for Nuclear Research", *Nature*, **162** (1948) 102–103.
41. Powell C. F. and Occhialini G. P. S. *Nuclear Physics in Photographs. Tracks of Charged Particles in Photographic Emulsions* (Clarendon Press, Oxford) 1947.

Occhialini and the Université Libre de Bruxelles. An Interview by L. Gariboldi

Alberto Bonetti and Leonardo Gariboldi

Prof. Bonetti, Occhialini left Bristol in 1948 to begin a new scientific adventure at the Centre de Physique Nucléaire of the Free University of Brussels, a group led by Max Cosyns. You joined this group a few months later.

Bonetti: Yes, I did. In spring 1948 Giuseppe Occhialini returned to Italy after almost ten years of absence. He paid a visit to his father Augusto Occhialini in Genoa, where the latter was restoring good teaching and research conditions after the difficulties of the war. It was then that I met for the first time Beppino, not yet Beppo for me. He asked his father if there was a young physicist willing to join him in Brussels. He added that the young man should go there with a microscope. At the time I knew nothing about microscopes and nuclear plates (I was playing with ultrasounds and had seen only a few photos of tracks of charged particles which Beppo sent his father as greeting cards!). I accepted the proposal of my professor Augusto, and with some naïveté I accepted also to take with me the century-old brass-bright monocular making a fine display with its immersion objective in a cabinet among other less obsolete optical instruments (I wonder if my prof intended somehow to test me out). I arrived in Brussels, Université Libre, Groupe de Physique Nucléaire, on September 1st 1948, and was put right away in front of my microscope to look through some small plates. I did not see anything else than ill-formed blobs of grains, but everybody was happy: "Those are electron tracks!", a first attempt to electron-sensitive emulsions.

A. Bonetti
Università di Firenze, Florence, Italy

L. Gariboldi (✉)
Istituto di Fisica Generale Applicata, Sezione di Storia della Fisica,
Università degli Studi di Milano, Milano, Italy
e-mail: leonardo.gariboldi@unimi.it

L. Gariboldi et al. (eds.), *The Scientific Legacy of Beppo Occhialini*,
https://doi.org/10.1007/978-3-031-37034-2_11

How was the situation of the researches in physics in Brussels at that time?

Bonetti: I discovered there something new for me, open to the world. I went on trying to learn quickly what I had to do and to understand the way of life in that centre of research, not big, but full of enthusiasm and ambition—and also of pressure. So I have only a vague recollection of what was going on in other parts of the University building, and also in other parts of the country. In the Physics Department, apart from Max Cosyns, I remember Auguste Piccard preparing something about a next expedition, Madame Piccard helping her husband, and Paul Kipfer (a nice person full of recollections and anecdotes, as the one about himself explaining the principles of thermodynamics to an absent-minded Wolfgang Pauli just before this latter had to give an undergraduate lecture). The professor of Theoretical Physics was Jules Géhéniau, who had a part in the direction of the Centre. His wife, Madame Géhéniau, was the very helpful and charming secretary of the Group. Max Cosyns was an able director, ready to satisfy Beppo's requirements in setting up the Group, starting with microscopes and scanners. In particular he was able to provide some money for extra people, as was the case for me.

Which were the main researches at the Centre?

Bonetti: The science was Cosmic Rays and Radioactivity, but much effort was put in the quest for a reliable processing technique of nuclear emulsions, which was at the time a crucial problem. I had a part in it under the leadership of Beppo and Constance (Connie) Dilworth, the two scientists who worked out the "temperature method". The method was optimised through a previous study of the geometrical stability and possible deformations of the sensitive layer during the various steps of the processing, which would result in a distorsion of the tracks. The study of the emulsion permeation and dilation in liquids of different *p*H was carried out by Edgard Picciotto, a geologist, and Adrienne Ficq, a biologist, both interested in the use of nuclear emulsions in their field of research. They succeeded in measuring precisely the thickness variation of emulsions soaked in solutions of different acidity and found that basic solutions produce a much larger swelling than acidic ones. Then Beppo found in an old photographic treatise that Amidol was the developer which allowed the lowest *p*H. Then again Connie was the one who got from the shelf a bottle of boric acid in order to prepare with sodium sulphite the buffer solution to stabilise the *p*H of the developer. Amidol with boric acid was a winning card of the Brussels group. The method was adopted by practically all the groups working with nuclear emulsions.

Was there a particular idea or contribution that led to the temperature method?

Bonetti: The basic idea of Beppo and Connie was that developers act very slowly at low temperatures; the second point was that the swelling time is much less dependent on temperature. I do not remember if Cosyns played a role in the method itself: while I was in Brussels he certainly contributed to the design and realisation of stainless-steel developing vessels of larger and larger dimensions and with temperature control devices. When "Poohnicastro", the first double-walled, quasi-optically flat, silver-plated, 45×45 cm vessel started working, Beppo told me: "Here it is! Now I have my own cyclotron." After Cosyns left the Group, the construction of a large-dimension developing plant went on in Brussels under the direction of Beppo: it was the largest at the time. The "temperature method" was instrumental in the use of thicker and thicker emulsions and in the management of stacks of stripped emulsions of large dimensions. Usually our Ilford thick emulsions looked a bit brownish after development. Kodak emulsions looked a bit reddish. In the 1957 meeting in Strasbourg devoted to the use of nuclear emulsions, a Russian physicist, Konstantin Sergeevich Bogomolov, presented his own emulsions: 400 μ thick, Amidol-developed, they were crystal clear. But the amount produced was small and the thickness inconvenient. I do not know what happened later with Russian emulsions.

How was the collaboration of Brussels with Waller of Ilford?

Bonetti: Waller was the man who manufactured reliable, high-quality emulsions with almost perfect reproducibility, independently of the amount requested by experimenters. I could hardly understand a single word of his spoken English, so Connie revealed to me a secret of British society: "Waller's family does not belong to the establishment. He is a self-made man, no public schools, so he speaks working-class English." Waller was a valuable person and very correct in his relationship with his customers. The Brussels group had a very kind and open relation with him. I remember when one day Connie stood up from the microscope shrieking: "Fungi! Fungi!". There was a sudden sensation of scare. She phoned Waller: "We have fungi!". The answer was: "Tymol in the washing water". We never had fungi again. The important step was the massive production of minimum-ionisation–sensitive emulsions which led to the discovery of the μ-e decay and of the K-meson by Bristol and of the Hyperons by Brussels-Genoa-Milan. A trick suggested by Bates and Occhialini [1] (still in Bristol) was to have the emulsions mounted on 1 mm thick glass plates: with a suitable objective it was possible to turn the plate and observe through the glass and even make two-plates sandwiches. Occhialini asked the British firm Cooke to produce a long-working-distance immersion objective to that effect. Later also Leitz and the Italian Koristka produced those objectives with magnification up to 100×. All these tricks turned out to be very useful with the advent of stripped emulsions, first proposed by Shapiro and Stiller of the Naval Research Laboratory, and then used massively by Bristol and the other nuclear emulsions groups. As it happened, I was the one who discovered Koristka because their rather rough binocular microscopes were cheap enough that we could buy them in Genoa at the very beginning, when I was back from Brussels, summer 1949. That serendipitous choice had unexpected consequences, because the important person in Koristka was Claudio Cantù, an old acquaintance and friend of Beppo at the time of Arcetri, twenty years before! The collaboration of Occhialini and Cantù led to the production of the outstanding MS-2 and MS-3 microscopes which made the glory of emulsions for more than ten years...

Which was Occhialini's particular contribution to the development of electron-sensitive emulsions?

Bonetti: No doubt that Beppo stimulated Ilford to improve the C3 emulsions which were the endpoint of a generation of emulsions with which Lattes, Occhialini and Powell obtained such important results as the π-μ decay. Beppo was not alone in the action of stimulus to Waller, but I believe that he played an important role in convincing Ilford of the soundness of producing the next-generation emulsions sensitive to minimum-ionisation particles. The first G5 were available in Brussels in spring 1949. Before going back to Genoa, I studied with Yves Goldschmidt-Clermont the range-energy relation for slow electrons. Then Giovanna Tomasini and I carried out a reliable statistics of the μ-meson decay in plates exposed to Cosmic Rays and developed by Brussels. We were dumb enough as to lose the possible identification of a K-meson decay in those plates. A K-meson was observed for the first time in Italy by Levi-Setti and Lovati in plates lent by Brussels.

How was the problem of emulsion distortion confronted?

Bonetti: A first step was the 1% glycerol bath which gave the processed emulsion very nearly the same thickness as the unprocessed one. Then Beppo invented the guard-rings method to dry emulsions with a minimum amount of distortions. In a drying room where there was a temperature-controlled air draft, the glycerol-treated emulsions were surrounded with strips of gelatine-coated, water-soaked plates whose thickness was brought to be much the same as that of the emulsions to be dried. This avoided the quick drying of the borders of the emulsions, which was the main source of distortions. Bristol proposed a nice geometric method to correct the false curvature of tracks introduced by quasi-regular shears in emulsions during drying.

The Brussels group also studied the magnetic deflection of tracks. What was Occhialini's role in it?

Bonetti: The seminal work by Bates and Occhialini [1] contained the suggestion of using the gap between sandwiched plates for magnetic deflection measurements. The first experiment was carried out by Yves Goldschmidt-Clermont and others in Bristol, and repeated by Merlin, Someda and others at the Cancano Lake in Valtellina (Italy). Goldschmidt-Clermont was a very attractive person. He had met Beppo in Bristol and played an important role in Brussels for his cleverness and for the kindness of his nature.

How was the relation with the producers of microscopes?

Bonetti: Beppo was in touch with Cooke, Troughton & Simms when he was in Bristol and obtained from them a first long-distance immersion objective. But he asked other firms to produce those objectives and also to produce microscope tables with precision movements. Leitz produced very good long-distance objectives, while the ones produced by Koristka were not perfect, but less costly, and contributed to a lot of good work. But the interesting result of the collaboration with Koristka (*i.e.* Claudio Cantù) was the design and production of the large-table, high-precision microscopes for multiple-scattering measurements already referred to. The mathematics of the MS-2 spring movement, based on the principle of certain seismographs (I can't remember who had the idea), was worked out by a good friend of Beppo's, J. Plainevaux.

How many microscopes were in use in Brussels?

Bonetti: When I first arrived in Brussels there were perhaps six or eight, some Cooke microscopes, others of Belgian make, all conventional binocular microscopes. A critical point was the quality of the movement of the table in view of measurements of angles and of lengths, both important in the measure of the mean angle of "multiple scattering" of fast particles. When the collaboration with Cantù began, there was a kind of competition between the method of the sagitta, measured with a micrometer at regular distances along the track, and that of the angle between successive segments of the track, measured by means of an ocular goniometer. I think Beppo felt that the method of the scattering angle was not promising, while Cosyns insisted on it. Cantù even produced a maquette following Cosyns' ideas. At that time Riccardo Levi-Setti went to Brussels and Beppo assigned to him the task of comparing the sagitta and the angle method. In that work Levi-Setti showed his understanding, his critical sense and the skill of his hands. At the end he went to Beppo and was worried because he had found that the sagitta method was the better one. Beppo found the way to tell the truth without insulting anyone. Since then the angle method was abandoned and Cantù started designing microscopes optimised for sagitta measurements: a first attempt was the MS-1, which was soon dwarfed by the MS-2 and MS-3. For these ones Cantù built also an interferometric gadget for the control of the quality of the movement of the table.

Where did you usually expose your plates?

Bonetti: In the first years mainly on the Pic-du-Midi, sometimes on the Jungfraujoch. Then there were the first balloon flights started by Bristol in England, if I am not wrong, and followed soon by flights in Italy, Sardinia and Po Valley. It was the beginning of collaborations among several Italian groups, Genoa, Milan, Padua, in association with Brussels where the plates were processed. Some of the balloons were even produced in Padua by Gianni Quareni, one of the valuable young physicists whom Michelangelo Merlin (who had spent several months in Brussels in 1949) was leading to success in the emulsion group he set up in Padua. Physicists like Marcello Ceccarelli, Gianni Quareni, Milla Baldo Ceolin, began their career in Merlin's group.

Can you describe the Brussels laboratory?

Bonetti: For some reason my memories are vague as far the structure of the place, less vague about people and the spirit of the place. There was a common room which looked like a kitchen with a cowl, shelves with chemistry vessels, a sink to wash everything including one's hands, gas bottles. We spent our spare time there, having coffee and discussing our work in an often casual way. We used to speak French, sometimes English with Connie. There were also fervent discussions on subjects unrelated to physics. I understood certain peculiarities, I dare say oddities, of Beppo's behaviour thanks to his discussions with a psychologist, the husband of one of the girl-physicists: they indulged in arguing heatedly about the psychological traits of the Nazi leaders on a Freudian basis (the Nurenberg trials were a big issue at the time). The talk was also about music and literature. I consider myself lucky in my life because I found that many of the physicists I met in Brussels and elsewhere were not only good in their field, but also carriers of a "humanist" culture. Beppo was one of them. He had a touching knowledge of Shakespeare and enjoyed quoting by heart the most notorious passages, generally with a direct reference to what was going on. I mentioned already Madame Géhéniau, the nice and efficient secretary of the group. Madame Cosyns came sometimes to help her husband in the preparation of an expedition whose scope I don't remember. There was Picciotto with his bottles for the *p*H studies and his penetrating judgements about science —and people. There was Aldo Igiuni who was to become a good friend and a perfect fellow worker. There were two dark rooms, in one of which there was my microscope. In the microscope room two microscopes were for Beppo and Connie. There they observed a notorious disintegration star with big nuclear fragments. With the after wit, one of those fragments might have been a hyper-nucleus. I think it was the last important work done with C3 emulsions. I remember a large room divided into two parts. The upper part gave way to the dark rooms; the lower part, among other things, contained the processing apparatus before the construction of the "Poohnicastro". The apparatus was quite cheap: stainless-steel tanks from military kitchens and simple thermostatic systems with a contact thermometer and an agitator. The temperature method was obtained by passing the nuclear plates, contained with the relevant chemicals in sealed tubes, from a cold tank to a warm tank and viceversa. There was no cafeteria in the building. Sometimes we brought a sandwich from home, but there were reasonably cheap restaurants nearby. Merlin and I went often there.

Why was a new journal created, the "Bulletin du Centre de Physique Nucléaire?

Bonetti: As usual it was a suggestion by Beppo, who asked Cosyns a mechanism to made known in a short time the relevant information concerning the technical aspects of nuclear emulsions handling and their applications. Probably Beppo had in mind the habit promoted in Arcetri by his "maestro" Garbasso who encouraged a quick publication of results. The Brussels Bulletin was a lucky initiative because it efficiently reached all those interested in the use of nuclear emulsions.

How many people worked in the laboratory?

Bonetti: When I arrived there no more than ten people including Beppo, Connie and Yves Goldschmidt-Clermont: some of them were guest researchers with temporary appointments, some of them students engaged in their graduation research. As far as I remember, Aldo Igiuni was the only attendant before becoming an outstanding technician. There were the microscope scanners, typically underpaid girls, following Bristol model promptly imitated by other emulsion groups in Italy and elsewhere

(Connie began her career as underpaid microscope scanner!). The atmosphere was warm and open. Everyone who wished to work in the laboratory was heartily welcomed. I remember when my sister, a biologist, came to Brussels to try to use nuclear emulsions in biological experiments. Beppo provided a place and the material she needed, Connie showed her the use of liquid emulsions. Louis Vermaesen, a clever young physicist, came from Gand: Occhialini associated him to the "temperature development" work. I do not know why he left the group and put an end to his scientific career. Very little I remember of Stephen J. Goldsack from England: he worked with Connie and later was for sometime in Milan. Guy Vanderhaege had been a very proficient student and was a member of the group from 1949, I believe. He became the senior researcher in later years, when the group was engaged in collaborations at European level. In the last years of the collaboration of Milan with Brussels the senior researcher was brilliant and attractive Monique René.

How was the scientific relation between Occhialini and Connie Dilworth?

Bonetti: Connie wrote to me, in August 1948, a very formal invitation letter in French, dictated to her by some Belgian colleague not without amusement, as I discovered later. When a few days later I met her in the common room in Brussels I found that Connie was not only not formal but also very attractive: I liked her direct and clearly intelligent expression. She proved to be an exceedingly talented woman with an inbuilt intellectual authoritativeness, facing scientific issues with uncommon clarity and rigour. Connie had the ability to elaborate quickly and precisely Beppo's ideas, as was the case when they were working out the "temperature development". But Connie's mind was also highly creative, I only quote here the constant sagitta method for the measurement of the mass of particles at the end of their range, and the first idea of the Lunar Occultation Satellite (HELOS).

Mario Schönberg, the renowned Brazilian theoretician, was also a member of the group.

Bonetti: Schönberg joined the Brussels Group for many months. He used to work by night, and the lights of his room were on until late in a cloud of smoke. Among the problems he treated I remember the ionisation-energy relation of high-energy charged particles, a work done partly in collaboration with the French theoretician Louis Michel. They studied in particular the ionisation rising after the minimum, a problem relevant to fast-particle measurements in nuclear emulsions. The problem troubled many people, with various solutions, on both sides of the Atlantic Ocean. Schönberg solution was similar to that of Fermi's. Schönberg's presence in Brussels was important, because he was a reference for students interested in theoretical physics. He was one of those physicists endowed with a "humanist" culture, whom I referred to in the above.

When Occhialini was in Bristol, he called Lattes and Camerini from Brazil. Did he call other physicists from Brazil once in Brussels?

Bonetti: No, as for as I remember. The interaction with Brazil was the UNESCO mission of Beppo to Chacaltaya.

A noteworthy member was Aldo Igiuni. Why did it happen that an assistant began to work as a researcher?

Bonetti: Igiuni had no scientific training at all, but he was a particularly talented young man. He learnt quickly and with full understanding the techniques in use in the laboratory. He started his training helping Beppo in the tests on the processing of large plates, both normal and stripped. He never used a microscope, because that was not his task, but he was there as long as nuclear emulsions were used both in Brussels and in Milan. I do not remember exactly when Beppo persuaded Aldo to accept an appointment as technician in the Milan section of INFN (the National Institute of Nuclear Physics). The INFN was open-minded and ready to provide highly qualified technicians with an administrative and financial position similar to that of researchers. Aldo Igiuni was one of them. After almost ten years of nuclear emulsions Aldo followed Beppo (and Connie) in the space adventure, contributing to the design and realisation of balloon and space instrumentation.

Was the activity in the laboratory strictly ruled?

Bonetti: Our activity was regulated by our commitment and common sense. We spent the daytime in the laboratory, except those who had worked until late in the evening. Nobody took notice at what time any of us went to work.

How was your life outside the laboratory?

Bonetti: I mentioned already that Beppo had a remarkable literary culture and strong interests in politics, so he enjoyed meeting personalities visiting Brussels. The Institute of Italian Culture offered the occasion of encounters often at dinner time with writers such as Elio Vittorini or Giuseppe Raimondi and the discussions on political and/or literary issue were heated.

What about the financial support to the laboratory?

Bonetti: As far as I know, the laboratory was supported by the Free University. The director was responsible for the financial needs of the Centre. Beppo's requests were wise and sensible—and were granted to him.

How were the relations of the Brussels group with the Italian groups led by Occhialini? Let us begin with Genoa.

Bonetti: Beppo had just won, 1949, the competition for a chair in Italy when he was offered the direction of a laboratory at the Tata Institute, in India. He even wrote to his father that he would accept the offer. But in the end Beppo accepted the position in Genoa, and Bernard Peters went to the Tata in his place. Beppo kept his chair in Genoa for two years till the death of his father, 1951, and supported strongly the emulsion group which I put together following Brussels model. The microscope scanners were undergraduate students making their graduation research on plates lent by Brussels. Some of those students were good and enthusiastic. I mentioned already Giovanna Tomasini who came from a family of farmers and enrolled in the University of Genoa to become a secondary school teacher. When Beppo moved from Genoa to Milan in 1952, his successor Ettore Pancini appreciated the courage and commitment of Giovanna and appointed her as leader of the emulsion group, maintaining the collaboration with the group of Milan and Brussels which was still the provider of the plates. The work of Giovanna continued to the end of her career in important collaborative experiments at the CERN accelerators. The more prominent person emerging from Genoa in those years was Livio Scarsi, the first graduated student of Beppo with a thesis on particle pairs production in emulsions by primary cosmic rays. Livio followed Beppo to Milan, working first in particle physics and cosmic-ray interactions in emulsions, then in space physics. He was faithful to cosmic-ray research to the end of his life.

Occhialini had been invited to Milan by Polvani and Caldirola. On the other hand the physicists in Milan had already worked with emulsions.

Bonetti: Polvani's assistant Antonio Lovati had even invited the Canadian physicist Pierre Demers to learn from him the technique of producing nuclear emulsions in the laboratory. The attempt was stopped when Beppo, keeping his position in Brussels with the agreement of both sides, shifted from Genoa to Milan. Lovati, Levi-Setti and Martina Panetti were there, Livio Scarsi joined them, and so did I one year later. Connie came from Brussels with her daughter in 1953. So the Milan group took its shape for the next eight years, Beppo being the link with Brussels.

When Occhialini moved to Milan, in 1952, there was also the tragic death of Loubens caused by a broken winch during a speleologic excursion.

Bonetti: The death of Loubens was a tragedy which provoked a crisis in the Brussels Centre, also because of the responsibility of Cosyns in the design of the winch. The good relationship between Beppo and Cosyns came to an end. After much bitterness Cosyns decided to leave Brussels and went to Paris, and then to East Germany. The new director was Pierre Baudoux, who was able to largely heal the wounds and appease the climate to the point that the collaborations with the Milan group and at international level could go on until the late fifties. Let us close this sad chapter with a happy anecdote. When Beppo's 80th anniversary was celebrated in Arcetri, 1987, Jacques Labeyrie, a member of the excursion at the Pierre Saint Martin, told the story of how Beppo, who had been rescued at great risk after two days spent at the bottom of the cave, solved the problem of the defective winch. Beppo made with a suitable rope a seamen knot with which it was possible to pull up the cable of the lift letting free the damaged wheel. The winch could be repaired satisfactorily enough to save the life of the trapped speleologists. Once again, Beppo.

How were the relations of the Brussel group with the English groups?

Bonetti: Beppo was the link with Bristol and Manchester, thanks to his personal relations with Powell and Blackett. It was the time of the discovery of "new" particles and we were interested in knowing what the Manchester cloud chamber was producing at the Pic-du-Midi. The same for Bristol: great rivalry and high esteem in the quest of new objects. The result was the collaboration which led to the G-Stack balloon experiment through the commitment in particular of Michelangelo Merlin who, after his stay in Brussels, joined for some time the Bristol group.

And what about a relation with French groups?

Bonetti: We had a correct relationship with the group led by Louis Leprince-Ringuet in Paris. Of course, we were interested in the results obtained with their beautiful multiplate cloud chamber. There were also lasting friendly relations with his junior researchers. The friendship of Charles Peyrou with Beppo and Connie lasted to the end of their lives, after the scientific clash about the mass of what came to be known as the K_{μ}. The French had found the notorious 918 mass, at variance with the value 960 from the mass of the τ-meson and the average value, again about 960, of the mass obtained by Diworth, Occhialini and Scarsi (1954) through a compilation of measurements from many groups with nuclear emulsions and Wilson chambers. The G-Stack solved the problem.

The Brussels groups was part of the Free University. How important was it for you to work in a "free" university?

Bonetti: I happened to be educated without any reference to religious beliefs, in the spirit of 19th century "libre pensée". So to live in a place where professors and students had to sign the "declaration de libre pensée" was to me a rewarding intellectual and emotional experience. I enjoyed Picciotto explaining to me some amusing aspects of such an ambience. The persons I met were used to a rigorous and critical attitude towards both traditional and novel ideas. A challenging case was that of Lysenko. All over Europe the communist press presented reports on Lysenko's "extraordinary discoveries" and the communist parties were under strong pressure, to say the least. The Belgian communist party was a small elite party, mainly of intellectuals. Among them Jean Brachet, the son of the embryologist Albert Brachet and himself biologist at the Université Libre, travelled to USSR to ascertain facts with his own eyes. Back to Brussels he exposed publicly the preposterousness of Lysenko's ideas and the indecency of the Soviet government in supporting amateur Lysenko against well-known geneticist Vavilov. The matter was discussed with perfect freedom and open-mindedness, a not at all obvious outcome.

Occhialini was awarded a doctorate "honoris causa" by the Free University.

Bonetti: It was a recognition of Occhialini's work on the occasion in which the Université Libre invited him to join Cosyns in the foundation of a promising research centre. From a rather eccentric personality one might expect some eccentricity. Indeed at the ceremony Beppo had to wear a toga, something which he did not care of, and carelessly he wore it, availing himself of the informal climate of the Université Libre. I reported to his father the ceremony describing Beppo, to his amusement, as "that short and rebellious Italian"...

A centre of nuclear physics in the immediate after-war is somehow associated to the atomic bomb. Were you conscious of such a problem?

Bonetti: I do not remember if we ever talked explicitly about the bomb. I cannot say why it was so, perhaps we felt we were short of arguments to rationally discuss such an issue, besides the emotional side of it. The cold war was at its beginning and many of the group (most were proud of having been resistant against nazis) were militant gauchistes, giving strong support to the movement of the "partisans de la paix".

The Brussels groups, such as other emulsion groups, were large groups if compared to those in the '30s. Scientific papers in the '30s were signed by one or two scientists. Lattes, Muirhead, Occhialini and Powell's 1948 paper was noticed also because it was signed by four scientists. A whole page was needed for the names of the scientists signing the G-Stack paper. Did you notice that something was changing?

Bonetti: The Bristol experience shows that Powell favoured the presence of the signatures of all those who had a part in an experiment. I wonder how much this was due to the influence of Occhialini, who was very careful in the recognition of the role played by each one and in the use of the alphabetic order of names. I think it is correct to say that the advent of visual particle detectors stimulated the formation of groups of researchers engaged in long-term projects. Nuclear plates were a model case: being a relatively cheap instrument, they required a lot of man power to get significant results in a fairly short time. Scanners were often clever people, but it was up to researchers to control the results of the observations, and especially the unusual events, with a systematic cross-checking. This favoured the formation of relatively large groups of researchers in a work which required not so much specialisation as efficiency. The next step was the coordination of several groups in large collaborations. The experience of nuclear plates passed first to the bubble chamber projects and then to the mammoth detecting devices of the big accelerating machines: they engage a lot of people not only because of the dimensions and the huge amount of data, but also because of the variety of specialisations requested by such complex projects.

We usually refer to physical studies made with the big accelerating machines as a "Big Science". Can we say that also in your case there was a transition towards a "big" science, even if of a different kind?

Bonetti: As already remarked, nuclear emulsions stimulated the formation first of large groups and then of large collaborations. In this sense the G-Stack was a first, if minor, version of Big Science.

Which was the importance of the G-Stack for all of you?

Bonetti: Ending the 1953 meeting on new particles and cosmic rays in Bagnères-de Bigorre, Leprince-Ringuet described particle physicists as belonging to two factions, the accelerating machines faction, striving to realise in the laboratory higher and higher accelerating energies, and the cosmic ray faction, striving to reach higher and higher altitudes to exploit the high-energy tail of the cosmic-ray spectrum beyond the reach of laboratory machines. This was the attitude of those who in 1954 joined the efforts of six emulsion groups to attack the still extant problem of the decay scheme of charged K-mesons by means of a many-litres stack of stripped emulsions flown at balloon altitude. Also the G-Stack did not supply more than a few tens of useful events, but the sample was good enough as to give a substantially correct answer.

Was the Brussels group aware, from the beginning, of the importance of the researches they were doing?

Bonetti: I would say yes, everybody felt to be on the way of important results.

During the years Occhialini spent in Brussels, both Blackett and Powell were awarded the Nobel Prize, respectively in 1948 and 1950. How did Occhialini react to these awards and his possible exclusion?

Bonetti: Blackett was the only person, I believe, for whom Occhialini felt a kind of veneration. Ten years older, Blackett exerted a great influence on Beppo, both on the scientific and the human side. Beppo was in touch with Blackett when he was en enemy alien in Brazil during the war and was trying to go to England to join the war effort against fascism. Thanks to Blackett he succeeded in reaching England, but was rejected from direct involvement in the war effort (he was a mistrusted Italian!). Always through Blackett he joined Cecil Powell who, being one of the few communists in England, worked in isolation in Bristol. Beppo's intellectual and human esteem for Blackett was such that he accepted Blackett's prize as one accepts a welcome thing, something concerning a teacher and a friend. I never heard from Beppo a single jealous word about Blackett who, in his Nobel Lecture, acknowledged officially, I should say warmly, Beppo's contribution. That was not the case with Powell. Beppo's contribution to the discovery of the pion and more generally to the technique of nuclear emulsions was at least as recognisable as his contribution to the discovery of electron pair production and to the control of Wilson chamber. When Powell was awarded the prize, Beppo expected at least an acknowledgement of his part in a joint endeavour. The absence of a mention of Occhialini in Powell's Nobel Lecture was blatant. The sadness was great, to say the least, even disregarding the fact that Occhialini himself was not awarded a deserved recognition. It took a long time to overcome the bitterness, but in the end it happened, and this, in my opinion, was a part of the success of the G-Stack. After several years, Beppo once told me: "I met Powell. He feels his death (sente la morte)". Powell actually died a short time later, prematurely. Beppo succeeded in having a bench with Powell's name on it on the hills near Como, a cherished English habit.

Although Occhialini was formally a member of the Centre de Physique Nucléaire up to 1959, in november 1949 he already moved to Genoa as a professor, and divided his working time between Italy and Belgium. Why did he come back to Italy?

Bonetti: Beppo felt he could not disappoint his father after the long years of absence from Italy. Among other things, Augusto Occhialini had avoided major damages to the Institute of Physics in Genoa during the war, also against the attempts of Germans to take away the valuable instrumentation—he certainly longed to leave his place to his son. This is also why Beppo did not accept the invitation by the Tata Institute. But this is not all. Having been openly and strongly antifascist, Occhialini felt he had some kind of duty towards his country, and this attitude was in consonance with that of the other prominent Italian physicists with whom he was familiar from the time when Arcetri in Florence and via Panisperna in Rome gave rise to the "Italian School of Physics". As it happened, just after the death of Augusto Occhialini, between 1951 and 1952 Amaldi and Bernardini in Rome, Deaglio and Wataghin in Turin, Polvani and Caldirola in Milan, Rostagni and Dallaporta in Padua promoted the formation of the Istituto Nazionale di Fisica Nucleare. The INFN was a highly successful structure around which the Italian research in Physics was organised in a modern way for many tens of years, outliving largely the founder-members. Occhialini accepted promptly the offer of Caldirola and Polvani (who was an ancient pupil of Augusto in Pisa before the 1st World War!) and joined the Milan section of INFN. Beppo's emulsion group in Milan played an important role in particle physics in the fifties and sixties. And then Livio Scarsi and Aldo Igiuni, and several younger members, followed Beppo and Connie in the new adventure of Space Physics after their 1960 sabbatical year with Bruno Rossi, the founder with Gilberto Bernardini of the group of Arcetri and the initiator of cosmic ray research in Italy.

References

1. Bates W. J. and Occhialini G. P. S. "Applications of the Reflecting Microscope to the Nuclear Plates Technique", *Nature*, **161** (1948) 473.
2. Dilworth C. C., Occhialini G. P. S. and Payne R. M. "Processing Thick Emulsions for Nuclear Research", *Nature*, **162** (1948) 102–103.
3. Dilworth C. C., Occhialini G. P. S. and Samuel E. "Eclaircissement des plaques photographiques nucléaires", *Bulletin du Centre de Physique Nucléaire de l'Université Libre de Bruxelles*, **2** (1948).
4. Cosyns M., Dilworth C. C. and Occhialini G. P. S. "Obturateur thermique pour plaques nucléaires", *Bulletin du Centre de Physique Nucléaire de l'Université Libre de Bruxelles*, **6** (1949).
5. Cosyns M., Dilworth C. C., Occhialini G. P. S. and Schönberg M. "Double Stars with Relativistic Particles from Cosmic Rays", *Nature*, **164** (1949) 129–131.
6. Cosyns M. G. E., Dilworth C. C., Occhialini G. P. S., Schönberg M. and Page N. "The Decay and Capture of μ-Mesons in Photographic Emulsions", *Proceedings of the Physical Society*, **62**, No.12 (1949) 801–805.
7. Occhialini G. P. S. "On the identification of high energy particles in electron sensitive plates", *Suppl. Nuovo Cimento*, **6**, Ser. IX, No. 3 (1949) 413–428.

8. Dilworth C. C., Occhialini G. P. S. and Vermaesen L. "On Processing Nuclear Emulsions. Part I. Concerning Temperature Development", *Bulletin du Centre de Physique Nucléaire de l'Université Libre de Bruxelles*, **13a** (1950).
9. Bonetti A., Dilworth C. C. and Occhialini G. P. S. "On Processing Nuclear Emulsions. Part II. After Development Technique", *Bulletin du Centre de Physique Nucléaire de l'Université Libre de Bruxelles*, **13b** (1951).
10. Occhialini G. P. S. "Technique des plaques nucléaires", *Colloque sur la sensibilité des cristaux et des émulsions photographiques*, in *Revue d'Optique théorique et instrumentale*, **23A** (1951) 296–300.
11. Bonetti A. and Occhialini G. P. S. "Technique of Nuclear Emulsions", *Suppl. Nuovo Cimento*, **11**, Ser. IX, No. 2 (1964) 222–227.
12. Bonetti A., Dilworth C. C., Ladu M. and Occhialini G. P. S. "Misure in lastre nucleari", *Rendiconti dell'Accademia dei Lincei*, **17** (1954) 311–314.

Elementary-Particle Physics at the University of Milan 1951–1956

Riccardo Levi-Setti

If history was made in the development of strange-particle Physics, under the guidance of Giuseppe Occhialini (Beppo), by the nuclear emulsion researchers at the Universities of Genoa and Milan, scant record is to be found in the literature under Beppo's name. Only toward the end of 1955 does Beppo's name appears, in the reports of the "G-Stack Collaboration" that he orchestrated. Although he would have been fully entitled, as well as his wife Mrs. Connie Occhialini-Dilworth, at first as Miss Dilworth, to co-author most of the publications of our group, neither of them did. In this brief summary of the new Physics we were fortunate to uncover, I take the liberty of relating the excitement of our discoveries also on behalf of my closest collaborators, A. Bonetti, M. Di Corato, B. Locatelli, M. Panetti, L. Scarsi, and G. Tomasini, the Genoa-Milan group. The list of publications of the group, covering the period 1951–1956 and reproduced here, was taken in its entirety from my CV. Thus Beppo's legacy is bound to transpire through the work of his pupils and apprentices, being myself one of them. The manner by which his guidance of our work manifested itself deserves a separate memoir of mine, to accompany this writing.

Cosmic rays were our only source of high-energy particles, and the higher the altitude of exposure of the emulsion stacks used in our work, the higher the chance of encountering new members of the particle zoo, due to the energy degradation suffered by the primary cosmic radiation in penetrating the atmosphere. How to expose the stacks at high altitude, by mountaineering and balloon flight expeditions became part of our education. We all learned to collaborate with small and large teams of colleagues, in the field and in our laboratories. Beppo taught all of us how to appreciate and endure the experience.

The story begins with my apprenticeship with Beppo at the Brussels' Centre de Physique Nucleaire during the summer of 1950. There I learned how nuclear emulsions were processed, to yield uniformly developed and undistorted ionizing particle

R. Levi-Setti (✉)
The Enrico Fermi Institute and Department of Physics, The University of Chicago,
Chicago, IL, USA

L. Gariboldi et al. (eds.), *The Scientific Legacy of Beppo Occhialini*,
https://doi.org/10.1007/978-3-031-37034-2_12

tracks throughout their considerable thickness (up to 1.2 mm), a prerequisite for the Physics to be. I was also introduced to the methods of particle identification by the measurement of pβ from the multiple scattering of their tracks [1, 2] in conjunction with that of the rate of energy loss dE/dx revealed by various methods of track ionization density measurements. In the latter endeavour I benefited greatly from the help of Miss Dilworth and Yves Goldschmidt-Clermont, experienced members of Beppo's Brussels team, and like Beppo, of Bristol extraction.

By the time my leave of absence from the University of Pavia expired, at the end of the summer, Beppo was already casting the foundations of what became to be known as the Genoa-Milan group. How this materialized involved the introduction of new characters accompanied by a musical-chair like reshuffling of the cast affiliations. While still at Pavia, I was introduced by Beppo, who held a professorship at the University of Genoa, to Alberto Bonetti and Giovanna Tomasini, recently initiated by him to the art of nuclear emulsions. Beppo arranged for me to collaborate with Tomasini in a study of the energy spectrum of μ-meson decay electrons by multiple-scattering measurements [3], on plates exposed at the Pic du Midy in the Pirenées, at 3000 m, on loan from the Brussels lab. While this collaboration was in progress, I was offered a position at the University of Milan, which I joined in mid-1951. There, under the auspices of the Chair of the Department of Physics, Prof. Giovanni Polvani, and the support of the INFN (Istituto Nazionale di Fisica Nucleare), led in Milan by Prof. Piero Caldirola, I began to organize a nuclear-emulsion scanning and measuring lab, and took steps to get Alberto Bonetti to join me, followed soon afterwards by Beppo and by then Mrs. Connie Occhialini-Dilworth. By early 1952, the Genoa-Milan nuclear emulsion group had taken shape, joined by Mrs. M. Panetti, wife of a Physics faculty colleague, already present at Milan.

While the above infrastructure was under development, Tomasini and I were struggling to compete with our counterparts at Bristol and Dublin in the study of the μ-meson decay spectrum. Our nemesis at Dublin was C. O'Ceallaigh (the joke, in Italian, at the time was that they have Occhiali, we have Occhialini) and the Bristol group who reported some evidence of a high-energy tail in the μ-meson decay spectrum, extending further than the tail we found. As much as Prof. M. Schönberg convinced us that the tail we observed was still consistent with a spectrum cut-off at 55 MeV, and to be expected due the large errors in our multiple-scattering measurements, we were left with a lingering doubt. The emulsions used by our competitors, flown at balloon altitude, may have contained decaying particles heavier than μ-mesons, not present or less frequent in our plates exposed at mountain altitude. To dispel our apprehension, Beppo provided us promptly with plates flown for two hours at 20000 m, on loan from the Brussels Centre, to continue our μ-decay study. This move paid off handsomely, as we shall see shortly. At the same time Beppo undertook to organize a balloon launching expedition which took place in Sardinia during the summer of 1952, as a collaboration between the University of Bristol, the Brussels Centre, a number of Italian Universities, other European laboratories, and the Italian Navy and Air Force. Numerous stacks of nuclear emulsions were flown over many hours at an altitude of 27 km in egg-shaped gondolas hanging from giant polyethylene balloons, under the guidance of the Bristol group headed by Cecil F. Powell, and tracked by radiotelemetry (to which I attended, in extreme summer temperatures). Some of the gondolas were recovered at sea in adventurous circumstances (e.g. amidst floating sea tortoises). Our labs became awash with plates to scan and discoveries to be made.

The newly acquired Sardinian plates yielded promptly our first discovery that K-mesons originated directly from strong interactions, rather than from the decay of even heavier particles. In a paper entitled *"Slow Heavy Mesons from Cosmic Ray Stars"* [4], G. Tomasini and myself described our respective findings. Shortly afterward, the Genoa-Milan group described [5] the first observation (by G. Tomasini, in the plates borrowed for the μ-decay spectrum) of a particle of "transprotonic" mass, called J, decaying into a fast secondary (we now would call it a $\Sigma^+ \rightarrow n + \pi^+$ decay), and the paper was cautiously entitled *"Observation of the Decay of a Heavy Particle"*. Back to scanning the 1952 Sardinian plates, on a lonely 1952 Christmas day spent at the lab, I had the venture of finding the first example of a $\Sigma^+ \rightarrow p + \pi^0$ decay. After struggling to demonstrate the direction of motion of both primary and secondary particles coming to rest, to exclude a dp reaction, the event (called J_p at that time) was accepted as a bona-fide alternative mode of decay of a Σ^+ hyperon, yielding a mass a few MeV away from the now accepted mass of this hyperon. (Within the uncertainty of the range-energy relation the energy of the stopping decay proton could be reliably determined, and so the Q-value of the assumed two-body decay reaction) In the spring 1953, Beppo arranged for me to be invited to present our evidence to the Royal Society in London, and I will never forget my emotion at speaking from that well-worn lectern (most likely of Newton's vintage) in a chapel-like, mahogany-paneled lecture room. In July 1953, our events were presented at the Cosmic Ray Conference of Bagnères de Bigorre, where our hyperprotons were rebaptized (fortunately temporarily, it took Gell-Mann in 1955 to call them Σ) Ω_p and Ω_π and finally published in the Fall 1953 [6] in a comparison with additional evidence from other laboratories. This paper, entitled *"On the Existence of Unstable Charged Particles of Hyperprotonic Mass"* was awarded the "City of Como" 1953 prize by the Italian Physical Society.

In the summer of 1953, another International Expedition to Sardinia took place, and before the end of the year another discovery was made, the first two-body pion decay of a Λ-hypernucleus, in fact, an example of the decay reaction ${}_{\Lambda}H^3 \rightarrow \pi^- + He^{3++}$ [7]. Although the experimental evidence was unquestionable, our paper carried implicitly Beppo's cautious approach by being entitled *"On the Possible Ejection of a Meson-Active Triton from a Nuclear Disintegration"*. In a subsequent note [8], the Λ binding energy was estimated at ~ 1 MeV, to be compared with that of the last neutron in H^3, equal to 6.24 MeV. In the summer of 1954, I attended the Varenna Summer School, where Fermi gave a lecture series on pion physics (tragically his last summer). With much trepidation, I also gave a lecture on hypernuclear physics and noticed that Enrico was intrigued by my mention of a $\sim$1 MeV Λ-binding in ${}_{\Lambda}H^3$. After my talk, he came over to me and told me: "*That is extraordinary, the Λ must have a wave function as large as this room....*"

At the beginning of 1954, several nuclear emulsion groups, primarily the Genoa-Milan, the Dublin-Bristol, and the Padua groups, were spearheading the study of strange particles and their interactions (they were yet to be called strange). It was pretty well established that we had to deal with charged K-mesons and Σ-hyperons which were produced in the interaction of high-energy cosmic rays with the emulsion nuclei, and the only access we had to neutral strange particles was confined to the Λ when bound in hypernuclei. While the knowledge of the mass and decay modes of the Σ hyperons was settled relatively quickly [6, 9] (an event almost identical to our first $\Sigma^+ \rightarrow p + \pi^0$ decay, reported by the Padua group in 1954, confirmed the two-body decay assumption) the numerous decay modes of the K-mesons could only be attributed to one and only charged K after a protracted struggle. Thus each decay mode was at first assumed to originate from K-mesons of different identities, and the latter given different names. The first K-meson standard was provided by the decay $\tau^+ \rightarrow \pi^+ + \pi^+ + \pi^-$ reported earlier by the Bristol group and fully identified through its decay at rest of the three pions. Then came a number of K's decaying into one charged secondary and one or more neutrals. A two-body decay $K_\mu \rightarrow \mu + \nu$ was proposed by the cloud-chamber group of Ecole Polytechnique, another two-body decay $\chi \rightarrow \pi + \pi^0$ by the Bristol group, and finally the three-body decays. These included the $\tau^+ \rightarrow \pi^+ + (\pi^0 + \pi^0)$, first envisaged by Pais and Dalitz and observed at Rochester, the $K^+ \rightarrow \mu^+ + \pi^0 + \nu$ proposed by O'Ceallaigh at Dublin, and the $K \rightarrow \beta + \pi^0 + \nu$ first observed at Bristol.

The paramount difficulty, even using thick emulsions or emulsion sandwiches, was that the geometry of the decay events very seldom provided sufficient track length for an unambiguous identification of either the secondary energetic pions, muons and electrons by multiple scattering and g^* measurements, or an accurate mass determination by multiple scattering *versus* range of stopping primaries. Although the latter measurements hovered by a few hundred electron masses around the known mass of the τ they were seldom accurate enough to establish that all the K-meson decay zoo represented alternative decay modes of the same particle.

This struggle and frustration is portrayed well enough in a number of reports by the Genoa-Milan group [10–12], describing successively increasing numbers of K-mesons, some of which could be attributed to specific decay schemes, but whose measured masses were not accurate enough to resolve the above impasse. Together with Beppo, we came to the realization that a radical solution of the impasse was called for, that of bringing to rest or to their fate in the emulsion all K-meson secondaries. This would allow their positive identification, the accurate measurement of their energies, and thus the precise determination of the decay Q-values, and ultimately of the K-meson masses. For this to be possible, the most energetic secondaries, the muons from the two-body K-meson decay (~20 cm long) would have to be traced through many emulsions of a very large emulsion stack. This was the motivation for an experiment never attempted before, the G-Stack (G for giant) Collaboration [13, 14]. We owe to Beppo the orchestration of this major effort, which involved the coordinated participation of 36 physicists from 6 laboratories, an absolute first at the time. In the words of Cecil F. Powell, (the master of the spoken word, as referred to by Beppo) "*There are, of course, difficulties in establishing a successful collaboration. It takes time for standard procedures to be worked out, and to be put into general use. But, we have found that the advantages greatly outweigh the difficulties. In particular, we have not found the collaboration onerous or that it tends to inhibit the flow of new ideas. On the contrary, the friendly mutual support has seemed to us to result in an easy relation between the different laboratories which has tended to release and encourage individual initiative*".

The above was then the spirit in which we undertook the daunting experiment. In the Fall 1954 a 63 kg stack of 250 emulsion sheets 600 μm thick and 37 × 27 cm in size, was flown at a mean altitude of 27000 m for six hours over Northern Italy. Mobilizing an amazing number of Italian Ministries, Armed Forces. airport and communication facilities and organizations. Once again, I found myself in charge of tracking the balloon by radiotelemetry from the Milan airport. Due to a parachute mishap, the stack-carrying gondola, weighing a total of 140 kg, descended somewhat too rapidly over the Apennines, and about 10% of the emulsions were damaged in the impact with the ground. This however did not affect significantly the outcome of the experiment, and enough stack subsets were distributed among the collaborating laboratories. The fun began when a long K-meson secondary exited one stack subset and had to be expertly traced to one of the adjacent subset often abroad. Difficulties notwithstanding, the experiment was an outstanding success and contributed significantly to establishing the unique parent identity of the K-mesons involved in the zoo of six decay modes. These results were reported first in a comprehensive paper entitled: *"Observations on Heavy Mesons Secondaries (G-Stack Collaboration)"* presented at the 1955 Pisa International Congress [13] and further elaborated in a paper entitled *"On the Masses and Modes of Decay of Heavy Mesons Produced by Cosmic Radiation (G-Stack Collaboration)"* [14]. As previously mentioned, these were the only publications where Beppo's name appeared among the authors of the Genoa-Milan group. Together with another discovery by our group, that the nuclear capture of negative K-mesons gave rise to the production of Σ- and Λ-hyperons (the latter bound in hypernuclei) [15], the G-Stack Collaboration results gave impetus to Gell-Mann's establishing the strangeness scheme and the "eightfold way". This took place in momentous conversations between Beppo and Murray Gell-Mann at the Pisa Congress in June 1955.

On the front of hypernuclei, I became involved with reviewing the existing evidence twice [16, 17], and more events from the Genoa-Milan group were described [18]. Another collaboration of the Genoa-Milan group, with Saclay and Brussels, led to a very precise measurement of the τ Q-value and hence of the K^+ mass [19]. Finally, the energy spectrum of the muon-decay electrons, reported earlier [3] from 278 measured events, was re-examined after the statistics were augmented to a total of 506 events [20]. In a detailed comparison with the theoretical spectra calculated by Michel for positive muons and by Porter and Primakoff for a mix of positive and negative muons, inclusive of radiative corrections, a fairly accurate value of the theoretical parameter ρ was obtained. This was regarded as a feat of the multiple scattering method of measuring particle pβ's in nuclear emulsions, to the point of allowing a rather sophisticated comparison of measurement with theory.

My association with the University of Milan, with Connie and Beppo,and my productive collaboration with the colleagues of the Genoa-Milan group, came to an end on July 1st, 1956, when I joined the University of Chicago. There, I could rekindle at first my interest in hypernuclear physics with emulsions exposed to separated K-beams from the Berkeley Bevatron and the CERN PS. It became a lot easier than cosmic-rays exposures to find hyperfragments. Beppo's legacy guided my carrier since that day.

References

1. Levi-Setti R. "Contribution to the Methods of Measurement of Scattering in the Photographic Plate", *Nuovo Cimento*8195196.
2. Levi-Setti R. "Causes of Error in Scattering Measurements in Nuclear Emulsions", *Proceedings of the International Summer School of Physics, Varenna 1953, Course I, Suppl. Nuovo Cimento*, **11**, Ser. IX, No. 2 (1954) 207.
3. Levi-Setti R. and Tomasini G. "On the Decay of μ-Mesons", *Nuovo Cimento*81951994.
4. Levi-Setti R. and Tomasini G. "Slow Heavy Mesons from Cosmic Ray Stars", *Nuovo Cimento* 919521244.
5. Bonetti A., Levi-Setti R., Panetti M., and Tomasini G. "Observation of the Decay at Rest of a Heavy Particle", *Nuovo Cimento*101953345.
6. Bonetti A., Levi-Setti R., Panetti M., and Tomasini G. "On the Existence of Unstable Charged Particles of Hyperprotonic Mass", *Nuovo Cimento*1019531736.
7. Bonetti A., Levi-Setti R., Panetti M., Scarsi L., and Tomasini G. "On the Possible Ejection of a Meson Active Triton for a Nuclear Disintegration", *Nuovo Cimento*111953210.
8. Bonetti A., Levi-Setti R., Panetti M., and Tomasini G. "Remarks on the Decay of Meson Active Triton", *Nuovo Cimento*111953330.
9. Bonetti A., Levi-Setti R., Locatelli B., and Tomasini G. "Mass and Decay of Charged Hyperons", *Suppl. Nuovo Cimento*, **12**, Ser. IX, No. 2 (1954) 292.
10. Bonetti A., Levi-Setti R., Locatelli B., and Tomasini G. "Mass and Decay of K Particles", *Suppl. Nuovo Cimento*, **12**, Ser. IX, No. 2 (1954) 227.
11. Bonetti A., Levi-Setti R., Panetti M., and Tomasini G. "Some Slow K Particles, Their Decay at Rest and Production in Cosmic Ray Stars", *Proceedings of the Royal Society* A2211954318.
12. Bonetti A., Levi-Setti R., Locatelli B., and Tomasini G. "Results on Some Secondary Particles of K Mesons from the Stacks Flown in Sardinia in 1953", *Nuovo Cimento*11955904.
13. Davies J. H., Evans D., Fowler P. H., Francois P. E., Friedlander M. W., Hillier R., Iredale P., Keefe D., Menon M. G. K., Perkins D. H., Powell C. F., Crane L., Johnston R. H. W. and O'Ceallaigh C., Anderson F., Lawlor G., Nevin T. E., Alvial A., Bonetti M., Di Corato C., Dilworth C., Levi-Setti R., Milone A., Occhialini G., Scarsi L., Tomasini G., Ceccarelli M., Grilli M., Merlin G., Salandin G. and Sechi B. "Observations on Heavy Meson Secondaries. (G-Stack Collaboration)", *Proceedings of Pisa International Congress, June 1955, Suppl. Nuovo Cimento*, **4**, Ser. X, No. 2 (1956) 398.
14. Davies J. H., Evans D., Francois P. E., Friedlander M. W., Hillier R., Iredale P., Keefe D., Menon M. G. K., Perkins D. H., Powell C. F., Bøggild J., Brene N., Fowler P. H., Hooper J., Ortel W. C. G., Scharff M., Crane L., Johnston R. H. W., O'Ceallaigh C., Anderson F., Lawlor G., Nevin T. E., Alvial A., Bonetti M., Di Corato C., Dilworth C., Levi-Setti R., Milone A., Occhialini G., Scarsi L., Tomasini G., Ceccarelli M., Grilli M., Merlin G., Salandin G., and Sechi B. "On the Masses and Modes of Decay of the Heavy Mesons Produced by Cosmic Radiation" *Nuovo Cimento*219551063.
15. Bacchella L., Di Corato M., Ladu M., Levi-Setti R., Scarsi L., and Tomasini G. "Heavy Mesons and Hyperons Giving Rise to Capture Stars", *Proceedings of Pisa International Congress, June 1955, Suppl. Nuovo Cimento*, **4**, Ser. X, No. 2 (1956) 465.

16. Grilli M., and Levi-Setti R. "Report of the Committee on Unstable Fragments", *Proceedings of Padua Congress, April 1954, Suppl. Nuovo Cimento*, **12**, Ser. IX, No. 2 (1954) 466.
17. Levi-Setti R. "Unstable Fragments", *Suppl. Nuovo Cimento*, **2**, Ser. X, No. 1 (1955) 263.
18. Ladu M., Levi-Setti R., Scarsi L., and Tomasini G. "Three Possible Hyper- fragments", *Proceedings of Pisa International Congress, June 1955, Suppl. Nuovo Cimento*, **4**, Ser. X, No. 2 (1956) 621.
19. Bacchella G. L., Berthelot A., Di Corato M., Goussu O., Levi-Setti R., Rene M., D. Revel, L. Scarsi, G. Tomasini, and Vanderhaege G. "On the Q-Value of the Tau-Decay", *Nuovo Cimento*419561529.
20. Bonetti A., Levi-Setti R., Panetti M., Ross G., and Tomasini G. "Lo spettro di energia degli elettroni di decadimento dei mesoni μ in emulsione nucleare", *Nuovo CimentoNuovo Cimento*3195633.

Personal Remembrances

Bertram Stiller

Shortly after their arrival in Brussels, Belgium, to take up professorships at the Université Libre, the Occhialinis, Beppo and Connie, organized the first International Conference on Cosmic Ray Research with Nuclear Emulsions, Fig. 1. It was my good fortune to receive an invitation to this conference.

During the conference, I was invited to their home for dinner and, because my hosts had planned to see a John Ford movie that night, they asked me to baby-sit their infant daughter, Connie Pooh, today known as Etra. Although somewhat surprised, I nevertheless agreed, since I had already experienced the infancy of my two sons. The evening went well, and upon their return, they asked me to spend the night since it was 1951 and taxis were still difficult to find. I readily agreed before I realized that there was only one bedroom, and that I would bed down on the living room floor. In the morning, I was awakened by a familiar chattering which turned out to be coming from a group of amused young women, who were the Occhialinis' scanning team. There was no room at the lab, as yet, and so microscopes had been set up in their living room. We had used them the previous evening, but they failed to mention that the scanning team would be in residence in the morning.

With this unexpected and startling experience began a friendship between us that lasted for well over 50 years. Although we shared a strong interest in cosmic rays, elementary particles, space physics, and in the use of nuclear emulsions as particle detectors, we never had the opportunity to work or publish together.

In July, 1957, while we were both attending the First International Symposium on Nuclear Emulsions in Strasbourg, France, Beppo invited me to visit his cosmic-ray research group at the University of Milan. Later that month, I went to Milan and spent a few days visiting his laboratory. Beppo asked one of his microscopists, who knew some English, to act as kind of a mentor during my visit, since I could not speak Italian. This was Mila Puppo, who was such a wonderful mentor that she became

B. Stiller (✉)
Washington DC, USA
e-mail: bstiller@mailaps.org

L. Gariboldi et al. (eds.), *The Scientific Legacy of Beppo Occhialini*,
https://doi.org/10.1007/978-3-031-37034-2_13

Fig. 1 Beppo and Connie, Université Libre, Brussels, Belgium, 1951

my wife in 1960, and this event has made the years since then as happy and joyful as one could possibly expect. This accidental and totally unexpected occurrence was something that the four of us would marvel at, on many occasions.

During 1959–1961, while Beppo, Connie and Connie Pooh were at MIT, at the invitation of Bruno Rossi, I spent a number of months in the Physics Department of the University of Milan, on Via Saldini, working with Beppo's graduate students. Beppo, after having visited our group at the Naval Research Laboratory in Washington, D.C., asked me to introduce his students to some of our ideas in automating track measurements and emulsion processing equipment that we had developed at NRL. I was keen to do this because it would give me an opportunity to become better acquainted with Mila. When we decided to marry in 1960, the Occhialinis were kind enough to offer us their apartment on Viale Argonne, during their stay in the States, which offer we were happy to accept because of the high cost of rental housing in Milan. Shortly after moving in, we learned that Freddy and Elizabeth Herz were also joining Beppo's group, and so we invited them to stay with us. Thus the Occhialini apartment became host to a "*mélange-à-quatre*" of newlyweds.

Since Mila, my wife, had all her family in Italy, it became customary for us to spend our summers in Italy. Thus it was, that we began to spend part of each summer visiting with Beppo and Connie. During the week that we generally spent with them, we managed to cover not only the latest turns in Italian and U.S. politics but also comings and goings of their many friends in science. In particular, the physicists F. Houtermans and B. Pontecorvo were closely followed, not only because of their research, but also because of the unusual political positions that they had both chosen for themselves. Of course, Beppo's experiences with two Nobel Prize winners, P. Blackett and C. Powell, were also topics that we returned to with fair regularity.

The research using a counter-controlled cloud chamber embedded in a magnetic field—a system which Beppo and Blackett developed together—led to their confirming the existence of the positron, a positively charged electron, and to their discovery that electron-positron pairs were produced by gamma rays. In 1948, Blackett received the Nobel Prize and the award was, in part, for the research done with Beppo. Blackett acknowledged this debt by opening several paragraphs in his Nobel Address with the phrase, "Occhialini and I did…". Beppo always spoke of Blackett with great warmth and admiration.

When Beppo joined Powell's group at the University of Bristol, he realized almost immediately that the nuclear emulsions being used to study cosmic rays needed to be increased in sensitivity, so that they would be able to register tracks made by high-energy charged particles. He induced C. Waller, a chemist with Ilford Ltd, to attack this problem, and after some intense efforts, a much increased level of sensitivity was achieved. Beppo then sent stacks of these sensitive emulsions to be exposed on mountains tops, at the Pic du Midi and at the Andes. These exposures then led to the discovery of the π-μ decay which, as is well known, was a fundamental discovery for high-energy particle physics. Powell was awarded the Nobel Prize in 1950 for the development of sensitive nuclear emulsions and their use in cosmic-ray physics. Powell failed to mention Beppo in his Nobel Address. If ever there as a poor award of a Nobel Prize, this is the prime example. Of course, Beppo remained disappointed about this for the rest of his days.

In 1968, Beppo and Connie acquired a summer cottage in the town of Cinquale on the Italian Riviera, some kilometers north of Viareggio. It was an A-frame house, of small dimensions, sitting on a flat plot of ground.

Here, they began their first ventures into gardening and landscaping. There was much discussion about where to plant trees, fencing, shrubbery and so on, in addition to the modifications to the cottage that Beppo planned. Mila, having spent the war years in a small village, Ponte Organasco, in the Apennines, was much more acquainted with gardens than I, and so she became the advisor to Connie. My role, because of the well-known *fai da te* knowledge of Americans, developed into discussions and disagreements with Beppo, about modifications that he wanted to make to the house. Beppo had grown enamored with power tools that he found in the U.S. and so one of his dreams was to parallel-wire the cottage for 110 V a.c. so that he could operate U.S. power tools. This we accomplished without too much difficulty.

The Superstrada A-12 had only been built from Genoa to La Spezia at the time that Connie and Beppo were summering in Cinquale. When the extension to Pisa was designed, the plan put the highway almost into their backyard. This was the end of summers on the Italian Riviera. It was a real disappointment for them because of all the effort they had put into the cottage. There was another possibility; a mountain

Fig. 2 Beppo, San Pellegrino

cottage that Beppo had inherited from his father. Unfortunately, Connie was not a mountain person and so this had no great appeal for her. On the other hand, Beppo, having spent some years in the Andes during the war, was quite keen on rehabilitating this long-abandoned house, Fig. 2.

It was located about a kilometer from the village of San Pellegrino, in the middle of the Alpi Apuane, at an altitude of about 1500 m. Everything such as food, structural materials, tools, etc., had to be carried from the town along a rough path through fields, brush and woods to reach the house. After much hard work, Beppo decided that a wagon would be an ideal alternative to carting everything on one's back. Pulling the cart over the path turned out to be almost as tiring as carting things on one's back. At any rate, here we began an even larger set of renovations than on the cottage in Cinquale. After all, that had been a livable home to start with. Here there was no heat and even in summer, at that altitude, nights are not very warm. But here, Beppo invited in friends from Milan, since I generally spent only a week or two with him, and there was an ongoing series of renovations that he needed help with. Beyond our endless political discussions, there was also culinary activity, since we had to feed ourselves, lacking any feminine associates. But most memorable and amusing, were the car trips that he and I took to the nearest large town, Castelnuovo di Garfagnana. In those days, neither Beppo nor Connie had learned to drive. Beppo had a very strong fear of car accidents, and so until he became accustomed to a friend's driving ability, he would carry on with a steady stream of advice to the driver. In my case, being a driver with U.S. driving habits, he never did cease from advising me. For example, I drove with the car's headlights always on. This really drove him "up the wall" because at that time, this was not the custom on Italian roads. The statement that this was a matter of safety, was always countered with the statement that it was distracting to Italian drivers. In addition, he claimed that I was given to driving as if I were in the *Mille Miglia*. The road from San Pellegrino to Castelnuovo, Fig. 3, although not 1000 miles, was as enjoyable as driving in the *Mille Miglia*.

Usually, by the time we reached Castelnuovo, Beppo was not speaking to me. All this was done in a 1969 Peugeot 400, which, although it was as good a handling car as I wanted, was not a Ferrari. However, after hunting around in a *ferramenta* store for some time, we would both be ready for an *espresso* to prepare ourselves for going back up the mountains. Usually, it was near evening and so the trip up was done much more sedately.

In 1979, the Wolf Foundation of Israel awarded Beppo the Wolf Prize. This award, which has also been given to a number of Physics Nobelists, came as recognition of the importance to physics of the research that Beppo had done, not only with the two Nobelists I mentioned earlier, but also in recognition of the discoveries that he and his group had made since the time of his return to Italy. It was the first and only occasion on which I received mail from Beppo. A postcard with two words; "mission accomplished". The monetary part of this award made it possible for Beppo and Connie to seek a home suitable for their retirement. They found a grand old Tuscan farmhouse, named Cannarecchi, near the village of Marcialla, in Tuscany, about midway between Florence and Siena. They gave up their apartment in Milan and took up residence at Cannarecchi. This led to some changes in their lifestyles. Connie learned to drive because they were no longer within walking distance of shops. Beppo purchased a wonderful small motorbike. And using the two vehicles, they began to deal with a new life as displaced urbanites. Of course, the new house needed more than the usual renovations. It was sufficiently large and old, consisting of two stories plus a partial basement, and a large barn, that there were many projects to plan. Connie now pitched into these projects in much greater measure than she had before. We updated our transport to a splendid 1975 Alfa Romeo Giulia Super Berlina and so we were able to run down to Marcialla more easily than before. In fact, shortly after they settled into Cannarecchi, Beppo was invited back to Israel, as were other recent Wolf Prize winners, for the presentation to the next winner. Mila and I went down to house-sit while they went to Israel, and we had a great time looking after the

Fig. 3 Our *Mille Miglia* route from San Pellegrino to Castelnuovo

house, their cat and dog, and Connie's garden. Although I never drove my Alfa down from San Pellegrino to Castelnuovo, the road between Marcialla and Cannerecchi has a fine, steeply descending S curve which was always a high speed challenge.

As the summers ticked by, Cannerecchi was slowly fitted with some modern conveniences but most of the interior was left untouched. A beautiful front door was fitted. A pond appeared one year, which Beppo said was required in the event of fire. In truth, the total volume of water it held would hardly have dealt with a serious house or brush fire. Of course, the inevitable 110 V a.c. was installed, along with a fine workshop area, which was usually too cluttered to be able to find workspace. Best of all perhaps, was the apartment that Etra, now an architect, designed and had built into the former pig sty, which was part of the ground level of the house. Summers being what they are in Tuscany, this became the most desirable sleeping room in the summertime. Connie's vegetable garden was always in need of some additional spading, and there was never a dearth of physical activity when we visited them. An unfortunately cold winter led to the death of many of their trees, and this became a reforestation project with the aid of the Italian Forestry Service. Cannerecchi became a permanent home for Beppo and Connie, and there was a steady stream of company coming and going. Reaching there without a car was done with some difficulty. A bus route existed from Florence but it only went through Tavarnelle, on its way to Siena. This meant that someone had to meet visitors who came on the bus, since taxis were not easily found. Beppo's bike was not up to carrying passengers, and so Connie and her VW Beetle served as a "taxi". Tavarnelle had many shops so there would always be something to buy, making such taxi service useful for the household.

After Beppo passed away in 1993, we felt even closer to Connie and tried to visit more than once each summer. Mila's parents began to need our help, and so I would visit Connie by myself, taking the train and bus instead of driving alone. These visits involved helping her with her bee hives, her wine cellar, vegetable garden, and her distillation of lavender flowers to make lavender essence. Also, it was time for nostalgia, after the many years that had been shared. But inevitably, Italian and U.S. politics kept us up far into the night. We made our last visit in the summer of 2000, when it became clear to me that Mila's deteriorating health would prevent us from visiting Cannerecchi again. It took Etra and myself a lot of scolding of Connie to obtain a computer and learn to use it. Thereafter, we kept in touch by means of ubiquitous e-mail messages until Connie entered a retirement home on the Italian Riviera. And now there are just Connie Pooh and me, to reminisce about Beppo and Connie, and that night long ago in Brussels.

Working with Beppo: Personal Recollections

Riccardo Levi-Setti

My association with Beppo lasted about six years, from the Summer of 1950 to the Spring of 1956. It was an intensive, rigorous training in many areas of scientific, historical, literary, inventive, political, and, yes, even artistic human endeavour.

My first encounter with Beppo's personality dealt with our interpersonal communication. Beppo's thinking was extremely fast, and often overlapping thoughts were expressed in incomplete sentences. It took long hours spent in the darkroom with Beppo, while monitoring the processing of thick nuclear emulsions when I first joined him in Brussels, for me to begin to follow his rapidly changing lines of thought so that I could answer in sync. This phase was eventually overcome, and in the process, I was exposed to fascinating Beppo's recollections, ranging from his adventures in the Amazon and Mato Grosso in Brazil, to his work with Patrick M. S. Blackett at Cambridge and Cecil F. Powell at Bristol. I soon came to the realization that some of the scientific attributes of these mentors may have rubbed off on Beppo, and, indirectly, I benefited from this exposure myself. Here I was, confronted with the "deus ex machina" who was largely responsible for the discoveries which brought the Nobel prize to his two tutors, telling me what he learned from his association with them, without expressing any bitterness from having been shunned from sharing the deserved reward. I was deeply moved.

Beppo's inventiveness transpired immediately in the training I was fortunate to have been exposed to. There was a three-dimensional way of thinking about designing laboratory equipment, from the plumbing needed in the specialized vats needed for processing photoemulsions, to the delicate devices to be constructed for improving the precision of microscope measurements which I was assigned to perform. I remember clearly how several of his inventions were immediately incorporated into custom-built microscopes with the aid of Ing. Cantù of the Koristka microscope factory in Milan.

R. Levi-Setti (✉)
The Enrico Fermi Institute and Department of Physics, The University of Chicago, Chicago, IL, USA

L. Gariboldi et al. (eds.), *The Scientific Legacy of Beppo Occhialini*,
https://doi.org/10.1007/978-3-031-37034-2_14

The rigours of the scientific method were imprinted by Beppo in my mind and that of my coworkers every time something new was found in our lab, and our results had to be conveyed to the scientific community in a convincing manner. This process usually took place at night, and I mean sometimes through the entire night, after we presented Beppo with a timid draft of one of our papers. There we were exposed to Beppo's dialectic abilities which generally threw to shreds our data and presentations, and often sent us back to the lab. At times, even if Beppo was (secretly) in agreement with our logic, he enjoyed taking an adverse, opposite viewpoint, and we had to struggle to defend our conclusions. And then, there was the process of finding an appropriate title for our papers, which was to be suggestive of a discovery, but had to preserve an escape route if we were carried away by our excitement. A delicate balance was to be attained, and that is why the titles of our papers contained subtle nuances manifested through the use of the adjectives "possible" or "probable". As a result, our papers carried a great deal of Beppo's guidance, and it was his generosity toward us that prevented him from adding his name among the authors. The same can be said of Connie (Occhialini-Dilworth), who not only helped us with measurements and critical insights, but who was always in charge of making our English intelligible.

The period in which I worked with Beppo marked the transition between individual experimentation and collaborative work with a number of colleagues, often residing at different institutions. This transition required at first an education with which we were not familiar. I learned my lesson when I tried to present a new finding of mine, and I was reminded that this finding was not entirely mine but belonged as well to distant researchers with whom I was tied by a collaboration agreement. I soon benefited by this ethic when the reverse situation occurred, and step-by-step we all became involved in collaborating successfully with 36 co-authors from six laboratories. This was the beginning of the communal efforts which became the norm in high-energy physics research, to reach enormous proportions at the modern TeV-region particle accelerators.

Besides having developed and perfected the technology of processing thick nuclear photographic emulsions, Beppo excelled in ordinary black and white photography, both from the technical to the artistic viewpoint. Still vivid in my mind is a day spent in Verona with Beppo, taking pictures of the 24 romanic bas-reliefs on the doors of the San Zeno cathedral at various times of the day, to let the shadows from the sun vary the moods of our photographs. And then, the action continued in the darkroom, where the principles of temperature development to which we were accustomed in dealing with nuclear emulsions were transferred to the printing of photographs. Thus, I learned how to modulate light and shadows while printing, to correct unwanted exposure gradients, by locally warming the paper in the developer bath, and many other tricks of the trade which could make photography akin to painting. I definitely owe it to Beppo to have inspired in me a passion for photography at both levels mentioned above, so that my career after Beppo always pivoted around photographic methods of physical research. After nuclear emulsions, I dealt with bubble chamber pictures of elementary particle phenomena, and, in a major switch of interests, with the development of high resolution scanning ion microscopy and imaging secondary ion mass spectrometry, and its applications to materials science and biology. Three-dimensional thinking, exercised with Beppo, was of great help in the design and construction of my ion microprobe. The use of this novel instrument compelled me to deal with another switch, from film photography to digitally-acquired images and computer-based image processing and printing. Even with this new tool, some of the canons of artistic photography inherited from Beppo, such as the importance of composition, have become an instinctive guidance. This can be detected now in my analytical maps of chromosomes (my presently sponsored biological research), and in my pictures of trilobites (my palaeontological avocation).

In conclusion, working with Beppo, although not always easy, left a profound imprint on my scientific personality, and I have no doubt that this has been the case with all my early collaborators who had the fortune of sharing my experience.

G. P. S. Occhialini: One of My Masters

Sergio P. Ratti

I think I had four "masters" in my life, over and above my parents: Giovanni Polvani, Beppo Occhialini, Marcello Conversi and Martin M. Block: four persons so different, for their character, culture, personality; mostly for their approach to younger physicists and students.

Giovanni Polvani was my "master of life", he hired me as "assistente incaricato" no more then 1 month after my graduation, November 14th, 1957; he hired me in a permanent position as assistant professor on February 2nd, 1959. When I got married he said to my wife: "young lady remember: 80% of your husband's career depends upon your patience, understanding and cooperation". She did that for over half a century now. Incidentally, Beppo wanted to call up a meeting the day of my wedding and insisted that I should postpone the ceremony not to interfere with the scientific program. I did not do that and he came anyhow to the wedding party telling his proverbial jokes. Polvani was the "Professor" with a capital P. He talked very rarely to me; in my career I visited his apartment only once; I always addressed him as a third person as usual in the Italian language when formally talking to a person. I was assisting him during the exams. For at least 6 years, over and above my teaching assignments, I held classes in his place while he was elected Rector of the University and later President of the National Research Council—CNR (he used to say: "my son, don't touch thermodynamics, just tell me when you get to the point". Close to the point—thermodynamics having been postponed to acoustics and waves—he would say: "my son, don't touch the II Principle. I want to cover that, it is too important". And only too close to the end of the term he would apologize for not having time to go personally into the classroom and give me permission to cover II and III Principles. For the benefit of the reader, the II Principle sounded as "the trace left by Joule's eddy experiment is indelible!" proving a very personal view of the subject). On the other hand I had maximum freedom to organize my own research. When he died, I felt the loss of my second father.

S. P. Ratti (✉)
University of Pavia, Pavia, Italy

L. Gariboldi et al. (eds.), *The Scientific Legacy of Beppo Occhialini*,
https://doi.org/10.1007/978-3-031-37034-2_15

Marcello Conversi—while at the University of Pisa— starting around 1955-56, was the co-leader, together with Carlo Succi the senior assistant professor to Polvani, of the experiment in which Carlo Rubbia and I did our thesis work in a small mountain laboratory in Val Formazza, northern Italy. When Carlo got the Nobel Prize in 1984, Marcello, in an article written on one of the major Italian magazines, was so kind to quote my name as one of his students. Many years later, Marcello, then at the University of Rome, joined an experiment I proposed at the Triga Mark II reactor of the University of Pavia—where I was appointed as full professor in 1972—to search for neutron-antineutron oscillations, predicted by one version of $SU(n)$ symmetries and strongly supported by Bob Marshak, a friend of mine, past President of the American Physical Society, President of the City University of New York, the inventor of the $V - A$ theory of weak interactions. With him, intense was the discussion of several physics issues, ranging from diffraction dissociation, resonance classification, Regge trajectory role in hadron physics, to cosmic ray physics and future development of accelerator technology. He asked me once to move to the University of Rome, in an ambiance where Enrico Persico, Giorgio Salvini, Giorgio Parisi, Nicola Cabibbo, Carlo Bernardini, Giorgio Careri, among other, where the leading scientists. He wanted me to go there one day to discuss seriously the possibility. I took the plane, went there and at the Alitalia Air Terminal (at that time into the same building of Stazione Termini), tried to get a taxi. It was a nightmare: people fighting to jump into the cars, shouting aloud. I was so discouraged that, as soon as I saw Marcello I told him I would rather continue to live into the provincial, parochial,... sleepy Pavia. We remained friends, I visited him rather regularly till he passed away many years ago now.

Martin Block was my master independent of the physics projects in which we were and are involved. I spent the academic year 1961–62 with him at Northwestern on sabbatical thanks to Beppo (I'll go back to it later) working on hyperfragment physics detected in a liquid helium bubble chamber that I still consider superb: it was a little larger than a children shoe box 8cm $\times$ 10cm $\times$ 20cm. During my sabbatical I wrote as many as 7 papers in a year, including one of the first measurements of the Λ^0-hyperon lifetime; the measurement of the Λ^0-K^- relative parity; the (still unique) measurement of the s-wave contribution to the totally neutral decay $\Lambda^0 \rightarrow n\,\pi^0$; the measurement of the Michel ρ parameter for the β-decay of negative muons, still the only experimental proof that the $V - A$ weak interaction basic parameters ($\rho = 3/4$ is the predicted value, $\rho = 0.751 \pm 0.035$ the measured value) must be the same for particle and antiparticle. During that year we almost destroyed a desk by kicking it with our feet in excited discussions on how to invert error matrices into statistical weighting factors in a certain computer program. His character is well known for not being among the mildest in the world; however, after 45 years we are still very good friends. He continuously invites me to visit him in Aspen, Colorado, were he now lives, but I could not (yet) find the time to do so, as I got plenty of things to do here in Italy.

Beppo: G. P. S. Occhialini was quite different. I never published a paper with him, I never worked in any of his experiments. Nonetheless he has been one of my "masters" (on an occasion I'll mention later, Bruno Pontecorvo said: "I suggest to all of you to collaborate with Beppo Occhialini, he won't get a Nobel Prize, but you would have a good chance...". Is it a sign of destiny?). When I was a student, I never had the privilege to listen to a lecture delivered by him in person. Alberto Bonetti and Livio Scarsi did that for him. However I learnt later that he was not "absent"; he used to meet his assistant professors in his house till very late at night to discuss all details of the lecture, to find out the best way to propose a new subject. I remember when I took the exam "Fisica Superiore" (actually, since 1991 till the new reform I held that chair in Pavia): I studied that exam together with Carlo Rubbia mostly in the laboratory up in the mountain. As "poor laureandi" we used to spend 25 days at the mountain laboratory and one week home; 25 days up, one week down. All that for about one and a half year. We were asking and answering each other the funniest questions and make calculations, discussing whether or not a π was proper in a formula or rather a 4π or even an 8π (due to the 2 spin projections of an electron or the 2 polarization states of electromagnetic waves). I felt very confident when I got to the table in a classroom located in the basement of the Physics Institute in via Saldini, Milan.

There were Beppo, Livio Scarsi and Alberto Bonetti; the exam lasted—as usual—about an hour and had nothing to do with the program written in the book and covered in all the 40 lectures of the course. At the end, after a brief private consultation, they called me back in the classroom and said: "Ok, you got an A (30/30)". My face did not look "that" happy and they did ask me why. In Italy there is the possibility of getting an A^+ (laudem-lode) and I thought that, after all they did not ask any "normal" questions and that I could have forgotten all the program and they would not know. Therefore I spoke out, gently but very plainly, my opinion. Beppo said: "Ok, let me then ask you another question in the program". At that point Alberto and Livio both said with almost an unique voice: "No Beppo, the student is right, we discussed physics with him and asked questions proper for an exam to appoint an assistant professor. It was not a normal exam. He is close to being a professional young researcher" and they added a "lode" to my vote. I felt very proud in that particular instant; furthermore I had the impression that from that moment Beppo looked at me smiling with more affection.

He was the first "professor" who allowed me to address him using the confidential Italian second person. He was on my thesis examining committee; my thesis advisor was Giovanni Polvani with a substantial help from Carlo Succi (and a minor contribution by Riccardo Giacconi at the beginning, very few months before he left to the United States).

The arrival of Occhialini in Milan encouraged the formation of a group to build Wilson chambers to be triggered—since that was the technique with which he did in England the search that did not gain him the Nobel Prize—and exposed to cosmic rays. Actually in the late fifties Carlo Succi and his group built perhaps the largest Wilson chamber (about 1 cubic meter sensitive volume) ever used in a cosmic-ray experiment.

In present terms my dissertation would have been preparing an apparatus to search for the τ lepton, then known "only" as heavy muon, based on the use of 2 triggered Wilson chambers. At that time a Russian group (or rather Georgian?) led by Alikha-

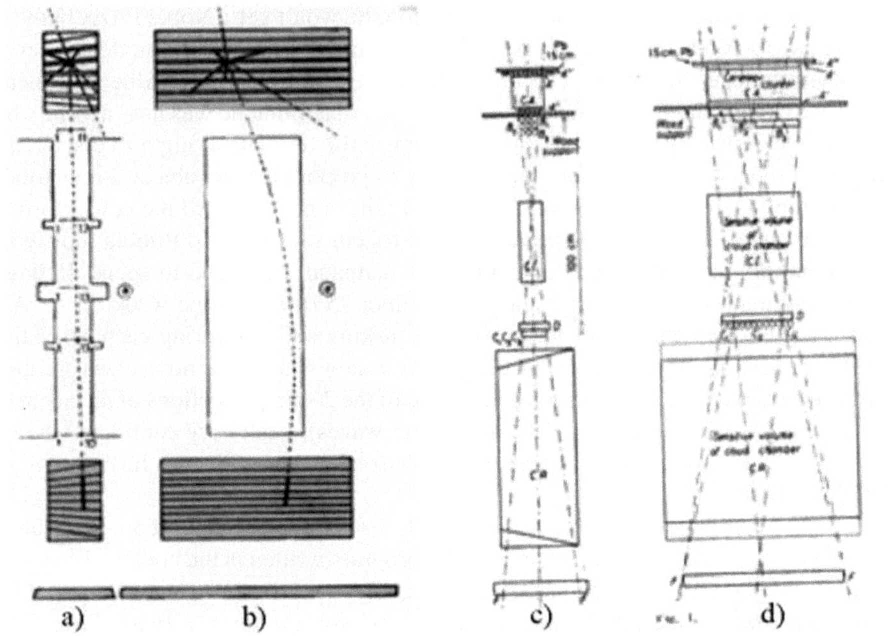

Fig. 1 **a, b** the apparatus of the Russian experiment, measuring the mass of particles produced into the upper chamber and stopped into the bottom plate-chamber measuring the mass by means of momentum and range; **c, d** the apparatus of the Milan-Pisa experiment measuring the mass of a particle by measuring the relativistic velocity through the ionization left in the upper CI chamber and the light produced into the Cherenkov counter, plus the range in the bottom CR chamber. At the very bottom an anticoincidence counter rejects the through tracks

nian (A. Alikhanian, A. Alihanov, H. Weissenberg, V. Morozov, G. Muschriswili, A. Dadaian, N. Shostakovich, A. Akaian and many others) in cosmic-ray experiments, exploited from 1948 till 1957, detected a number of non interacting particles, similar to the muons, generated in a upper Wilson chamber, stopping in a set of carbon plates 1cm thick of a second Wilson chamber separated by a horizontal magnetic field. The mass was the result of the contemporary measurement of momentum and range. Figure 1a, b shows their setup. Those experiments claimed the detection of particles having a mass $m = 550\,m_e$, m_e being the electron mass, i.e. about 280 MeV/c^2. The title of my thesis was *"Design of an experiment to search for a particle having mass 550 electronic masses"*; unfortunately it is now known that the mass of the τ-lepton is around 1.8 GeV/c^2. By measuring the relativistic velocity β (from ionization with a "top" helium Wilson chamber and/or with a water Cherenkov counter —Fig. 1c, d and range in a "bottom" cloud chamber, equipped with a set of carbon plates 0.3 cm thick, we did not detect any of such particles but rather exactly the predicted number of nuclear interactions ending into nuclear stars delivering fragments which would not cross more than 3 plates, i.e. fragments which would be "not" visible in the gas gaps outside 1 cm thick carbon plates.

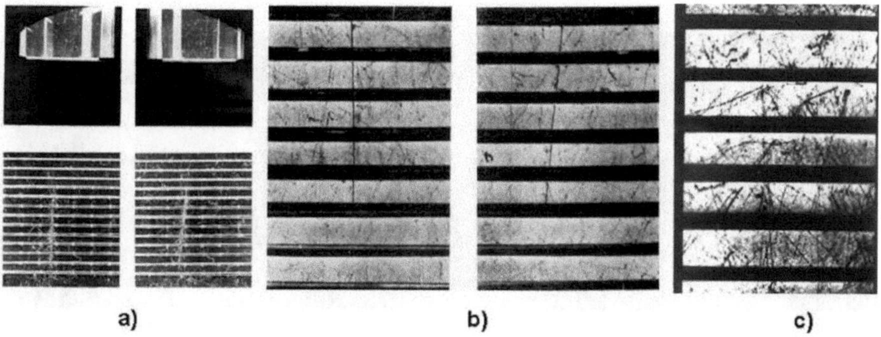

Fig. 2 **a** Polaroid picture showing a particle crossing the ionization chamber and producing a high-energy interaction in the 5th plate 0.3 cm thick, whose secondary particles are absorbed by the remaining plates, thus not triggering the veto counter; **b** stereo-pictures of a stopping muon; **c** an enlarged view of a neutral interaction in the second visible gas gap, with a produced pion plus secondary fragments stopped within 1 cm of carbon (3 consecutive plates)

Figure 2a shows a Polaroid picture taken to check the performance of the 2 chambers containing an incoming cosmic ray in the ionization chamber (where the electric field separates positive and negative ions condensed into 2 separate parallel tracks); Fig. 2b shows a stopping muon in plate 7, while Fig. 2c shows a neutron star in the second gas gap, producing fragments all stopped within 3 plates, i.e. stopped by less than 1 cm of carbon.

Figure 2a shows a Polaroid picture taken to check the performance of the 2 chambers containing an incoming cosmic ray in the ionization chamber (where the electric field separates positive and negative ions condensed into 2 separate parallel tracks, while Fig. 2b shows a neutron star in the second gas gap, producing fragments all stopped within 3 plates, i.e. stopped by less than 1 cm of carbon.

I was told Occhialini appreciated my thesis work. I think he had a role in having assigned to my dissertation the "Giovanni Gentile jr." Prize for the best thesis 1957.

I still have the picture (Fig. 3a) of the major professors in the examining committee—Giovanni Polvani, Piero Caldirola, Giuseppe Occhialini, Carlo Salvetti, Guido Tagliaferri, Antonio Lovati—taken the day of my thesis defense. Among the "laureandi", that day were Guido Vegni and Elisabetta Abate (who married a German physicist and lives now in Bochum), a skinny and short student whose name is Skof (known by us all confidentially as Skoffino—I do not remember his name) who was later appointed associate professor at the University of Bari, plus other students whose names do not come to my mind after half a century (Fig. 3b).

In 1952, the Istituto Nazionale di Fisica Nulceare (INFN) was founded in Italy as a branch of Comitato Nazionale per l'Energia Nucleare (CNEN). The first director was Piero Caldirola and the second Beppo Occhialini. His direction was not free from some "extravaganza". To move from one town to another, from Milan to the mountain Lab, or to CERN that was being built at fast speed, all of us had to fill in a form: where to go, when to go, how to go. Among the possible transportation

Fig. 3 **a** From left: Beppo Occhialini, Giovanni Polvani, Carlo Salvetti, Piero Caldirola (in the background, from left: Guido Tagliaferri, Antonio Lovati and Elisabetta Abate); **b** the students getting the title: from left: a skinny Guido Vegni, not identified, E. Abate; not identified, Skoff, myself, not identified

means the form contempled: train, airplane, mule, sleigh, roller skates, camel (just in case an experiment would happen to be done in the dunes of the desert). Milan was the only INFN Section having listed all those transportation possibilities. I found it no more than "peculiar". Some year after graduation Beppo decided to move to cosmic and space physics with Livio Scarsi and Alberto Bonetti who, a bit later, got a full professorship in Florence. Guido Vegni—together with Mariella Di Corato and Adele Sichirollo—were fully immersed in experiments with nuclear emulsions: the G-stack collaboration was —perhaps—the very first worldwide collaboration set up in high-energy physics, even after CERN entered into full operation. Ettore Fiorini and myself—both very young indeed—wanted to move to bubble chamber physics. We were introduced to the CERN community and joined the analysis of the heavy liquid bubble chamber BP3 built by André Lagarrigue at the Ecole Politechnique.

The adventure to CERN was the result of several discussions with Beppo. He used to invite us in his apartment for after dinner meetings. In his family there was a strict hierarchy established all along: the little dog Dynah a dachshund, was the boss; then came Connie Pooh, then Connie and—he used to say—after the janitor, there was Beppo, ready to say always "yes" (but I do not believe a word of all that!). My personal believe is that the real boss was Connie Dilworth, authoritarian enough to manage the family with Beppo delegating to her all possible family affairs, but Connie Pooh whom he loved immensely, I believe. For the little girl it must have not been soo easy to deal with such an extraordinary father, living thirty years ahead of everybody around.

Beppo's character was not that easy; he liked a lot to discuss by paradoxical arguments; as of a sudden changing subject for a long while, then going back to the original point without an apparent reason. On the contrary he was miles ahead of all the others; jumping from an issue to another was a kind of joke for him, jokes built up with words. As Emilio Segré said once: "he is a living example that appearance is often misleading: he seems to be a little crazy but is full of good sense: he seems to be naïve in physics, while he knows a lot; he speaks 4 or 5 different languages but it is very difficult to understand him in any of them". His office looked very similar to being a real mess; appearing totally disorganized he would be able to put together a worldwide collaboration covering 3 to 4 continents. He was so full of ideas that not all of them were within intellectual and technical reach.

During several house meetings we investigated the possibility of using our cloud chamber at CERN. Honestly bubble chambers could collect data at a much faster rate; they could be filled with hydrogen or deuterium, both very neat and nice nuclei indeed; the dead time of the cloud chamber was too long and it could not recover adequate sensitivity to meet the fast rate of the accelerator beam. One night, Beppo came up with one of his "paradoxical" proposals: "why don't we build a dozen cloud chambers or so" he said, "organize them in such a way that they can be exposed, one after the other, matching the pace of the accelerator beam, for instance moving them on a round, circular rail, say some 15 m in diameter, thus letting enough time for the Wilson chambers to recover their sensitivity?". The idea was not unreasonable at all. We took it seriously but soon we had to realize that—over and above the "bureaucratic burden" of going through the CERN establishment—we had neither the manpower nor the financial resources to build a dozen 2m $\times$ 2m $\times$ 0.8m cloud chambers in our department. Thus the idea was abandoned. In parallel to the CERN adventure, in 1960, Beppo made a phone call to Gianni Puppi at the University of Bologna asking for help to introduce the group (Ettore Fiorini, and me) into the bubble chamber community of the USA. At that time Martin Block was visiting the Universities of Trieste and Bologna on sabbatical from Duke University; he had built a tiny neat helium bubble chamber; Puppi was too a famous professor and we could never address him directly! Beppo made arrangement so that Ettore and I could go to Bologna and work for a side analysis on pictures of the only liquid-helium bubble chamber that ever produced prominent physics, doing scanning and measurements from 8 pm till 8 am every night; the Bologna member of that mini-group was Giorgio Giacomelli who just went back to Bologna after receiving a PhD in physics at the University of Rochester, N.Y. The chamber was exposed to the Berkeley Bevatron to a stopping negative K-meson beam, to take advantage of the s-wave states in the K^- capture by the very symmetrical helium nucleus with the scope of studying nuclear hyperfragments generated in K^--^{4}He well-defined states. The pictures detected quite a lot of stopping muons as unavoidable background. During that analysis Ettore Fiorini moved for a sabbatical to Duke University where Martin was full professor. I remember shipping there an aluminum briefcase full of punched cards to join the measurements done at Duke. Working overnight was not that exciting, nonetheless I found that arrangement very charming because of the nice and cheap restaurants easy to find in Bologna. Meanwhile, Martin moved from Duke to Northwestern University, in Evanston Illinois and the research was concluded while I was there spending my sabbatical year 1961–62. I am still proud to have written in a scientific paper that we corrected for all systematic errors. Indeed: "*...to demonstrate explicitly our a priori knowledge in this experiment, concerning our systematic errors...*" we quoted the value with the statistical error only. Of the activities at NWU, I already wrote few lines at the beginning of this contribution. Coming back from Northwestern (or maybe, was it a little before my sabbatical in the

USA?), I found that Beppo appointed Alfred Freddy Herz, an Australian physicist as our group leader in place of Connie Dilworth who got tired to "grow up a bunch of kids" who wanted to get involved in research that was not inspiring her (when I became member of the CERN Track Chamber Committee, Charlie Peyrou was so kind to let me read the letter through which I was introduced to him by Connie. It was a very "peculiar" letter indeed. I need not to recall the details but at the very end Charlie and I ended the episode with a good laugh). Freddy was a guy of the "iron sergeant" style. We managed to come along and we wrote the paper containing the discovery of the A_1-particle produced in diffraction dissociation off complex nuclei. I presented the paper to the Rochester conference in Berkeley in 1966. In his review talk Arthur Rosenfeld stated: "the A_2 is a genuine resonance, the A_1 is a European effect". Nonetheless, we discovered the A_1, finally acknowledged by the Data Particle Group.

Beppo came back to Italy from Bruxelles in 1949, after the well-known peregrinations in South America and Europe (someone else has written about this point in this book). In 1968 the Institute of Physics celebrated the twenties anniversary of his going back to our country (and his sixtieth birthday) with a very spontaneous and touching ceremony. It is interesting to go through the signatures of the participant that I reproduce in Figs. 4, 5 and 6. Some 130 persons participated into the event, from all over Europe and the Americas. If you read carefully the names you may find the signature of 3 Nobel Prize Laureates (Cecil F. Powell, P. M. S. Blackett and Emilio Segré) but also the signature of two young men who much later would go and shake hand with the King of Sweden, i.e. Riccardo Giacconi (in the middle of the sixth row, first page: Fig. 4) and Carlo Rubbia (in the middle of the seventh row, second page: Fig. 5), not to mention Bruno Pontecorvo (first signature in the fifth row, first page: Fig. 4) who didn't obtain a Nobel Prize but would have deserved one as much as Beppo Occhialini. Polvani wanted to be among the last to sign (right signature, second line from the bottom: Fig. 6).

Concerning Bruno Pontecorvo signature there is kind of a mystery. At that time, the time of the iron curtain, the visas to leave the USSR were very difficult to obtain, even for a member of the Russian Academy of Sciences like Bruno, most of all for those considered as "refugees" who wanted to go back to their original countries. In the first page of signatures, Bruno Pontecorvo clearly appears on the fifth line, first column, but in the written contribution to the Symposium, Bruno writes that he regrets not having been able to attend in person. It must be that he gave permission to insert a label with his signature on it into the list. I cannot see any alternative explanation for the puzzle.

At that symposium I presented a paper that I found appropriate to the man, by the title: "*Some speculations in the framework of Regge Poles*", an issue that at that time was quite popular. The opening words are: "*Having in mind the extremely dialectical games that Beppo plays, from time to time, with his disciples and collaborators, I am tempted to propose a numerical game. More correctly I should say a speculation on experimental numbers. I learnt from him that experimentalists provide numbers and that theorist propose interpretations, but that both should think about the measured numbers which basically constitute the real facts. Since all in the audience are friend*

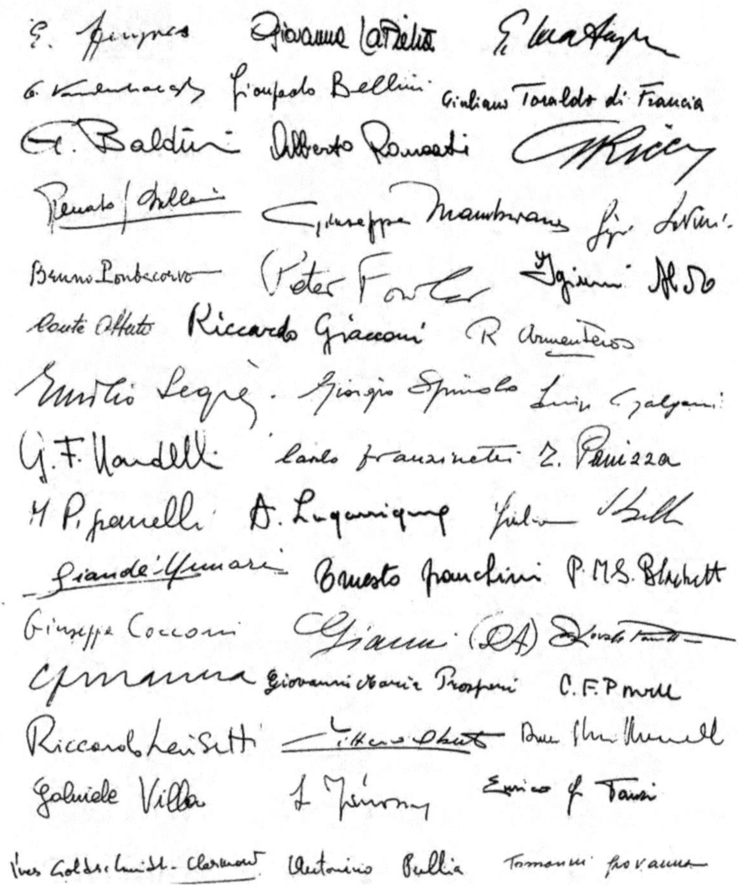

Fig. 4 First page of the signatures

of him, all know how he is able to play with words; perfectly capable to reach the conclusion he likes by using your words".

I presented an analysis of a set of 2-body and 3-body reaction produced in $\pi^{\pm}$p and pp collision at lab. momentum between 4GeV$/c$ and 12GeV$/c$, analyzed in terms of Regge pole exchange to prove the approximate validity of amplitude factorization: a questionable but valid interpretation of the data.

At that time I was collaborating to a daily newspaper: "Il Giorno", published in Milan, trying to spread scientific knowledge to ordinary people. On that occasion I wrote an article that appeared in the issue of October 15th by the title "*G. Occhialini contestatore dell'universo* ("G. Occhialini protester of the Universe", Fig. 7).

Indeed 1968 was the "big year" of the student protest, the occupation of the University buildings and it was a pretty hard time for all faculty members. Beppo was very uneasy with all that; he basically understood what the students wanted,

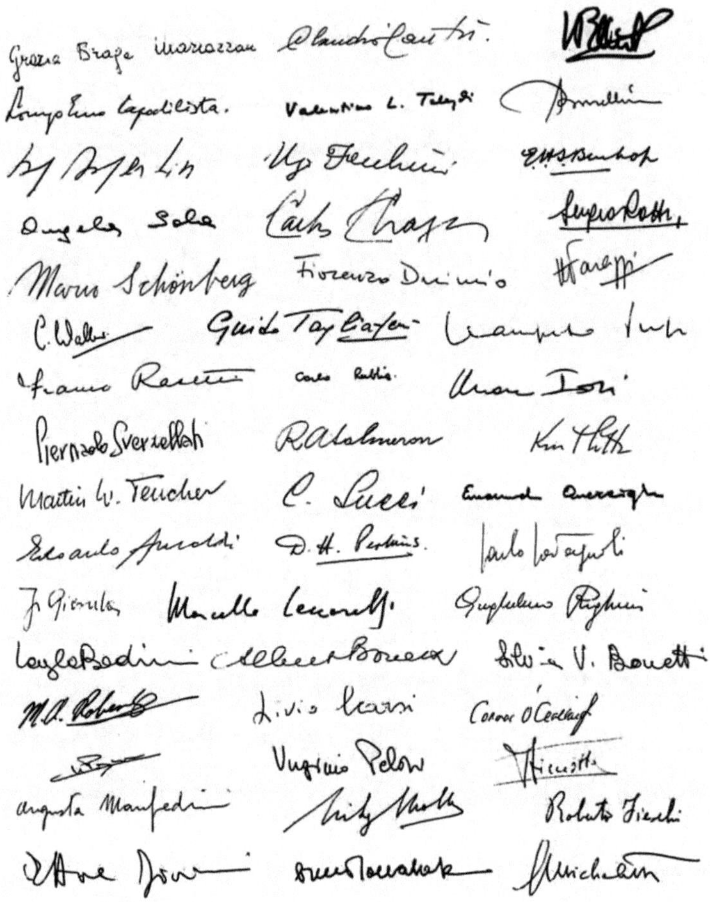

Fig. 5 Second page of the signatures

their issues, their instances, but did not approve the method, the violence, the arrogant approach. However he could not find the right way to address the issue. The permanent assembly meetings of over 400 shouting students squeezed like sardines into the large auditorium were not for him the place to stand at the podium and talk; it was not the kind of situation he would have liked to approach. During more than one period of occupation he first locked himself into his own office and finally took off the wall the plate with his name written on it. At the very end of the symposium to celebrate Beppo's coming back to Italy a couple dozen students asked to be admitted to the symposium and questions to the conveners. They wanted to know about the role of Science into the Human Society; when, how and where the scientists fought their fights to impose the right use of science and technology. There was then a natural appendix to the meeting that lasted about 2 hours. Beppo did not talk but answers came

Fig. 6 Third page of the signatures

from several foreign participants. Among the contributers to the symposium there were jewish like Emilio Segré who left the Country to the USA, due to the prosecution of the fascist regime, spanish physicists like Raphael Armenteros whose family had to expatriate from Spain due to the dictatorship of the Caudillio general Franco, Erwin Fenyves, a Hungarian physicist still under a soviet domination. Thus the answers came, either in English or in French or in a shaky Italian, but with a series of clean and concise argumentations, keen to capture the real message of a renovated spirit, the embryo of a society that might have been a new one. They *explained* why certain "difficult and painful" decisions were taken to avoid the happening of catastrophic events: how and why movements encouraging free thinking spread around, all over the World; how and why scientist can never avoid assuming personal responsibilities. Most of these people, although scientists, did not dare to speak out aloud against the "nazi" domination, against violence, against maccarthism, fundamentalism and blind capitalism.

Fig. 7 The title of an article written to celebrate the 60th birthday and the 20th year since the return of Beppo to Italy

There were, in summary, several eminent personalities who had protested twenty, thirty or even forty years before, who had been deeply anti-establishment although senior members of the "academy", who acted with strict coherence during all their lives, who in that moment were facing, in a new form, a living evidence of the logical continuation of their own juvenile protest. Beppo listened to all these talks with apparent satisfaction and all that turned out to be an event within the event.

Three years after the 1968 symposium I was appointed full professor at the University of Pavia, where I still teach and work so that I lost "direct" and constant contact with Beppo Occhialini, meeting him from time to time either at the Annual Conference of the Italian Physical Society, or at Conferences or at the Accademia dei

Lincei Meetings, where he was a distinguished effective member and I was nothing but a simple invited auditor.

My research field moved from bubble chambers to faster detectors. Searched —unsuccessfully at the Triga Mark II reactor of the University of Pavia—for neutron-antineutron oscillations, experiment by which I recovered the collaboration with the group lead by Marcello Conversi and in which I used flash chambers and Resistive Plate Chamber RPC for the first time. Thereafter, I moved to systematic search for charmed particles, both mesons and baryons, discovering a charmed baryon that would be called "primordial neutron", a [css] state (Ω_c^0) containing none of the ordinary quarks that has the shortest lifetime ever directly measured, i.e. $\tau = [72 \pm 11(\text{stat}) \pm 8(\text{syst})] \times 10^{-15}$ s. Got interested in fractal geometry and complexity; got interested in the application of the RPC detectors to the Positron Emitting Thomography (PET). In brief , I got along on my own for about 35 years.

I was shocked when I learnt—it was in Rome at a workshop of the Accademia dei Lincei, in the late nineties—that Beppo gave up smoking; I could not imagine him without a cigarette in his hand, without a cloud of smoke around his body, without the smell of burnt tobacco around him. Beppo, always smoking a cigarette, using its last quarter inch to light the next one.

My scientific career was deeply affected by the influence of professor G. P. S. Occhialini, although I am nor sure I know what the initials P. S. stand for. Somebody told me they are his second and third names Paolo and Sebastiano. But I am not sure.

When he died I was in the USA and could not attend his funeral. Since when I left Milan, I miss him a lot: the man who always stimulated my curiosity, and let my passion for physics grow to the limit of my intellectual capability. When I close my eyes, I still see him, hear his voice and his provocative words, arguing about physics. But this is probably a sign of my age, which is growing fast. I feel I was unable to fully collect his scientific inheritance.

G. P. S. Occhialini vu par un de ses amis

Jacques Labeyrie

Giuseppe Occhialini, "Beppo" pour ses amis, naquit le 5 septembre 1907 à Fossombrone, dans les Marches (Italie). Etant jeune, il rêvait de s'adonner à la poésie ou à la peinture, mais son père, professeur de physique, le persuada de se diriger vers cette science. Il commença ses études à Pise, puis à l'Université de Florence, où il fut nommé assistant de recherches en 1930. Il eut aussi une jeunesse sportive: il avait constitué avec quelques amis un groupe de spéléologie. Dans la Montagne de Marbre, au-dessus de Carrare, ils découvrirent un très grand gouffre, le Corquia, et l'explorèrent jusqu'à plus de 500 m de profondeur (le record du monde à l'époque). Il disait souvent: "la spéléologie, c'est l'exploration du pauvre". Voulait-il dire par là qu'il faisait partie des pauvres?

C'est à Florence qu'il se lia de grande amitié avec un autre jeune physicien qui devait lui-aussi devenir célèbre: Bruno Rossi. Celui-ci avait déjà commencé à étudier les "rayons cosmiques", et dès lors, ce sujet gouverna leur activité de chercheurs. De ces rayons cosmiques, on savait seulement qu'ils étaient très pénétrants, et donc porteurs d'une énergie immense. C'est sans doute les propriétés inconnues de leur énergie immense qui attirèrent Beppo. Il faut se rappeler qu'à cette époque on ne connaissait, en fait de particules nucléaires (on ne disait pas encore "élémentaires") que les "rayons" alpha, bêta, et gamma des radio-éléments naturels. On commençait tout juste à essayer de construire les premiers accélérateurs de protons, et Chadwick n'avait pas encore découvert le neutron.

En 1931, Beppo reçut une bourse d'étude de son gouvernement pour séjourner un an au fameux laboratoire Cavendish, à Cambridge. Il y resta, en fait, plus de deux ans auprès de Patrick Blackett, un physicien de la jeune science nucléaire, et qui étudiait celle-ci au moyen de la Chambre de Wilson. Beppo lui apporta de Toscane la pratique des compteurs de Geiger, et des circuits électroniques associés à ceux-ci. Je crois que c'est Beppo qui eut l'idée de marier les deux techniques, pour déclencher la Chambre de Wilson par "coïncidence" au moment du passage d'une particule cosmique, au

J. Labeyrie (✉)
Gif sur Yvette, France

L. Gariboldi et al. (eds.), *The Scientific Legacy of Beppo Occhialini*,
https://doi.org/10.1007/978-3-031-37034-2_16

lieu de la déclencher au hasard, comme cela se faisait jusque là. On gagnait ainsi énormément en efficacité pour détecter ces particules et leurs éventuelles réactions; cette astuce les amena presqu'aussitôt à la découverte des "cascades" et "gerbes" produites par l'action des rayons cosmiques sur la matière. Surtout, dès 1932, elle leur permit d'observer pour la première fois la création d'antimatière, sous la forme d'une paire d'électrons, négatif et positif. Blackett et Occhialini confirmaient ainsi les photos que venait d'obtenir en Californie Carl Anderson, également à la Chambre de Wilson, et qui montraient des traces de particules positives dues à l'action des rayons cosmiques (celui-ci les baptisa "positrons"); un peu plus tard, il découvrit des trajectoires d'une sorte d'électron lourd, le méson μ. Anderson reçut le Prix Nobel en 1936 "pour la découverte de l'électron positif". Blackett reçut le Prix Nobel en 1948 "pour le développement de la chambre à brouillard de Wilson"; quant à Occhialini, il fut oublié. Mais il resta quand même en très bons termes avec Blackett.

En 1937 Beppo fut invité comme Professeur à l'Université de São Paulo. Il y était encore, en 1942, lorsque le Brésil se rangea aux côtés des Alliés. Devenu ainsi malgré lui un "ennemi", il fut obligé de quitter l'Université et alla vivre chichement non loin de là, comme guide pour de rares touristes, dans la forêt de montagne d'Itatiaia, dans la région de São Paulo.

En 1944, c'est le retour en Grande-Bretagne, et Blackett l'aide alors à trouver une place au laboratoire Wills, à Bristol. Sous la direction de Cecil Powell une petite équipe y étudiait les réactions de physique nucléaire au moyen des émulsions photographiques. Beppo, toujours attiré par l'étude des rayons cosmiques, fut séduit par l'apparente simplicité de ces détecteurs. De plus, leur substance active, l'émulsion de gélatine et de bromure d'argent, avait l'avantage d'une densité plus de mille fois supérieure à celle des gaz de la Chambre de Wilson. Pour accroître le volume utile des plaques, et donc leurs chances d'interaction avec les rayons cosmiques, il travailla avec le directeur de la firme Ilford, Waller, à réaliser des émulsions qui étaient beaucoup plus épaisses que celles que l'on fabriquait jusque-là: cent cinquante microns contre une vingtaine de microns. En outre on avait accru leur concentration en bromure d'argent; on augmentait ainsi leur sensibilité, ce qui permettait de voir les traces dues aux particules très rapides, donc très peu ionisantes, et qui étaient invisibles jusque là. Beppo fit de plus un travail considérable pour transformer ces émulsions épaisses, molles et déformables, en un véritable appareil de mesure, permettant non seulement de détecter toutes les particules ionisantes, mais aussi d'identifier leur masse, leur charge et leur énergie.

Il se trouve qu'après son retour du Brésil, Beppo était tombé sous le charme de Max Cosyns, physicien à l'Université Libre de Bruxelles et collègue d'Auguste Piccard avec lequel il avait réalisé les premiers ballons stratosphériques, puis le Bathyscaphe. Cet amoureux de la verticale s'était aussi spécialisé dans la recherche et l'exploration des grands gouffres pyrénéens. Pendant les vacances de 1946, Beppo l'avait accompagné, emmenant à tout hasard dans ses poches quelques exemplaires de ses nouvelles émulsions.

Hélas! cet été-là, il pleuvait sans cesse, ce qui confinait nos explorateurs dans une petite auberge du Pays Basque. Ils profitèrent de ce contretemps pour se faire emmener dans la Jeep de l'aubergiste à l'observatoire du Pic du Midi, à 2883 m d 'altitude. C'était la place idéale pour y déposer durant quelques semaines un "stack" (un paquet) de ces premières émulsions épaisses dans l'espoir d'y capturer la trace bien visible de quelque interaction cosmique inconnue, d'une "étoile" comme Beppo devait les appeler plus tard.

De retour à Bristol il développa les plaques, et les analysa au microscope à immersion, aidé de Cesar Lattès et de Hugh Muirhead, et des "scanneuses" du laboratoire: dans l'une de ces plaques on voyait pour la première fois la trace d'une trajectoire de particule très rapide, dont la masse apparaissait inférieure à celle du proton, mais supérieure à celle du méson μ, et d'où partait, une longue trajectoire, celle d'un méson μ, se terminant par sa désintégration en électron et neutrino celui-ci étant invisible. C'est ainsi, dans l'été 1946, grâce à l'utilisation de ces nouvelles émulsions épaisses et denses, que fut découvert le méson π, ou "pion". C'était la particule, très recherchée, de l'interaction "forte", qui liait entre eux les nucléons et qui avait été prédite par Yukawa dès 1935, mais que personne n'avait pu encore observer.

Powell eut le Prix Nobel, en 1950, "pour le développement de la méthode photographique pour étudier les processus nucléaires". Une fois de plus, Occhialini fut oublié (et même deux fois: pour la découverte du méson π et pour la transformation des émulsions en un instrument de détection de très grande qualité). Entre temps, dès 1948, il avait quitté le laboratoire de Bristol, et rejoint celui de Bruxelles, auprès de Cosyns, où il exerça bientôt ses talents d'animateur scientifique incomparable. En été, avec quelques collègues ils continuaient leurs explorations spéléologiques. C'est ainsi que le dernier jour des vacances de 1950, Beppo fut parmi les trois spéléos qui découvrirent, l'entrée du formidable réseau souterrain de la Pierre Saint-Martin. L'exploration de celui-ci, deux ans plus tard, fut un des grands moments de sa vie (voir Fig. 1).

En 1950 il retourna en Italie, où il reprit la vie universitaire et fut nommé professeur à Gènes, puis à Milan, en 1952. En 1960 il y eut un court intermède d'un an à Boston, où il alla comme professeur invité auprès de son ami Bruno Rossi.

En Europe cependant, comme aux Etats-Unis, la recherche sur les particules élémentaires avait bien changé: c'en était fini des expériences astucieuses dans des laboratoires universitaires, à la façon qu'aimait Beppo et qui était aussi celle de Blackett, de Chadwick ou des Joliot à Paris. La pluie de rayons cosmiques, seule source jusque-là de ces particules élémentaires, était devenue bien trop faible au goût des physiciens et des administratifs de la Science. On vit alors, dès la fin de la Seconde Guerre mondiale, des accélérateurs géants se construire ici et là, fournissant sous des énergies de plus en plus élevées des flux de protons de plus en plus grands, et qui étaient servis par des équipes de plus en plus nombreuses: la "Physique des Particules" était née. Mais Beppo, homme aux goûts simples, n'aime pas trop cela, et pour continuer la recherche sur ses chers rayons cosmiques il va s'évader dans un domaine qui vient de naître: l'espace extra-terrestre. Dès 1960 il va élargir sa petite équipe de Milan par une alliance avec la jeune équipe d'astrophysique de Saclay. Ensemble, elles développent l'utilisation d'un nouveau détecteur observable à distance, la Chambre à étincelles; embarquée sur les ballons stratosphériques du CNES, celle-ci va tout d'abord confirmer dans la haute atmosphère la présence en grand nombre d'électrons cosmiques primaires de très haute énergie (déjà aperçus

Fig. 1 Août 1950. Beppo en train d'assurer un de ses camarades lors de la remontée du gouffre d'Escurets dans les Pyrénées

en petit nombre deux ans auparavant par une équipe américaine avec une Chambre de Wilson embarquée). Ensuite c'est la découverte de la première source galactique de rayons gamma (le pulsar du Crabe). Entre temps, Beppo organise un monde nouveau: celui des expériences sur les satellites scientifiques européens. Il est l'un des fondateurs et animateurs du "COS-Group" de l'ESRO (le groupe cosmique de l'Organisation Européenne de Recherche Spatiale, devenu ensuite l'ESA, pendant européen de la NASA), et étend la coopération des cosmiciens à quatre autres groupes, anglais, danois, hollandais et allemand.

Jusqu'en 1982, aidée par leurs moyens nationaux respectifs, en plus de l'ESRO, de la NASA et le l'INTERCOSMOS soviétique, cette association amicale des six laboratoires va réaliser une douzaine de grosses expériences spatiales pour découvrir et étudier les rayonnements à très haute énergie qui arrivent sur notre planète.

Je crois que c'est dans cette période là de sa vie que Beppo reçut le Prix Wolf qui lui fit certainement un grand plaisir.

Ces travaux demandaient une longue patience. Ainsi il fallut plus de dix ans pour achever le plus significatif d'entre eux, l'obtention de la première carte de notre Galaxie par son rayonnement gamma de très haute énergie (entre 70 MeV et 5 GeV) à partir du satellite COS-B. Mais on retombait ainsi dans une nouvelle sorte de "Big Science" et Beppo la quitta vers 1982 pour intégrer le CFR (Centre des Faibles Radioactivités) à Gif sur Yvette, et y participer avec un très grand talent à une nouvelle méthode de datation des roches par les traces de fission spontanée: il était enfin revenu à ce travail de laboratoire qu'il aimait tant. Beppo mourut à Paris le 30 Décembre 1993, entouré de l'affection de quelques amis.

Un mot encore, on n'a pas parlé d'une de ses qualités les plus frappantes qui était la modestie.

Appendix
List of Scientific Publications of G. Occhialini

1–Occhialini G., "Uno spettrografo magnetico per raggi β emessi da sostanze debolmente radioattive" *Rendiconti della Reale Accademia Nazionale dei Lincei*, **14** (1931) 103–107.

2–Blackett P. M. S. and Occhialini G., "Photography of Penetrating Corpuscular Radiation" *Nature*, **130** (1932) 363.

3–Blackett P. M. S. and Occhialini G. P. S., "Some Photographs of the Tracks of Penetrating Radiation" *Proceedings of the Royal Society of London Ser. A*, **139** (1933) 699–727.

4–Chadwick J., Blackett P. M. S. and Occhialini G., "New Evidence for the Positive Electron" *Nature*, **131** (1933) 473.

5–Occhialini G., "Le recenti ricerche intorno all'elettrone positivo" *La Ricerca Scientifica*, **1** (1933) 372–373.

6–Chadwick J., Blackett P. M. S. and Occhialini G. P. S., "Some Experiments on the Production of Positive Electrons" *Proceedings of the Royal Society of London Ser. A*, **144** (1934) 235–249.

7–Occhialini G., "Il Positrone" in *Enciclopedia Italiana* (Istituto Treccani, Roma) 1934.

8–Bernardini G. and Occhialini G., "Il Congresso di Fisica Nucleare a Zurigo" *La Ricerca Scientifica* (1936) 426–434.

L. Gariboldi et al. (eds.), *The Scientific Legacy of Beppo Occhialini*,
https://doi.org/10.1007/978-3-031-37034-2

9–Occhialini G., "La radiazione gamma del Polonio-Berillio" *Rendiconti della Reale Accademia dei Lincei*, **25** (1937) 188–194.

10–Occhialini G., "Diffusion des rayons gamma du thorium C''" *Réunion internationale de Physique-Chimie-Biologie, Paris, octobre 1937* (Hermann et C., Paris) 1938.

11–Occhialini G. P. S., "A Simple Type of Non-Ohmic Resistance for Use with Geiger-Müller Counters" *Journal of Scientific Instruments*, **15** (1938) 97–99.

12–Occhialini G., "Mesures de l'effet de latitude pour les gerbes" *Comptes Rendus Hebdomadaires des Séances de l'Académie des Sciences*, **208** (1939) 101–103.

13–Occhialini G. P. S. and Schönberg M., "Sobre uma componente ultra molle da radiação cósmica (I)" *Annaes da Academia Brasileira de Ciências*, **11**, No. 4 (1939) 351–355.

14–Occhialini G., "Contributo allo studio dell'effetto di latitudine per gli sciami" *Annaes da Academia Brasileira de Ciências*, **12**, No. 1 (1940) 39–44.

15–Occhialini G., "Sull'effetto di latitudine degli sciami" *La Ricerca Scientifica*, **11**, aprile (1940) 231–234.

16–Monteux Y. and Occhialini G. P. S., "Sur un nouveau type de compteurs plans (I)" *Annaes da Academia Brasileira de Ciências*, **12**, No. 2 (1940) 125–129.

17–Occhialini G., "Sur la radioactivité beta du rubidium" *Annaes da Academia Brasileira de Ciências*, **12**, No. 1 (1940) 155–158.

18–Occhialini G. and Damy De Souza Santos M., "Effetto dell'eclissi totale di sole del 1° ottobre sull'intensità della radiazione cosmica" *La Ricerca Scientifica*, **11**, ottobre (1940) 792.

19–Occhialini G. P. S., Pompéia P. A. and Saboya J. A. R., "Nota sobre a estabilização de tensão em corrente alternada" *Annaes da Academia Brasileira de Ciências*, **12**, No. 4 (1940) 349–352.

20–Occhialini G. P. S. and Schönberg M., "Sobre uma componente ultra molle da radiação cósmica (II)" *Annaes da Academia Brasileira de Ciências*, **12**, No. 3 (1939) 195–202.

21–Occhialini G., "Contributo allo studio della componente ultramolle della radiazione cosmica" *La Ricerca Scientifica*, **12**, novembre (1941) 1193–1195.

22–Occhialini G. and Damy De Souza Santos M., "On a Method of Recording Random Events" *Annaes da Academia Brasileira de Ciências*, **13**, No. 1 (1941) 57–62.

23–Occhialini G., "Metodo per la stabilizzazione di alte tensioni" *La Ricerca Scientifica*, **13**, giugno-luglio (1942) 319–321.

24–Occhialini G. and Damy De Souza Santos M., "Two Useful Gadgets for Controlled Wilson Chambers", in *Symposium sôbre Raios Cósmicos, agôsto 4–8, 1941, Academia Brasileira de Ciências, Rio de Janeiro 1943* (Imprensa Nacional) 1948, pp. 165–168.

25–Powell C. F., Occhialini G. P. S., Livesey D. L. and Chilton L. V., "A New Photographic Emulsion for the Detection of Fast Charged Particles" *Journal of Scientific Instruments*, **23** (1946) 102–106.

26–Occhialini G. P. S. and Powell C. F., "Multiple Disintegration Processes Produced by Cosmic Rays" *Nature*, **159** (1947) 93–94.

27–Occhialini G. P. S. and Powell C. F., "Nuclear Disintegrations Produced by Slow Charged Particles of Small Mass" *Nature*, **159** (1947) 186–190.

28–Lattes C. M. G. and Occhialini G. P. S., "Determination of the Energy and Momentum of Fast Neutrons in Cosmic Rays" *Nature*, **159** (1947) 331–332.

29–Lattes C. M. G., Muirhead H., Occhialini G. P. S. and Powell C. F., "Processes Involving Charged Mesons" *Nature*, **159** (1947) 694–697.

30–Lattes C. M. G., Occhialini G. P. S. and Powell C. F., "Observations on the Tracks of Slow Mesons in Photographic Emulsions" *Nature*, **160** (1947) 453–456; 486–492.

31–Bates W. J. and Occhialini G. P. S., "Applications of the Reflecting Microscope to the Nuclear Plates Technique" *Nature*, **161** (1948) 473.

32–Powell C. F. and Occhialini G. P. S., *Nuclear Physics in Photographs: Tracks of Charged Particles in Photographic Emulsions* (Clarendon, Oxford) 1947.

33–Occhialini G. P. S. and Powell C. F., "The Scattering of Fast Neutrons by Protons" *Philosophical Society of Cambridge Conference Report* (1947) 150.

34–Occhialini G. P. S. and Powell C. F., "The Artificial Production of Mesons" *Nature*, **161** (1948) 551–552.

35–Lattes C. M. G., Occhialini G. P. S. and Powell C. F., "A Determination of the Ratio of the Masses of π- and μ-Mesons by the Method of Grain-counting" *Proceedings of the Physical Society*, **61** (1948) 173–183.

36–Dilworth C. C., Occhialini G. P. S. and Payne R. M., "Processing Thick Emulsions for Nuclear Research" *Nature*, **162** (1948) 102–103.

37–Occhialini G. P. S. and Powell C. F., "Observations on the Production of Mesons by Cosmic Radiation" *Nature*, **162** (1948) 168–173.

38–Dilworth C., Occhialini G. and Samuel E., "Eclaircissement des plaques photographiques" *Bullettin du Centre de Physique Nucléaire de l'Université Libre de Bruxelles*, **2** (1948).

39–Cosyns M., Dilworth C. and Occhialini G., "Temperature Shutter for Nuclear Research Emulsions" in *Cosmic Radiation: Colston Papers* (Butterworth) 1948.

40–Cosyns M., Dilworth C. and Occhialini G., "Obturateur thermique pour plaques nucléaires" *Bullettin du Centre de Physique Nucléaire de l'Université Libre de Bruxelles*, **6** (1949).

41–Occhialini G. P. S., "On the Identification of High Energy Particles in Electron Sensitive Plates" *Suppl. Nuovo Cimento*, **6**, Ser. IX, No. 3 (1949) 413–428.

42–Cosyns M., Dilworth C. C., Occhialini G. P. S. and Schönberg M., "Double Stars with Relativistic Particles from Cosmic Rays" *Nature*, **164** (1949) 129–131.

43–Cosyns M. G. E., Dilworth C. C., Occhialini G. P. S., Schönberg M. and Page N., "The Decay and Capture of μ-Mesons in Photographic Emulsions" *Proceedings of the Physical Society*, **62**, No. 12 (1949) 801–805.

44–Dilworth C. C., Occhialini G. and Vermaesen L., "On Processing Nuclear Emulsions, Part I. Concerning Temperature Development" *Bullettin du Centre de Physique Nucléaire de l'Université Libre de Bruxelles*, **13a** (1950).

45–MEULEMANS G., OCCHIALINI G. P. S. AND VINCENT A. M., "The Wire Method of Loading Nuclear Emulsions" *Il Nuovo Cimento*, **8** (1951) 341–344.

46–BONETTI A. AND OCCHIALINI G. P. S., "Cylindrical Emulsions" *Il Nuovo Cimento*, **8** (1951) 725–727.

47–BONETTI A., DILWORTH C. C. AND OCCHIALINI G. P. S., "On Processing Nuclear Emulsions, Part II. After Development Techniques" *Bullettin du Centre de Physique Nucléaire de l'Université Libre de Bruxelles*, **13b** (1951).

48–OCCHIALINI G., "Technique des plaques nucléaires" in *Colloque sur la sensibilité des cristaux et des émulsions photographiques, Revue d'optique théorique et instrumentale*, **23 A** (1951) 296–300.

49–DILWORTH C. C., OCCHIALINI G. P. S. AND SCARSI L., "Heavy Mesons" *Annual Review of Nuclear Science*, **4** (1954) 271–314.

50–BONETTI A., DILWORTH C., LADU M. AND OCCHIALINI G., "Misure in lastre nucleari" *Rendiconti dell'Accademia Nazionale dei Lincei*, **17** (1954) 311–314.

51–BONETTI A. AND OCCHIALINI G., "Technique of Nuclear Emulsions" *Suppl. Nuovo Cimento*, **11**, Ser. IX, No. 2 (1954) 222–227.

52–DAVIES J. H., EVANS D., FRANCOIS P. E., FRIEDLANDER M. W., HILLIER R., IREDALE P., KEEFE D., MENON M. G. K., PERKINS D. H., POWELL C. F., BGGILD J., BRENE N., FOWLER P. H., HOOPER J., ORTEL W. C. G., SCHARFF M., CRANE L., JOHNSTON R. H. W., O'CEALLAIGH C., ANDERSON F., LAWLOR G., NEVIN T. E., ALVIAL G., BONETTI A., DI CORATO M., DILWORTH C., LEVI SETTI R., MILONE A., OCCHIALINI G., SCARSI L., TOMASINI G., CECCARELLI M., GRILLI M., MERLIN M., SALANDIN G. AND SECHI B., "On the Masses and Modes of Decay of Heavy Mesons Produced by Cosmic Radiation" *Il Nuovo Cimento*, **2** (1955) 1063–1103.

53–ALVIAL G., BONETTI A., DILWORTH C., LADU M., MORGAN J. AND OCCHIALINI G., "Measurements of Ionization" *Suppl. Nuovo Cimento*, **6**, Ser. X, No. 2 (1956) 244–254.

54–G- STACK COLLABORATION, "Observations on Heavy Mesons and Hyperons" *Nature*, **175** (1955) 971–973.

55–DAVIES J. H., EVANS D., FOWLER P. H., FRANÇOIS P. E., FRIEDLANDER M. W., HILLIER R., IREDALE P., KEEFE D., MENON M. G. K., PERKINS D. H., POWELL C. F., CRANE L., JOHNSON R. H. W., O'CEALLAIGH C., ANDERSON F., LAWLOR G., NEVIN T. E., ALVIAL G., BONETTI A., DI CORATO M., DILWORTH C., LEVI- SETTI R., MILONE A., OCCHIALINI G., SCARSI L., TOMASINI G., CECCARELLI M., GRILLI M., MERLIN M., SALANDIN G. AND SECHI B., "Observations on Heavy Mesons Secondaries" *Suppl. Nuovo Cimento*, **4**, Ser. X, No. 2 (1956) 398–424.

56–OCCHIALINI G. P. S., "Possible Ways to Improve the Nuclear Emulsion Situation" *Premier Colloque de Photographie Corpusculaire* (Strassbourg) 1957, pp. 1–7.

57–IGIUNI A. AND OCCHIALINI G. P. S., "Operational Methods for the Processing of K.5 Emulsions with the Semi-Automatic Apparatus of Brussels" *Colloque de Photographie Corpusculaire* (Montréal) 1958.

58–BHOMWIK B., EVANS D., FALLA D., HASSAN F., KAMAL A. A., NAGPAUL K. K., PROWSE D. J., RENÉ M., ALEXANDER G., JOHNSTON R. H. W., O'CEALLAIGH C., KEEFE D., BURHOP E. H. S., DAVIS D. H., KUMAR R. C., LASICH W. B., SHAUKAT M. A., STANNARD F. R., BACCHELLA G., BONETTI A., DILWORTH C., OCCHIALINI G., SCARSI L., GRILLI M., GUERRIERO L., VON LINDERN L., MERLIN M. AND SALANDIN A., "The Interaction and Decay of K^- Mesons in Photographic Emulsion. Part I. General Characteristics of K^--Interactions and Analysis of Events in which a Charged π-Meson is Emitted" *Il Nuovo Cimento*, **13** (1959) 690–729.

59–BHOMWIK B., EVANS D., FALLA D., HASSAN F., KAMAL A. A., NAGPAUL K. K., PROWSE D. J. J., RENÉ M., ALEXANDER G., JOHNSTON R. H. W., O'CEALLAIGH C., KEEFE D., BURHOP E. H. S., DAVIS D. H., KUMAR R. C., LASICH W. B., SHAUKAT M. A., STANNARD F. R., BACCHELLA M., BONETTI A., DILWORTH C., OCCHIALINI G., SCARSI L., GRILLI M., GUERRIERO L., VON LINDERN L., MERLIN M., SALANDIN A., "The Interaction of K^--Mesons with Photographic Emulsion Nuclei. Part II. The Emission of Hyperons from K^--Interactions at Rest" *Il Nuovo Cimento*, **14** (1959) 315–364.

60–DILWORTH C. C. AND OCCHIALINI G., "Satelliti scientifici europei per astrofisica stellare" *Annuario della EST*, **6** (1973) 127–138.

61–OCCHIALINI G., "Occhialini, Giuseppe" in *Scienziati e Tecnologi Contemporanei*, Vol. II (Mondadori, Milano) 1974, pp. 322–324.

62–OCCHIALINI G. P. S., contribution to the "Memorial Meeting for Lord Blackett, O.M., C.H., F.R.S. at the Royal Society on 31 October 1974" *Notes and Records of the Royal Society of London*, **29** (1975) 144–146.

63–OCCHIALINI B., "Cesar Lattes: The Bristol Years" in *Topics on Cosmic Rays. 60th aniversary of C. M. G. Lattes*, edited by BELLANDI FILHO J., CHINELLATO C. AND PEMMARAJU A., Vol. **1** (Unicamp, Campinas) 1984, pp. 6–8.

